D1327624

WITHDRAWN

THE
PAUL HAMLYN
LIBRARY

⸻ • ⸻

DONATED BY
THE PAUL HAMLYN
FOUNDATION
TO THE
BRITISH MUSEUM

⸻ • ⸻

opened December 2000

INSECTS AND OTHER INVERTEBRATES IN CLASSICAL ANTIQUITY

INSECTS AND OTHER INVERTEBRATES IN CLASSICAL ANTIQUITY

by
Ian C. Beavis

UNIVERSITY OF EXETER

592 093 8 BEA

First published 1988 by the University of Exeter
© 1988 Ian C. Beavis

ISBN 0 85989 284 0

Exeter University Publications
Reed Hall
Streatham Drive
Exeter, Devon
EX4 4QR.

Printed in Great Britain by
the Alden Press, Osney Mead, Oxford

CONTENTS

		Page
Preface		ix
Note on Abbreviations and Short Titles		xi
Introduction		xiii

I WORMS, LEECHES, CENTIPEDES, WOODLICE, etc.

ANNELIDA: LUMBRICIDAE (EARTHWORMS)
    1. *Ges enteron/Lumbricus* — 1

ANNELIDA: HIRUDINEA (LEECHES)
    2. *Bdella/Hirudo* — 4

ARTHROPODA: CHILOPODA (CENTIPEDES)
    3. *Skolopendra/Centipes* — 10

ARTHROPODA: DIPLOPODA AND ISOPODA (MILLIPEDES AND WOODLICE)
    4. *Ioulos, Oniskos/Multipeda* — 13

II SCORPIONS, SPIDERS, MITES AND TICKS

ARACHNIDA: SCORPIONES (SCORPIONS)
    5. *Skorpios/Scorpio* — 21

ARACHNIDA: PSEUDOSCORPIONES (FALSE SCORPIONS)
    6. *Skorpiodes* — 34

ARACHNIDA: ARANEAE (SPIDERS)
    7. *Arachne/Araneus* and *Phalangion* — 34
    7a. *Arachne/Araneus* (sensu stricto) — 35
    7b. *Phalangion* — 44

ARACHNIDA: ACARI (MITES AND TICKS)
    8. *Kroton/Ricinus* — 56
    9. *Akari* — 60

III GRASSHOPPERS, COCKROACHES, MANTIDS, MAYFLIES, etc.

INSECTA: COLLEMBOLA (SPRINGTAILS)
    10. *Petaurista* — 61

INSECTA: ORTHOPTERA (GRASSHOPPERS, LOCUSTS, CRICKETS)
    11. *Akris/Locusta* — 62

12. *Troxallis/Gryllus*   78

INSECTA: DICTYOPTERA (COCKROACHES AND MANTIDS)
13. *Silphe/Blatta*   80
14. *Mantis*   85

INSECTA: EPHEMEROPTERA (MAYFLIES)
15. *Ephemeron/Hemerobion*   88

IV   **CICADAS, BUGS AND LICE**

INSECTA: HEMIPTERA
16. *Tettix/Cicada*   91
17. *Koris/Cimex*   104
18. *Tiphe/Tippula*   106
19. *Kokkos/Coccus*   108
20. *Lakkos*   111
21. *Margarites chersaios*   111

INSECTA: ANOPLURA AND MALLOPHAGA (LICE)
22. *Phtheir/Pediculus*   112

V   **BUTTERFLIES, MOTHS AND WOOD-BORING LARVAE**

INSECTA: LEPIDOPTERA (BUTTERFLIES AND MOTHS)
23. *Psyche/Papilio*   121
24. *Phallaina*   129
25. *Ips/Convolvolus*   132
26. *Ses/Tinea*   136
27. *Bombyx*   140
28. *Pityokampe*   148
29. *Xylophoron*   148

INSECTA: LEPIDOPTERA AND COLEOPTERA
30. *Kerastes/Cossus* and *Karabos/Lucavus*   149

VI   **BEETLES**

INSECTA: COLEOPTERA (BEETLES)
31. *Kantharos/Scarabaeus*   157
32. *Melolonthe*   164
33. *Kantharis*   168
34. *Bouprestis*   173
35. *Lampyris/Cicindela*   175
36. *Kis/Curculio*   177
37. *Thrips/Tinea*   181

38. *Sphondyle* and *Staphylinos* 184
39. *Kleros* or *Skleros* 186

VII BEES, WASPS AND ANTS

INSECTA: HYMENOPTERA (BEES, WASPS AND ANTS)
40. *Sphex* and *Anthrene/Vespa* and *Crabro* 187
41. *Tenthredon, Anthredon, Pemphredon, Bembix* 195
42. *Bombylios* and *Bombykion* 197
43. *Seiren* 198
44. *Myrmex/Formica* 198
45. *Myrmex chrysorychos/Formica Indica* 209
46. *Psen/Culex ficarius* 212
47. *Gall wasps* 216

VIII FLIES AND FLEAS

INSECTA: DIPTERA (FLIES)
48. *Myia/Musca* 219
49. *Myops, Oistros/Asilus, Tabanus* 225
50. *Konops, Empis/Culex* 229
51. *Oinokonops/Bibio* 236
52. *Kynomyia* 238

INSECTA: DIPTERA OR ODONATA (DRAGONFLIES)
53. *Hippouros* 239
54. *Mulio* 239

INSECTA: SIPHONAPTERA (FLEAS)
55. *Psylla/Pulex* 240

IX UNIDENTIFIABLE AND FABULOUS INSECTS AND INVERTEBRATES

56. *Orsodakne* 243
57. *Prasokouris* 243
58. *Knips* and *Sknips* 245
59. *Skolex tes chionos* 247
60. *Pyrigonos* 247
61. *Myrmekoleon/Formicoleon* 249
62. *Serphos* 251
63. *Psylla/Pulex* 253
64. *Phtheir* 253
65. *Vermis caeruleus* 254
66. *Skolex leukos* 254
67. *Galba* 254

68. *Rauca* 254
69. *Biurus* 255
70. *Phryganion* 255
71. *Names listed by Greek lexicographers* 255
72. *Names listed by Polemius Silvius* 258

Index 259

PREFACE

The following work consists of a revised version of a thesis originally submitted to the University of Exeter for the degree of Doctor of Philosophy in September 1984.

Its appearance in print so shortly after that of Davies and Kathirithamby's *Greek Insects*, apparently covering the same field, perhaps requires some explanation and justification. The explanation is simply the coincidence of the respective authors being independently inspired by D'Arcy Thompson's glossaries of Greek birds and fishes to attempt something similar in the field of entomology. The justification is that the apparent duplication between the two works is actually more apparent than real, so that they may be seen as complementary.

As regards zoological scope, *Greek Insects* does not include the annelids, arachnids, myriapods and woodlice covered by the present work, while it includes the domestic honey bee, here omitted. In terms of literary coverage, it generally excludes Latin literature, with the exception of Pliny. It does, however, give attention to the role of the insect types in art, an area generally omitted from consideration here, and it includes some comparative material from post-classical literature and folklore. The Introduction to *Greek Insects* will be found to include a number of discussions of general topics not directly addressed by the present work; namely, the role of insects in Aesopic fable and Greek poetry, the achievements of Aristotle and his abilities as an entomologist, and the place of insects in Greek art.

Davies and Kathirithamby do not bring forward any fresh identifications of insect names and generally refrain from discussing the suggestions made by previous authors, referring their readers to the conclusions of D'Arcy Thompson in the notes to his translation of Aristotle's *Historia Animalium*. The present work, on the other hand, gives particular attention to this subject and seeks to provide a full analysis and revision of previous efforts at identification.

Although total comprehensiveness is never perfectly attainable, since some significant material will inevitably slip through the net of any survey, the present work does, within the limits set out in the Introduction, aspire in that direction; while the authors of *Greek Insects* deliberately disclaim an attempt at comprehensiveness, providing instead a broad treatment which refrains 'from quoting too many obscure and out of the way authors'. Therefore, even where the same field of classical evidence is under consideration by both works, there will be found a difference of approach and presentation.

In presenting this work to the public, I would wish to express my gratitude to my research supervisor in the Department of Classics at Exeter, Professor T. P. Wiseman, for his advice, help and encouragement from the very beginning, when this rather unusual thesis subject was first proposed, through the three years of research, and latterly in the preparation of the revised text for publication. I would also gratefully acknowledge the assistance given me by Dr. R. J. Wootton of the Department of Biological Sciences at Exeter, who read the thesis both in a draft form and in its completed version as submitted, and who offered many helpful suggestions on identifications and other matters of zoological fact. Thanks are due too to my external examiner, Professor E. K. Borthwick of the University of Edinburgh, for his helpful comments on the text of the thesis as submitted, and for giving me a preview of his unpublished article on the praying mantis, and to Mrs B. V. Mennell, the Publications Officer of the University of Exeter, for so efficiently seeing the project of publication through from its inception to its completion.

Publication has been made possible by grants from both the Royal Society and the Hugh Last Fund administered by the British School at Rome, and the generous assistance of these two institutions is gratefully recorded.

Tunbridge Wells   IAN C. BEAVIS
July 1987

NOTES ON ABBREVIATIONS
AND SHORT TITLES

Abbreviations of the titles of classical works are intended to be self-explanatory. Reference to chapter, section or page generally follow the numbering of the standard editions cited in Liddell-Scott-Jones *Greek-English Lexicon* (here abbreviated to LSJ) and the *Oxford Latin Dictionary*. The only significant exception is the works of Philo, where the section numbering is that of Colson and Whitaker's edition in the Loeb Classical Library. In the case of patristic literature, references are given to Migne's *Patrologia Graeca* (MPG) and *Latina* (MPL).

The abbreviations used for the titles of periodicals are those employed in *L'Année Philologique*. Th.L.L. denotes the *Thesaurus Linguae Latinae*, and PW Pauly-Wissowa's *Realencyclopaedie der Classischen Altertumswissenschaft*.

Short titles have been used to refer to the following frequently cited works:

H. AUBERT and F. WIMMER (ed. and trans.)	*Aristoteles Thierkunde*, Leipzig, 1868.
F. S BODENHEIMER	*Materialien zur Geschichte der Entomologie bis Linne*, Vol. 1, Berlin, 1928.
F. S. BODENHEIMER	*Animal and Man in Bible Lands*, Leiden, 1960.
P. CHANTRAINE	*Dictionnaire Étymologique de la Langue Grecque*, Paris, 1968–80.
M. DAVIES and J. KATHIRITHAMBY	*Greek Insects*, London, 1986.
A. ERNOUT and A. MEILLET	*Dictionnaire Étymologique de la Langue Latine*, ed. 4, Paris, 1959.
L. GIL FERNANDEZ	*Nombres de Insectos en Griego Antiguo*, Manuales y Anejos de *Emerita*, XVIII, Madrid, 1959.
G. FRÖHLICH and W. RODEWALD	*Pests and Diseases of Tropical Crops*, London, 1970.
H. GOSSEN	*Die Tiernamen in Aelians 17 Büchern Peri Zoon*, QGN, IV. 3, 1935.
H. GOSSEN	*Die Zoologischen Glossen im Lexicon des Hesychius*, QGN, VII, 1940.
A. S. F. GOW and A. F. SCHOLFIELD (ed. and trans.)	*Nicander: The Poems and Poetical Fragments*, Cambridge, 1953.
A. HORT (ed. and trans.)	*Theophrastus: Enquiry into Plants*, Loeb Classical Library, 1916–26.

Imms' General Textbook of Entomology, ed. 10, rev. O. W. Richards and R. G. Davies, London, 1977.

Z. KADAR   *Survivals of Greek Zoological Illuminations in Byzantine Manuscripts*, Budapest, 1978.

O. KELLER   *Die Antike Tierwelt*, Leipzig, 1909–13.

O. KELLER and IMHOOF–BLUMER   *Tier und Pflanzenbilder auf Münzen und Gemmen des Klassischen Altertums*, 1889.

H. LEITNER   *Zoologische Terminologie beim Alteren Plinius*, Hildesheim, 1972.

A. L. PECK (ed. and trans.)   *Aristotle: Generation of Animals*, Loeb Classical Library, 1942.

A. L. PECK (ed. and trans.)   *Aristotle: Historia Animalium*, Vols. 1–2, Loeb Classical Library, 1965–70.

H. RACKHAM and W. H. S. JONES   *Pliny: Natural History*, Loeb Classical Library.
(ed. and trans.)

A. F. SCHOLFIELD (ed. and trans.) *Aelian: On the Characteristics of Animals*, Loeb Classical Library, 1958–9.

K. G. V. SMITH (ed.)   *Insects and Other Arthropods of Medical Importance*, London, 1973.

E. J. L. SOULSBY   *Helminths Arthropods and Protozoa of Domesticated Animals*, London, 1968.

J. H. STAPLEY and F. C. GAYNER   *World Crop Protection*, Vol. 1: *Pests and Diseases*, London, 1969.

A. STEIER   *Aristoteles und Plinius: Studien zur Geschichte der Zoologie*, Wurzburg, 1913.

R. STRÖMBERG   *Griechische Wortstudien: Untersuchungen zur Benennung von Tieren Pflanzen . . .*, Göteborg, 1944.

C. J. SUNDEVALL   *Die Thierarten des Aristoteles*, Stockholm, 1863.

D'ARCY W. THOMPSON   *The Oxford Translation of Aristotle*, Vol. 4; *Historia Animalium*, Oxford, 1910.

INTRODUCTION

In the introduction to his valuable work on the etymology of Greek insect names, Gil Fernandez states that although it was not his intention to provide a complete natural history of the species known in classical antiquity, along the lines of D'Arcy Thompson's glossaries of Greek birds and fishes, he entertained the hope that his book might provide a basis for the compiling of such a history by some later author. It is the intention of this present work to provide just such a natural history as Fernandez envisages.

Classical entomology, to use a convenient if imprecise term to denote the natural history of terrestrial invertebrates as it was known in antiquity, is a subject which has been much neglected. And it is one which has more often been the domain of zoologists with some additional knowledge of classics than of classicists interested in entomology. D'Arcy Thompson, the most reliable worker across the whole field of classical zoology, although publishing some material on insects in the notes of his Oxford translation of Aristotle's *Historia Animalium*, did not provide a comprehensive account of this group of animals comparable to his works on birds and fishes (the latter including marine invertebrates).

The present study has been written with a twofold aim. In the first place, it seeks to establish, so far as it is possible to do so, the probable identifications of the numerous insect names recorded by classical sources. This question of establishing definitions is one which has engaged the attention of previous authors more than any other, with variable results. On the one hand, there has been a tendency to refrain from innovation, which has resulted in many identifications being simply passed down unquestioned from one writer to another. On the other hand, there has been a tendency in some cases, and in the work of one author in particular, to put forward highly specific conclusions that are in no way warranted by the very limited evidence upon which they are supposedly based. This study, therefore, in addition to its constructive role, has also of necessity something of the nature of a ground clearing exercise, considering the extent to which some previous suggestions are justified.

A word is perhaps in order here concerning the respective merits of those authors chiefly responsible for these identifications. As has been said, D'Arcy Thompson is the most authoritative although his work is restricted to the insects of Aristotle. Also restricted to Aristotle, although more comprehensive, is the important study by Sundevall, foundational in this field, which in turn has provided the basis for the glossary of Aristotelian animal names prefixed by Aubert and Wimmer to their two volume edition of the *Historia*. Sundevall's conclusions are generally modest and sensible, despite occasional aberrations, but his work is inevitably dated in many respects. Covering a wider field is Keller's *Antike Tierwelt*, which seeks to deal with all the invertebrates referred to by ancient authors. His conclusions, like those of Sundevall, are generally cautious, and do not make an over-ambitious attempt at precision; but his treatment is far from comprehensive, and a number of varieties of insect are omitted. A similar favourable judgement cannot, unfortunately, be passed on the conclusions of Gossen, author of comprehensive articles on the animals of Aelian, Hesychius, Athenaeus and the lyric poets, and of the majority of the entries under the insect names in Pauly-Wissowa's *Realencyclopaedie*. Gossen has put forward an enormous number of highly specific identifications which are for the most part purely speculative, and for which little or no explanation is given. His basic error seems to have been to ignore the fact that the ancients did not in any respect share our modern concept of a species as a clearly defined zoological entity firmly distinguished from its fellows by precise diagnostic criteria. Mention must also be made of the work to which the present author is above all indebted, Gil Fernandez' *Nombres de Insectos*, which has already been cited. As has been said, that work is primarily concerned with the etymology of insect names, rather than with classical entomology in general, and should therefore be regarded as complementary rather than parallel to the present study.

In the second place, this work seeks to present, by drawing together scattered details, a composite account of all that was known or believed about the habits and life histories of terrestrial invertebrates, and of their economic importance and their place in popular thought, insofar as this is expressed in literature. In doing so, it also exhibits the dependence of ancient authors upon one another, most notably the manner in which later compilists such as Pliny handled, or mishandled, the earlier foundational material provided for them by the Aristotelian corpus.

As regards the scope of this work, its primary concern is with the data to be found in literature. It does not, therefore, seek to treat comprehensively the evidence which is available from coins, inscribed gems, inscriptions, and other non-literary sources. Nonetheless, a good deal of such evidence has been considered where relevant, in particular where it has been employed by previous authors in support of identifications. It may be added that, so far as the particular aims of this study are concerned, non-literary sources do not for the most part appreciably supplement the picture that is available to us from literature.

AD 600 has been selected as a convenient chronological limit since it marks the *floruit* of Isidorus of Seville, a writer who more than any other in the field of natural science marks the transition in the Latin West from the classical to the mediæval world, summing up the surviving results of ancient scholarship, albeit in a sadly degenerate form, and passing them on to the encyclopaedists of later times. The same terminus has been adhered to for Greek literature, although here, with a tradition of natural history writing which continues throughout the life of the Byzantine state, there is really no such convenient stopping place. Byzantine compilations such as the *Geoponica*, which are essentially no more than compendia of material from classical sources, have been treated in full, but strictly post-classical works such as Manuel Philes' didactic poem on animals and Demetrius Pepagomenos' account of falconry, which have sometimes been included as witnesses to classical zoology, have been excluded. The designation 'classical literature' has been taken to include the Septuagint translation of the Old Testament, which has necessitated some consideration of the Hebrew original.

Within these chronological limits, all varieties of insect have been included, with the single exception of the domestic honey bee, which would have occupied a hugely disproportionate amount of space in relation to the rest. Classical apiculture is an extensive subject which belongs more to the realm of agriculture than to that of natural history, and which has in any case been well treated by earlier authorities (cf. Davies and Kathirithamby, *Greek Insects*, pp. 47–72). Of terrestrial invertebrates other than insects, internal parasites of man and animals have been excluded, as have snails, the latter being dealt with as shellfish in D'Arcy Thompson's *Glossary of Greek Fishes*. The animals treated have been arranged, so far as possible, in modern taxonomic order, each section dealing with a particular 'type' or group of types distinguished in antiquity. Named insects which are not classifiable in this way have been placed together at the end.

In cases where an insect name is only rarely referred to in literature, all examples of its appearance have been considered. In the case of those which occur more frequently, an effort has been made to include all references which contain significant statements of fact, or which illustrate popular attitudes, as opposed to those which simply name the insect in passing. Where there are frequently recurring motifs attached to certain very well known insects, for example cicadas and ants, it will be found that examples from the major poets and dramatists have been comprehensively listed along with a representative selection of parallel references from other types of literature.

I

WORMS, LEECHES, CENTIPEDES, WOODLICE, etc

Annelida: Lumbricidae (Earthworms)

1. *Ges enteron/Lumbricus*

1. Identification and nomenclature

Γῆς ἔντερον and *lumbricus* are the chief classical names for the terrestrial annelids which we know as earthworms.[1] The former term—literally 'earth's guts'—is vividly descriptive of the creatures' general appearance. They are normally spoken of collectively, and examples of the use of the singular are rare (for example, Aelian IX.3). The order of the two elements in the name tends to be fixed, though there are a few cases of the form *enteron ges* (for example, Numenius fr. 1) and one with the article between (Athenaeus 305a). The eventual evolution of the name into a compound is attested by the Latinised form *gesentera—us* (Cassius Felix p. 44; Dioscorides Latinus; *C.Gl.L.* III.495.61) and by survivals in some modern south European dialects.[2] Fernandez correctly points out that the distinction of meaning supposed in LSJ, between 'wormcasts' in Theophrastus and one or two other places and 'earthworms' elsewhere, is not supported by the evidence; in all cases it is the creatures themselves which are being referred to. A Latin translation *terrae intestinum* is recorded (Chiron 360).

In Greek usage the more general term σκώληξ, which covers also insect and invertebrate larvae, is often found with the meaning of 'earthworm' from the *Iliad* (XIII.654) onwards (for example, Arrian *Epict. Diss.* IV.11.31; Clement Alex. *Protr.* p. 75; Origen *c. Celsum* IV.23; *C.Gl.L.* II.434.37). There is a corresponding Latin usage with *vermis* unqualified. Also the term ἕλμινς normally used only for parasitic worms[3] is employed in this sense in the *Geoponica* (XX.32,24).

The Latin equivalent *lumbricus* was used both in popular parlance and among technical writers. By the latter it is employed interchangeably with *vermis terrenis* (or *vermiculus terrestris*) and often appears with the addition of distinguishing adjectives or nouns (i.e., *lumbricus terrestris* or *terrenis* in Marcellus and Q. Serenus, and *lumbricus terrae* in Q. Serenus only). Pliny prefers *vermis terrenis* and uses *lumbricus* only once (XI.140), while the later medical sources show no such consistency. The frequent addition of *terrestris*, etc., to

---

[1] Cf. Keller, *Ant. Tierwelt*, II, pp. 501–2.   [2] Fernandez, *Nombres*, pp. 44–5.
[3] Ibid., pp. 144–5.

the latter is due to the fact that *lumbricus* was also used to refer to various internally parasitic worms attacking man and domestic animals.

In addition to its usual and more specific meaning of 'earthworm', *ges enteron* seems also to have had a somewhat wider application, being used to refer to similar creatures found in marine and freshwater habitats. Aristotle (*HA* 570a 14 ff.) speaks of such animals being generated on the shores of rivers and marshes as well as by the sea; while Numenius (fr. 1 *ap.* Athenaeus 305a) refers under this name to the collection of worms for bait 'along the tops of hillocks by the shore'. In view of their stated habitat these could either be marine worms exposed at low tide or earthworms living in close proximity to the shore; most probably they are the latter. These worms are further described as ἴουλοι. . .μέλανες γαιηφάγοι ἔντερα γαίης. *Ioulos* generally means a woodlouse or millipede, so its appearance here is somewhat anomalous. Fernandez[4] suggests that μέλας is in this case being employed as a noun, but it seems unnecessary to suppose this.

Other rare synonyms of *ges enteron* are as follows:

(i) Γαφάγας. Hesychius defines this word as a Syracusan synonym of *ges enteron*. It refers to a popular belief that earthworms fed upon earth, and may be compared with the adjective γαιηφάγος applied to the creature by Numenius.[5]

(ii) Δρῖλος. Used only among surviving literature in *AP* XI.197, this term is discussed by Fernandez,[6] who relates it to the second element of κροκόδιλος.

(iii) Χαμαισκώληξ. Found only in Herodian *Tech.* I.46.

(iv) Ἀλακάται. Meaning elsewhere a distaff or object so shaped, this is given as a synonym of *ges entera* by Hesychius.

(v) Ἔμβρυλλαι. Defined as a synonym of *ges entera* by Hesychius.

2. Life history

Earthworms were believed, quite understandably, to take shape spontaneously in mud or damp soil (Aristotle *HA* 570a 16 ff.), particularly after heavy rain (Nicander *Ther.*388; Sextus Empiricus *Pyrrh.* I.41) (cf. Arnobius *adv.Nat.* II.7; Augustine MPL 32.1365). Lucretius, discussing the possibility of spontaneous generation, speaks on three occasions (II.871–2, 897–9, 928–9) of what he describes vaguely as *vermes* being brought forth by the putrefaction of soil and organic litter under the influence of rain.

In addition, the opinion was held by some that the earthworms so produced subsequently developed into eels. This was accepted even by Aristotle (*HA* 570a, 14 ff., *GA* 762b 27 ff.; cf. Plutarch *Mor.* 637 f.), who insists that the fish in question neither mate nor lay eggs but that they develop from a larval stage—he uses σκώληξ, his standard term for an insect larva—constituted by the spontaneously generated earthworms and their marine (seashore) and freshwater relatives. He adds, somewhat cryptically (in *HA*), that eels 'have been observed working themselves free out of these creatures' and also 'come to view when these are pulled apart or cut open' (ἐν διακνιζομένοις καὶ

---

[4] Ibid., pp. 100–1.   [5] Ibid., pp. 105–6.   [6] Ibid., p. 226.

διαιρουμένοις γίγνονται φανεραί, cf. GA ἐν οἷς ἐγγίνεται τὸ σῶμα τὸ τῶν ἐγχέλεων).
Peck[7] suggests that the *ges entera* spoken of here are not earthworms but rather the horse-hair worm (*Gordius*) which in modern folk belief has been credited with being the origin of eels.

D'Arcy Thompson,[8] on the other hand, considers that what were known as *ges entera* were the young eels known to us as elvers returning from their marine breeding grounds to fresh water: these often take short journeys overland in the course of their migration.

It may be added that classical sources were not unanimous in their views on the eel's life history, and that versions occur which do not involve earthworms (for example, Oppian *Hal.* I.513–21; Pliny IX.160; Athenaeus 298c).

3. Habits

Earthworms, as has been mentioned, were held to feed upon soil, the substance from which they were originally produced (Numenius fr. 1; Plautus *Cas.* 126–7) and are characterised as delighting in muddy and squalid localities (Arrian loc. cit.; Origen *c.Celsum* IV.23 ff.; Clement Alex. *Protr.* p. 75). Their emergence from their subterranean burrows is mentioned by Plautus (*Aul.* 628). Their mode of locomotion by movement of the whole body is described by Aristotle (*PrA* 705b 28, 709a 30).

4. Weather lore

According to the meteorological treatises of Aratus (*Phaen.* 958–9) and Theophrastus (*de Sign.* 42), followed by Pliny (XVIII.364), an unusual number of earthworms wandering about on the surface constitutes a sign of rain and storm. The scholiast on Aratus explains this behaviour as being due to changes in temperature below ground.

5. Non-medical uses

The use of earthworms as a fisherman's bait is mentioned in a few other places besides Numenius fr. 1 (*Geoponica* XX.32,24; Ausonius *Epist.* XIV.57). We also hear of them being employed as a lure in bird traps (Plautus *Bacch.* 792).

6. Medicinal uses

Earthworms figured quite extensively in classical pharmacology, and we find them prescribed for a remarkably wide variety of unrelated conditions. We also find a variety of ways in which the creatures are to be used. They may be crushed and applied externally in that condition (for example, Marcellus XXXV.29, XXXVI.61, II.4), boiled up with or applied with oil, vinegar, honey, etc. (for example, Pliny XXX.23,80,106,110), burnt and their ash applied with honey or oil (for example, Marcellus XXXIV.16, XII.31, VII.8; Pliny XXX.76,115,119,134), or taken internally with wine or other liquids (for example, Galen *de Simp.Med.* XI.1.42; Pliny XXX.66,93,125). Pliny (XXIX.92) recommends that they be stored in honey for medicinal use.

[7] *Gen.An.*, p. 565.   [8] D'Arcy W. Thompson, *A Glossary of Greek Fishes*, Oxford, 1947, p. 59.

Used in these various ways, they are recommended for the treatment of damaged sinews (Dioscorides *DMM* II.67, *Eup.* 1.162; Galen *de Simp.Med* XI.1.42; Cyranides p. 105; Marcellus XXXV.22; Q.Serenus 968; Pliny XXX.115; Paulus Aegineta VII.3), pains in the sinews (Pliny XXX.110,125; Marcellus XXXV.29) and the repair of broken bones (Pliny XXX.119), for ear conditions (Dioscorides *DMM* loc. cit.; Cyranides p. 105; Cassius Felix p. 44; Marcellus IX.2,109,8; Q.Serenus 48; Pliny XXIX.135, XXX.23; Dioscorides *Eup.* 1.57) and toothache (Dioscorides *DMM* loc. cit.; Cyranides p. 105; Marcellus XII.31; Q.Serenus 243), for wounds (Q.Serenus 1087; Pliny XXX.115), bleeding at the nose (Cassius Felix p. 61) and various kinds of ulcers (Cyranides p. 106; Marcellus XXXIV.16; Pliny XXX.39), as well as for the diagnosis of *strumae* (Marcellus XV.55), for the removal of corns (Marcellus XXXIV.106; Pliny XXX.80), for gout (Cyranides p. 106; Marcellus XXXVI.58,61; Pliny XXX.76–7), jaundice (Pliny XXX.93) and *ignis sacer* (Cyranides p. 105; Q.Serenus 759; Pliny XXX.106), for tertian fevers and δυσουρία (Cyranides p. 288; Dioscorides *DMM* loc. cit.; Galen loc. cit.), for complaints of the head (Marcellus II.4), for asthma (Pliny XXX.66), for breast conditions (Cyranides p. 105; Q.Serenus 352; Pliny XXX.125) and to stimulate the flow of milk (Pliny XXX.125), for the bringing away of the afterbirth (Pliny XXX.125), and as an antidote to the stings of scorpions (Cyranides p. 105; Pliny XXIX.91). Pliny describes a Magian remedy (XXX.54; cf. Marcellus XXV.45) for sciatica, involving the placing of an earthworm upon a split wooden dish mended with iron, the pouring of water over the dish, the reburial of the worm in the place from which it was originally obtained, and finally the drinking of water from the dish.

Earthworms were also employed in veterinary medicine (Chiron 360).

Annelida: Hirudinea (Leeches)

2. *Bdella/Hirudo*

1. *Identification: general*

Identification of the creatures known to the ancients by the above names is straightforward. We are dealing with the blood-sucking annelids known to us as leeches, which although being feared as potentially harmful to man and his domestic animals were also recognised as being of considerable usefulness to doctors as a means for blood letting.[9] There are many species of leech inhabiting freshwater in Europe, but only two are of medical importance, and it is these which brought themselves to the attention of classical authors. In the first place we have the medicinal leech (*Hirudo medicinalis* Linn.) which was employed until relatively recent times for the purpose of blood letting and often cultured for that purpose. It is not entirely impossible that the ancients may have used other species as well, but we may safely assume that this was the one chiefly employed. The medicinal leech is not of any great importance as a parasite, but may occasionally affect animals.[10]

[9] Cf. Keller, op. cit, II. pp. 502–3; Fernandez, op. cit, p. 109.   [10] Soulsby, *Helminths*, p. 340.

In the second place we have *Limnatis nilotica* (Savigny), which is the species of primary importance as a potential parasite affecting man and a wide range of domestic animals. It inhabits freshwater pools containing plants in Europe and North Africa. The young leeches occur near the surface and are attracted by footsteps at the water's edge. Thus they are easily swallowed by animals or human beings coming to drink, whereupon they attach themselves in the throat and may remain there sucking blood for several days, causing severe discomfort in the process, and in the case of animals even death.[11]

1.2. Identification: nomenclature and descriptions

Among Latin medical and veterinary authors the term *sanguisuga* is often used as a synonym of *hirudo*. Pliny uses the former more often than the latter and at one point explicitly comments that the word was entering the language in his time (VIII.29, '*quam vulgo sanguisugam coepisse appellari adverto*'). In later Latin a variety of curious aberrant spellings of *hirudo* begin to appear (Viz. *herudo* in Marcellus XIV.25; *erudo* in Marcellus VII.11; *irudo* in C.Gl.L. V.25.30; *herugo* in C.Gl.L. IV.86.10; *erugo* in C.Gl.L. V.628.40). The appearance of *tinea* in the latter gloss is presumably due to confusion with the intestinal worms known by this name in late medical writings.

Physical descriptions of invertebrate animals are rare in classical sources, the author generally assuming that the reader will be able to visualise the creature he is considering. However, we do find references to the colour of leeches as κυανόχροα ἑρπετά (Oppian *Hal.* II.597) and to the fineness of their outer skin (Apuleius *Met.* VI.26). Pliny (XXXII.123) refers to a reddish form of leech known to doctors as particularly liable to leave its head in the wound if proper care is not taken in its removal from the patient. This would presumably be a distinctive colour variety of the medicinal leech or perhaps of *Limnatis*.

Hesychius lists four synonyms of *bdella*:

(i) *Βλέτυες* (pl.), which according to Fernandez[12] contains in its etymology the idea of voracity, a quality which leeches were regarded as typifying.

(ii) *Δρίλαξ*, said to be an Elean word and regarded by Fernandez[13] as allied to δρῖλος ('earthworm').

(iii) *Δεμβλεῖs* (pl.). This is compared by LSJ with a leech name δεμελέας (acc. pl.) attested only in an inscription (*IG* 4.951.98).

(iv) *Βαῖτυξ*. Found both in Hesychius (Latte reads *Βλίτυξ* here) and in Bekker *Anec. Gr.* I.199.

2. Life history

No information on the life history of leeches is given by any surviving classical author. The leech is one of the few invertebrate types not considered in detail by Aristotle. Isidorus (*Or.* XII.5.3) places it in his section *de Vermibus*, thus implying that he regarded it

---

[11] Ibid., p. 340–2.   [12] Op. cit., p. 105.   [13] Ibid., p. 226.

as spontaneously generated, but he does not explicitly state this. Pliny simply mentions in passing at one point (IX.162) that it is one of those creatures which are born (*gignuntur*, a word which implies spontaneous generation) at a fixed time of year, in this case in the spring.

3. Habitat

Leeches are often characterised by references to their distinctive freshwater habitats. It was observed that they were most commonly found in marshes (λιμνᾶτις βδέλλα, Theocritus II.56; Oppian *Hal.* II.597: Pliny XI.116) or stagnant pools (Q.Serenus 670), as well as in streams (Q.Serenus 407) containing luxurious growths of water plants (Nicander *Alex.* 496–7).

4. Habits

The feeding habits of leeches will be considered below (8.2) in the context of their use in medicine. Their mode of locomotion, by 'looping', is described by Aristotle (*PrA* 709a 30 ff.).

5. Popular attitudes

The attitude of the ancients toward leeches was inevitably somewhat ambivalent. They were regarded on the one hand as a threat to human health and to the health of domestic animals, and viewed with general distaste for their unpleasant feeding habits; but it was nevertheless recognised that they constituted an important item in the doctor's equipment and that when rightly used they were able to save lives. Because of their habit of sucking blood incessantly until they reached the limit of their capacity, swelling markedly in the process, leeches were seen as typifying greed and voracity (Cicero *ep. ad Att.* I.16.11; LXX *Prov.* 30.15 and John Chrysostom *ad loc.* MPG 64.734; Cratinus fr. 44 .). And because it was observed that they would not release their hold until fully engorged, they are also taken as typifying undesirable tenacity (Plautus *Epid.* 188; Horace *Ars* 475–6). Theocritus (II.56), with the idea of draining life-blood in view, compares love to the leech.

6. Leeches as human parasites and their treatment

The inadvertent swallowing of leeches by human beings was regarded by ancient medical writers as a not uncommon occurrence (Pliny XXVIII.160, XX.143, XXIII.55; Galen VIII.265, XIV.143, 440, 538 K; Dioscorides *Eup.* 138; Celsus V.27.12; Ps. Eustathius *Hex.* MPG 18.748; Scribonius Largus 199; Isidorus *Or.* XII.5.3). This could occur either by drinking water directly from pools or rivers, or by the creatures being accidentally transported into the home in water vessels (Nicander *Alex.* 495 ff.). Once swallowed, they would attach themselves to the interior surface of the throat, or alternatively they might find their way as far as the stomach and affix themselves there (Dioscorides loc cit.; Nicander loc. cit.; Marcellus XVI.tit.), And there, unless measures were taken to evict them, they would remain until fully fed, causing pain, obstruction of breathing and the bringing up of blood (Nicander loc. cit.; Galen VIII.265 K).

The aim, therefore, of a doctor treating those unfortunate enough to be suffering in this way was somehow to induce the creature to release its hold so that it could safely be ejected; and there were a variety of means at his disposal to achieve this. In the first place, substances calculated to cause detachment of the leeches could be introduced into the throat, those mentioned being herbal preparations—especially involving rue—(Pliny XX.143; Galen XIV.440 K; Dioscorides loc. cit.), vinegar (Pliny XXIII.55; Galen XIV.143, 538 K; Celsus loc. cit.; Marcellus XVI.96; Nicander loc. cit.; Dioscorides loc. cit., DMM V.13), brine (Galen XIV.143, 538 K; Nicander loc. cit.; Dioscorides Eup. loc. cit), ice or snow (Galen XIV.143 K; Nicander loc. cit.), nitre (Dioscorides loc. cit.), earth boiled up (Nicander loc. cit.), butter (Pliny XXVIII.160; Galen XIV.538 K; Marcellus XVI.96), and the sauce known as garos (Galen XIV.538 K). Or alternatively burning bedbugs could be used as a fumigant (Pliny XXIX.62; Galen XII.363, XIV.538 K; Marcellus XVI.95; Dioscorides DMM II.34; Geoponica XIII.14.6). A soft sponge soaked in cold water could be inserted into the throat so that the leech would transfer to it and thus be removed. Or as a fourth possibility the patient could be immersed up to the chin in warm water and in that position take cold water into the mouth: the leech was then said to move toward the area of lower temperature (Galen XIV.143, 538 K: Dioscorides Eup. loc. cit.).

The use of the bills of woodpeckers or owls as amulets supposed to protect the wearer against leeches is referred to by Pliny (XXIX.92).

On farms it was regarded as an advantageous feature if the bodies of water there were well stocked with creatures known to prey upon leeches, especially freshwater crabs and eels (Geoponica II.5.6,15).

7. Leeches as parasites of domestic animals

The taking up of leeches by domestic animals, especially cattle, horses and sheep, in their drinking water was considered by farmers as an important threat to their livestock, and methods of dealing with this occurrence are considered by most agricultural and all veterinary sources (Pliny XXIX.62; Columella *RR* VI.18; Vegetius *Vet.*IV.24; Pelagonius *Vet.* 0.3; Gargilius *cur.bov.* 11; Geoponica XIII.17, XVI.19, XVIII.17; *Hippiatrica Ber.* 88.1–5, *Par.* 529). In some cases the creatures might be content to attach themselves to the inside of the mouth or the under surface of the tongue (*Hipp.Ber.* 88), and in such situations they did not constitute a serious problem, since they could easily be manually removed or detached by means of a leaf or a twig. Or they could be persuaded to detach themselves by the application of crushed bedbugs (*Hipp.Ber.* 88.4; Gargilius 11). However, if the leeches affixed themselves inside the throat, or even found their way into the nasal passages (*Hipp.Ber.* 88.5) or down into the stomach (Columella loc. cit.; *Hipp.Ber.* 88.3), the consequences could well be fatal to the animal and more serious measures needed to be taken. The chief method of detaching and ejecting the parasites was to introduce warm oil into the animal's throat by means of a pipe, reed or horn

(Columella loc cit.; *Geoponica* XVI.19, XVIII.17.5; *Hipp.Ber.* 88.1–3, 5, *Par.* 529). Vinegar (Columella loc. cit.; *Geoponica* XVIII.17.5; Vegetius loc. cit.; *Hipp.Ber.* 88.2), salt (*Hipp.Ber.* 88.3,5) and a preparation of the herb birthwort are mentioned as being employed in the same way. According to Columella, hot vinegar is effective in killing leeches that have found their way to the stomach. Alternatively, it is recommended that the smoke from burning bedbugs be applied through a pipe (Columella loc. cit.; Vegetius loc. cit.; Pliny XXIX.62; *Geoponica* XIII.17, XVI.19; *Hipp.Ber.* 88.4). It is also mentioned (*Hipp.Ber.* 88.5), that after the application of warm oil or other substances the leeches may be removed with surgical forceps or cauterised with a heated iron implement.

8.1. Direct medicinal uses

Leeches allowed to rot in vinegar or their ashes applied with vinegar were believed to be effective both for the removal of unwanted hair (Cyranides p. 104; Pliny XXXII.76,136; Marcellus VIII.184; Q.Serenus 670–3; Theodorus Priscianus *Eup.* I.41) and for the colouring of hair (Pliny XXXII.67–8; Marcellus VII.11). Serenus (407) lists leeches among cures for complaints of the spleen; and Marcellus (XIV.25) states that a living specimen enclosed in a nutshell and hung round the neck as an amulet is effective against conditions of the palate.

The smoke from burning leeches is said to act as a deterrent fumigant against bedbugs (Pliny XXXII.124;136; *Geoponica* XIII.14.6).

8.2. The use of leeches in blood letting

A stock of leeches for blood letting was evidently a normal element in the ancient physician's customary equipment. Their employment was not without its risks, but it was considered that the experienced medic could effectively circumvent these. To use leeches properly was clearly something of an art, and careful instructions on the subject were included in medical treatises. The most detailed account is given by Antyllus (*ap.* Oribasius VII.21), the same information occurring in an abbreviated form in the first chapter of Galen's treatise *de Hirudinibus etc.* (XI.317–9 K; cf. also Soranus *Gyn.* II.11 and Cassius Felix pp. 10–11). According to this source, it is advisable, rather than continually collecting new specimens, to keep a supply of leeches in captivity, since those that have been used on several occasions become accustomed to what is required of them and readily attach themselves when placed on the patient, where newly captured ones may be reluctant to do so. If freshly caught specimens are to be used, these should be kept for a day in order to allow time for any toxic element within them to be dissipated. The part of the body to be bled should first be cleaned with nitre and then prepared by being smeared with animal's blood or clay or fomented or scratched with the nails; this is to persuade the creature more readily to affix itself. Once this has been done, the leeches—the account assumes throughout that more than one will be employed—may be taken with a sponge from the wide vessel of warm water in which they have been kept at the ready and applied to the spot. After they have settled themselves in position, the affected part should be immersed in warm water or at least have warm water poured over it. The leeches will not naturally detach themselves until they are full, so that if before this stage it is felt that sufficient blood has been taken they must be artificially induced to do so by sprinkling their mouthparts with salt, nitre or ash. If it should be necessary due to shortage of specimens to employ leeches that are already fully engorged, they may be made to continue feeding by clipping off their tails with scissors; blood will then pour out, but the creatures will nevertheless continue feeding, being now unable to reach satiety. When the leeches have finally been removed, a cupping glass should be employed if possible to draw off any toxic matter which the animals may have left behind.

It now remains to be seen to what extent the fragmentary details on blood letting found in other authors correspond with Antyllus' description. Pliny (XXXII.123) and Oppian (*Hal.* II.597 ff.) describe how after application the leech becomes swollen with blood and how when full it will drop off of its own accord (cf. also Horace *Ars* 475–6). The former author mentions the sprinkling of the mouthparts with salt to induce premature detachment, adding that since the leech imbeds its mouthparts in the patient's flesh (cf. also XI.116) it is of great importance to ensure that no portion of them is left behind through mishandling. If this should occur, he continues, the place will suppurate and consequences may be fatal; he cites an actual case where this occurred. Pliny also mentions the process of snipping off the ends of leeches with scissors, but seems to have become confused about its purpose; according to his account those so treated gradually deflate and die and thus can be safely removed without leaving their mouthparts behind in the wound.

With regard to the particular medical conditions for which treatment with leeches was considered appropriate, Antyllus states in general terms that when deciding if leeches should be employed the doctor must bear in mind that they do not draw blood from deep down but rather from that which lies close to the surface. Accordingly, they are especially suitable for infected wounds as well as for any parts of the body where a cupping glass cannot conveniently be used. Opian (loc. cit.) also specifically mentions their use for the treatment of wounds. Pliny (XXXII.123) and Q.Serenus (784) both recommend them for use in cases of gout, while Nicander (Ther.930) suggests that they be applied to the bites of venomous animals. The Cyranides (p. 104) describes them for dropsy and spleen and eye conditions, Soranus (*Gyn.* II.11,23) for gynaecological conditions, and Cassius Felix (pp. 3, 34, 162, 169) for head pains, epilepsy, abscesses and pleurisy; while Caelius Aurelianus (*Ac.* I.76, II.30, III.21, 183, *Chron.* I.13,91,160, II.22,67, 174, III.25,53,127, IV.97, V.11,36,75) recommends their use, often as part of a programme of treatment, for a wide variety of ailments including internal disorders, headache and insanity.

In all these references it may be assumed that it is the medicinal leech which is primarily in view, although none of our sources explicitly distinguishes it from harmful forms like *Limnatis*. None, that is, unless Pliny's reference to a 'reddish' variety (XXXII.123) particularly liable to leave its mouthparts behind is to be seen as separating *Limnatis* from its more medicinally reliable relative.

9. Leeches in veterinary medicine

The use of leeches in the blood letting of domestic animals is mentioned only in the *Hippiatrica* (*Cant.* 71.18), where they are recommended for treating bites caused by venomous creatures such as scorpions.

10. Exotic species

(i) Africa and India. According to Pliny (VIII.29), leeches are often swallowed with drinking water by elephants, whereupon they attach themselves within the animals' breathing passage, causing acute discomfort.

(ii) Egypt. Herodotus (II.68) and several later authors (Aelian III.11, XII.15; Apuleius *Ap*.8) describe how the jaws of crocodiles become infested with leeches and how the animals allow the Egyptian plover to enter and feed upon them. Other writers state that it is scraps of food which the birds remove (Plutarch *Mor.* 980d–e; Ps. Aristotle *HA* 612a 29 ff., *De Mir.Ausc.* 331a 11 ff.).

(iii) Libya. The monstrous leeches reported by Strabo (XVII.3.4) from a river in this region are probably, since they are credited with gills, to be regarded as freshwater lampreys described in highly exaggerated terms.[14]

Arthropoda: Chilopoda (Centipedes)

3. *Skolopendra/Centipes*

1. Identification: general

The σκολόπενδρα was known in antiquity primarily as a venomous animal, in which context it figures in all the major medical sources. Its identification is not difficult since the descriptions we have of it point clearly to the various species of centipedes (Chilopoda). The species primarily in view would not have been the ubiquitous Lithobiomorpha, such as our own *Lithobius forficatus* (Linn.),[15] but the more impressive Scolopendromorpha containing a number of brightly coloured southern European species. *Scolopendra morsitans* Linn., for example, the variety named by Aubert and Wimmer,[16] ranges from yellow to green, with darker markings. *S. cingulata* Latr., *S. subspinipes* Leach, and *S. valida* Lucas have also been mentioned.[17] Though not actually poisonous, the bites of these Scolopendromorphs can produce painful swelling of the affected area, so that the fear they evoked in classical times is quite understandable.[18]

---

[14] D'Arcy W. Thompson, op. cit., p. 29.

[15] So identified by Sundevall, *Thierarten*, p. 235.

[16] Aubert and Wimmer, *Thierkunde*, I, p. 165; cf. Keller, op. cit., II, p. 481.

[17] J. Scarborough, *Nicander's Toxicology, Pharmacy in History*, XXI, 1979, p. 19.

[18] Smith, *Insects*, p. 475; J. L. Cloudsley-Thompson, *Spiders, Scorpions, Centipedes and Mites*, London, 1958, pp. 43, 52–3.

1.2. Identification: physical descriptions

The centipede appears in Aristotle as an elongated many-legged creature (*HA* 489b 22, *Pr.A.* 708b 5 ff.) similar to the *ioulos* (*HA* 523b 18) which continues to live if cut up (*Pr.A.* 707a 30). Galen (III.177 K) describes it similarly. According to Pliny (XX.12) it is longish (*oblongam*), with hairy legs, and distinguishable from the *multipeda* or *milipeda* by the fact that it is smaller (a curious error) and more venomous, and that it does not arch its back as it moves (XXIX.137). Nicander (*Ther.* 812–14) speaks of its fast-moving limbs as resembling the oars of a ship, and also credits the creature with the possession of two heads; the scholiast on the passage notes the erroneous nature of this last statement, but explains it by saying, on the claimed authority of Aristotle (perhaps a misreading of *HA* 532a 1 ff.), that the centipede is capable of running backwards (which is equally erroneous).

However, the most detailed depiction of the *skolopendra* is that provided by Aelius Promotus (in *Rh.M.* XXVIII, p. 274), which renders certain the identification with the brightly coloured Scolopendromorphs referred to above. Aelius states that the creature is many-legged, broad and slender, with a hairy mouth—that is, one with many appendages. In colour it is marked with yellow on a dark background (τὴν χροιὰν μηλίναις κατάστικτον ἐκ μέλανος ἔχον στιγμαῖς), though it may sometimes be uniformly dark. It is to be found in damp places.

1.3. Identification: land and sea centipedes

In terms of ancient zoological nomenclature, the creature we are discussing is strictly speaking the 'land centipede' (σκολόπενδρα χερσαία, *scolopendra terrestris*), so called to distinguish it from the marine variety (σκολόπενδρα θαλάττια; *scolopendra marina*) that was believed to be essentially identical. Ancient authors normally specify which of the two they are dealing with. It has been generally recognised that the so-called 'sea centipede' is to be identified with various species of Polychaete worms, notably the ragworms or Nereidae, which resemble centipedes in their general shape and in many cases in their colouration also.[19] Aristotle (*HA* 505b 13 ff.) characterises them as being somewhat smaller than terrestrial centipedes, redder in colour, and with more numerous and slender limbs. The ἑρπῆλαι δολιχήποδες mentioned by Numenius (fr. 1 *ap.* Athenaeus 305a) as a fish bait are probably to be classed, as Gossen[20] suggests, as Polychaetes (cf. fr. 2 *ap.* Athenaeus 306c). But in view of the fact that he describes them as to be found 'where the sandy cliffs are washed at the topmost break of the surf'—in other words, not necessarily in a marine habitat, it is not impossible that terrestrial centipedes, some species of which inhabit the sea shore, are in view here.

1.4. Identification: nomenclature

(i) *Centipes*. Pliny gives the centipede its Latin name on only one occasion (IX.145), preferring elsewhere to use the Greek equivalent spelt as *scolopendra*. In the *Corpus*

---

[19] Cf. Keller, op. cit., II, pp. 481–2. [20] 'Zoologisches bei Athenaios', *QGN*, VII, 1940, p. 377.

Glossarum (II.99.32, 433.50, 546.64, 572.43, V.276.3) *centipes* (with variant spellings *centipedium* and *contifex*) is the standard rendering for the Greek σκολόπενδρα. Two of the above glosses define the word as a *genus serpentis*; this is paralleled by Isidorus' inclusion of what he calls *centupeda* in his snake section (*Or.* XII.4.33). The form *centipeda* is found in Arnobius *adv.Nat.* II.52.

(ii) *Seps.* In Greek authorities σήψ is always the name of a venomous snake, but Pliny uses it on two occasions for the centipede, claiming this to be a Greek usage (XXIX.137, XX.12).

(iii) 'Οφιοκτόνη. The curious name 'snake-killer' is found only in Pseudo-Dioscorides (*Peri Iobolon* 5), where it appears as a synonym of *skolopendra*. Its origin presumably lies, as Fernandez[21] suggests, in some unrecorded popular belief to the effect that centipedes preyed upon snakes.[22]

(iv) 'Αμφισδεσφάγανον. According to Hesychius, a synonym of *skolopendra*. The word is corrupt, but appears to have something to do with the belief found in Nicander that the centipede was two-headed. Fernandez[23] suggests the reading 'Αμφιφάσγανον.

2. Centipedes as venomous

As with other venomous and reputedly venomous invertebrates, centipedes were believed to produce severe symptoms by their bites and even to be capable of causing fatalities (Nicander *Ther.* 812–4). Descriptions of their effects are given by a number of authors (Philumenus *de Ven.An.* 32.3; Ps.-Dioscorides *Iob.* 5; Paulus Aegineta V.9). Prescribed treatments include the use of various herbal remedies (Dioscorides *DMM* V.109, *Eup.* II.128; Pliny XX.12,145,157,162,245, XXII.64,68, XXIII.55), the use of pitch or honey (Dioscorides loc. cit.) and urine (Pliny XXVIII.67), and the immersion of the affected place in warm salt water followed by application of salt and vinegar (Philumenus loc. cit.; Pliny XXIII.55, XXXI.99). Plutarch's account of a centipede being emitted from the human body sounds like an example of what is known today as pseudo-parasitism (*Mor.* 733c).

It is recorded by Theophrastus (*ap.* Pliny VIII.104) and Aelian (XI.28, XV.26) that the people of Rhoeteum on the Troad were compelled to evacuate their town on account of an invasion of centipedes. Stories about populations being driven out by animals of various kinds are common in ancient authors, though it may be noted that millipedes accompanied sometimes by centipedes and woodlice have been known to engage in large scale migrations.[24] In this connection mention may also be made of the report (Aelian XI.19) that just prior to the destruction of the town of Helike by earthquake all of its centipedes, snakes, mice and *sphondylai* were observed making their escape.

3. Centipedes as harmful to domestic animals

Centipedes are also recorded as constituting a danger to cattle (Pliny XX.12; *Hippiatrica*

[21] Op. cit., p. 139.   [22] Cf. the locust name ὀφιομάχης (11,1.3,xxiii).
[23] Op. cit., p. 87.   [24] J. L. Cloudsley-Thompson, op. cit., p. 23.

Can. 71.12). According to Pliny, their bites cause swelling and suppuration. The *Hippiatrica* prescribes that the animal should be given asphodel and wine as an antidote, or alternatively that an application of figs, wine and vetch should be made to the affected place.

4. Uses to man
Pliny (XXIX.64) records fumigation with burning centipedes as a possible method of combating infestations of the bed-bug or *cimex*.

Arthropoda: Diplopoda and Isopoda (millipedes and woodlice)

4. Ioulos, Oniskos/Multipeda

1. Identification: general
The creatures most commonly known to the Greeks as ἴουλοι, ὄνοι, ὀνίσκοι, and to the Romans as *multipedae* or *milipedae*, were closely associated with the centipede, together with which they comprised a distinctive group of invertebrates characterised by the possession of an unusual number of legs. In terms of modern zoology they fall into two taxonomic groups; firstly, the elongated millipedes (Diplopoda) and secondly the much shorter woodlice (Isopoda). Typical millipedes are thus very different from woodlice in their external appearance, but one form, represented by the common *Glomeris*, is remarkably similar to the Armadillidiidae, a well known and distinctive family of Isopods, both in shape and colour and in its habit of rolling into a ball when disturbed.

However, it is not at all easy to determine whether any of the considerable array of classical names—a phenomenon paralleled[25] in some modern languages—for this group of animals are in fact consistently used to refer either to millipedes or woodlice, or whether in normal usage all of them were more or less interchangeable. It has commonly been supposed[26] that in the original Greek usage, as typified by Aristotle, *ioulos* was the specific term for millipedes and *onos/oniskos* for woodlice plus *Glomeris*, and that it was authors such as Pliny who later confused the issue by synonymising all the Greek names and using the blanket term *multipeda/milipeda* for the entire group without discrimination. But in fact there is evidence for interchangeability of names in the Greek sources as well: the lexicographers tend to synonymise *onos/-iskos* and *ioulos*, and Theophrastus' and Aratus' weather prognostications involving *ioulos* and the epithet ἰουλόπεζος applied to ships both seem more applicable to woodlice than to millipedes.

1.2. Identification: the individual names

(i) Ἴουλος. The *ioulos* of Aristotle is nowhere given the full treatment devoted to most animals in the *Historia*, but is described in various brief mentions (*HA* 523b

[25] J. L. Cloudsley-Thompson, op. cit., p. 5.
[26] Aubert and Wimmer, op. cit., I pp. 165, 169; Keller, op. cit., II pp. 481–2; A Marx in PW II.1744–5; Leitner, *Zool. Terminologie*, pp. 168–9.

18, *PA* 682a5,682b 1 ff.) as an elongated creature with many legs, similar to the *skolopendra*, which continues to live even if cut up. This would appear to be the millipede. The term is defined by various lexicographers (Hesychius, *E.M.*, Suidas, Photius) and scholia (Apollonius I.972, Lycophron 23, Aratus 957) as referring to a ζῷον πολύπουν similar to the *skolopendra* or as being a synonym of *onos*. Dio Chrysostom (X.10) characterises it as a creature which, though having many legs, is the slowest of creeping things. In a Latinised form, *iulus*, it is synonymised by Pliny (XXIX.136) with *oniskos* and the Latin *milipeda/multipeda*.

Theophrastus (*de Sign.* 19) and Aratus (*Phaen.* 957–8) use *ioulos* for a creature whose activity is a sign of rain, and their observations are appropriate to the woodlouse. According to the scholiast on Apollonius (loc. cit.), Theophrastus (fr. 185 Wimmer) stated that *onos* was another name for the same animal. In literature ships with their projecting oars are sometimes compared with *iouloi* (Apollonius Rh. I.972: Lycophron 23) and here as well the analogy would seem to be with woodlice, which are more boat-shaped than millipedes. Nicander calls his venomous myriapod *ioulos*, which is rendered by Gow and Scholfield[27] as 'woodlouse'. Scarborough[28] prefers to identify Nicander's creature as a millipede, and indeed these might be thought more likely to be suspected of constituting a danger to man; although it should be noted that folk beliefs about animals are not notable for their rationality, and apparently woodlice have been considered poisonous in modern times.[29] The illustrations in the illuminated manuscripts of Nicander,[30] for what they are worth, support the view of Scarborough in depicting the *ioulos* as an elongated creature like the *skolopendra*. In the opinion of Fernandez[31] and others, the name *ioulos* was acquired because of the creatures' 'hairy' or 'fluffy' nature; that is, in reference to its numerous and close-packed limbs; but this seems clearly erroneous. The more correct explanation is that provided by Sundevall[32] and supported by Theophrastus *HP* III.5.5, to the effect that it is the botanical meaning of *ioulos* as 'catkin' which lies behind the zoological usage. A catkin is long, cylindrical in form, and apparently 'segmented', and thus bears some resemblance to a millipede or pill-woodlouse.

However, it is not only millipedes and pill-woodlice which are cylindrical and segmented, and so we find the word *ioulos* occasionally applied to other invertebrates as well. In Suidas we are given a subsidiary definition of the term as ὁ ἐν ταῖς ἀμπέλοις σκώληξ (= Photius).[33] In its normal meaning *ioulos* is sometimes described as a σκώληξ, though this term is more naturally applied to larval forms (Photius s.v. *Onos*; Sch. in Aratum 957; Sch. Nicander *Ther.* 805). Numenius[34] uses *ioulos* as another name for the earthworm.

(ii) Ὄνος. While *ioulos* seems to be used both for millipedes and for woodlice, this

---

[27] *Nicander*, pp. 186–7.   [28] 'Nicanders Toxicology', *Pharmacy in History*, XXI, 1979, p. 18.
[29] J. L. Cloudsley-Thompson, op. cit., p. 5.   [30] Kadar, *Zoological Illuminations*, p. 48.
[31] Op. cit., p. 39.   [32] Op. cit., p. 236.
[33] I.e. the ἴψ or *convolvolus* (cf. 25 below).   [34] Cf. 1.1 above.

second commonest Greek term appears in surviving sources to refer only to the latter, and more especially to those which roll themselves into a ball, the Armadillidiidae or pill-woodlice (plus the millipede *Glomeris*). It is used by Galen (XIII.111,113 K), the Cyranides (p. 271), Dioscorides (*DMM* II.35) and Paulus of Aegina (VII.3) to refer to the medicinal woodlouse. Galen does use the name *ioulos* once (III.177 K) but only in reproducing an Aristotelian observation. In all cases the appearance of the name is accompanied by an explanatory comment to the effect that the creature lives around water containers and rolls up, these observations constituting also the standard lexicographical definition of *onos* as given by Hesychius and Suidas. In lexica the word tends to be synonymised with *ioulos*. The name 'donkey' is evidently a folk appellation given to the creature in allusion to its general form and grey colour;[35] the applying of mammalian names to woodlice is paralleled in modern languages. A compound name ὄνος ἰσόσπριος—that is, the woodlouse which looks like a pea when rolled up—was, so we are told by the scholiast on Apollonius (I.972) employed in a simile—κυλισθεὶς ὥς τις ὄνος ἰσόσπριος—by Sophocles (fr. 363). Consequently explanatory definitions are given by Hesychius and Photius.[36]

(iii) Ὀνίσκος. A diminutive of *onos* used as a simple synonym by Galen on two occasions (XII.366,634 K). It appears as the Greek equivalent of the Latin *asellus* in the *Corpus Glossarum* (II.24.1,4, III.400.64, 439.72). In a Latinised form *oniscus* it is noted by Pliny (XXIX.136, XXX.53,68) as the standard Greek equivalent of *multipeda/milipeda*, and of *porcellio* by later medical writers (Cassius Felix p. 44; Theodorus Prisc. *Eup.* II.44; Caelius Aurelianus *Chron.* I.119,129).

(iv) Ὄνιννος. This equivalent of the above, explained by Strömberg[37] and Fernandez[38] as a compound of ὄνος and ἴννος but maybe just a variant diminutive of *onos*, is found only in Theophrastus (*HP* IV.6.8), where it is used not for a terrestrial woodlouse but for a marine creature similar in appearance (perhaps one of the seashore Isopods, e.g., *Ligia* spp))

(v) Κύαμος. According to Galen (XII.366 K), the pill-woodlouse *oniskos* was termed the 'bean' by some 'because they are dusky in colour and resemble edible beans when they roll into a ball'. The same analogy is seen in the name ὄνος ἰσόσπριος.

(vi) Πολύπους, Μυριόπους. The Greek equivalents of the Latin *multipeda/milipeda* are normally used only as adjectives, but in rare cases are found in an absolute sense. The first is so used by Marcellus (IX.33 *multipedes . . . quae polypodas Graeci appellant*) and appears in Hesychius defined as εἶδος φθειρῶν (presumably φθείρ is being used in a vague sense as in our English term 'wood*louse*'). Μυριόπους as a noun is found in the scholiast on Nicander's *Theriaca* 805 and in the *Hippiatrica* (O & H II.256).

---

[35] Fernandez, op. cit., pp. 49–50.   [36] Cf. ibid., pp. 40–1.
[37] *Gr. Wortstudien*, p. 11.   [38] Op. cit., p. 54.

(vii) *Κουβαρίς*. Another name for the *onos* found in the title to chapter 35 of Dioscorides *DMM* II. Regarded etymologically the name appears to contain an allusion to the creature's habit of rolling into a ball.[39]

(viii) *Κόβαρος*. A variant of the preceding found in Hesychius, where it is defined simply as *onos*.[40]

(ix) *Σηνίκη*. According to Hesychius, a word which refers, among other things, to a 'many-legged creature similar to the domestic *onoi*'.[41]

(x) *Multipeda*. This is Pliny's standard equivalent for the Greek *ἴουλος* and *ὀνίσκος*, regarded by him as synonymous, as he explicitly states in XXIX.136. His description in this same passage is somewhat confused; he calls the creature *animal . . . e vermibus terrae* and says that it is many-legged and hairy (which is either a misleading reference to the animal's legs or an inexplicable error), that it rolls up, and that it moves along '*arcuatim repens*'. Pliny uses the term both to refer to the medicinal woodlouse (XXIX.143, XXX.31,47,53,75,93,101,114) and to the venomous myriapod of Nicander (XX.12,41,257, XXII.122,130, XXIII.55, XXVIII.161). Later writers, Marcellus (IX.10.33, XXXII.43, XXV.21, XV.72) and Isidorus (*Or.* XII.5.6), use the form *multipes*. Augustine (MPL 32.1070) describes a *reptans bestiola multipes* which some of his students found lying on the ground *in opaco loco* and cut in half, only to be very much startled by the two halves running off rapidly in opposite directions.

(xi) *Milipeda*. Used by Pliny (XXIX.136, XXX.35,40,68,86) interchangeably with the above, though less frequently.

(xii) *Centipeda*. More correctly a name for the centipede, but given by Pliny (XXIX.136, cited above) as a synonym of *multipeda/milipeda*.

(xiii) *Cutio*. Found only in Marcellus, who uses it twice to refer to the medicinal pill-woodlouse characterised by its habit of rolling up (IX.33, XV.72). In the former of these passages he describes it as many-legged and as having a hard outer shell (from which it may derive its name).[42] In the third place (VIII.128), however, he uses the word with reference to a mysterious creature described as a 'hairy beetle' which is similar to the true beetle (*scarabaius*), yellowish (*pseudoflavo quasi leonino*) in colour and shining (*lucentes*), found in old stone boundary walls and ditches

(xiv) *Asellus*. Corresponding exactly to the Greek *ὀνίσκος*, this woodlouse name is found only in the *Corpus Glossarum* (see (iii) above) and in late medical writers such as Cassius Felix (p. 44).

(xv) *Porcellio*. Another name found only in late Latin, being the standard term for the

---

[39] Ibid., p. 36.   [40] Cf. ibid.

[41] The etymology is unclear and is discussed by Fernandez, op. cit., p. 34. The word *ἀκατίς*, found only in a scholion on Hippocrates, is claimed by Strömberg and Fernandez, op. cit., p. 34, as a woodlouse name on rather slender evidence.

[42] Ernout and Meillet, *Dict. Ét.*, p. 161.

medicinal pill-woodlouse in medical (Theodorus Priscianus *Eup.* II.44; Caelius *Chron.* I.119,129) and veterinary (Pelagonius 49) sources. It is found in the Latin translation of Dioscorides. The woodlouse evidently acquired the appellation of 'little pig' as it acquired the name *onos/oniskos*, by reason of its shape and colour. There are parallel names in modern languages, e.g., the English dialectical 'sow' or 'sow-bug'.[43]

2. Habits and habitat

The defensive reaction of the pill-woodlouse in rolling itself into a tight ball when touched was well known in antiquity and is universally mentioned as one of the two characteristic features of the *onos/oniskos* or *multipeda/milipeda* (Aristotle *PA* 682b 21 ff.— though unnamed here; Theophrastus *HP* IV.3.6; Sophocles fr. 363; Dioscorides *DMM* II.37; Galen XII.366,565,634 K; Paulus Aegineta VII.3; Hesychius, Photius and Suidas s.v. ὄνος and ἴουλος; Pliny XXIX.136; Marcellus IX.10.33, XV.72, XXXII.43, XXXV.21; Isidorus *Or.* XII.5.6). The fact that the majority of woodlice species do not have this habit may either have been ignored, or alternatively it is possible that it was only the distinctive pill-woodlice which were regarded as being of medicinal importance. Millipedes, it may be noted, have a somewhat similar mode of behaviour, but they coil themselves up like a snake and therefore would not fit the classical descriptions which use words like σφαιρούμενον and speak of the rolled up animal as resembling a pea or bean.

If phrases such as Dioscorides' σφαιρούμενοι κατὰ τὰς ἐπαφὰς τῶν χειρῶν are more or less universal wherever woodlice are mentioned, statements about the creatures' fondness for damp places are equally so. Woodlice require a humid environment in order to survive, and therefore congregate in numbers in suitable situations. Galen (XII.366 K) states that on farms they are to be found in large quantities 'underneath the water vessels in which the farm workers convey water from the springs and which they place by the hearth'. Elsewhere (XIII.111) he characterises them more briefly as οἱ ὑπὸ τὰς ὑδριὰς ὄνοι, as do Dioscorides (*DMM.* II.37), Hesychius (s.v. ἴουλος), Photius (*ditto*) and Suidas (*ditto*), or (XII.634 K) as ἐν τοῖς ὑδρηροῖς ἀγγείοις, paralleled by the entries under ὄνος in Hesychius and Suidas. Galen's mention of the water containers being indoors is the explanation for the epithet 'domestic' (κατοικίδιος) applied to the *onos* elsewhere in his works (XII.634 K).

The occurrence of woodlice under stones is referred to by Marcellus (XXXII.43, XXV.21). They are also noted as being associated with dung (Galen XII.634, XIII.113 K; Marcellus XV.72). No description of the life history of woodlice or millipedes has come down to us, but it would seem that, like most terrestrial invertebrates, they were believed to be spontaneously generated in their distinctive habitats. Galen (XII.366, 634 K), Marcellus (XV.72, XXXII.43), and the lexicographers use the verbs γίγνομαι and *nascor*, which when applied to invertebrates always imply spontaneous generation, in connection with their occurrence under water jars and about dung. Also Isidorus' phrase *vermis terrenus* (*Or.* XII.5.6) means in the context 'an animal generated from the earth'.

[43] Cf. the Byzantine Greek σκρόφα, Fernandez, op. cit., p. 235.

3. Weather lore

We read in the meteorological treatises of Theophrastus (de Sign. 19) and Aratus (Phaen. 957–8) that it was regarded as a sign of impending rain or storm if large numbers of iouloi were to be observed crawling up walls. This behaviour corresponds with the habits of woodlice.

4. Reputation as venomous

The appearance of the creatures we are considering on the classical list of venomous animals is certainly unjust and not readily explicable, whether millipedes or woodlice or both are concerned, since all are equally harmless and cannot even bite. Exactly what harm they were credited as capable of doing is not necessarily clear. Nicander (Ther. 811), our only Greek authority on the subject, gives no explanation, simply listing the ioulos in passing, and Pliny's mention of 'bites' could be simply the result of an erroneous analogy with the centipede. In more recent folk belief[44] woodlice were reputed to render poisonous articles of food which they might wander over. It may also be mentioned that, according to Galen (XII.366 K), medicinal woodlice could cause harm if not responsibly prescribed.

As regards antidotes for poisoning by these creatures, Nicander has one general list which serves for all invertebrates and is discussed below (7b, 3). Pliny prescribes leaves of gourd (XX.12), fennel root (XX.257), onion (XX.41), barley or millet meal (XXII.122,130), vinegar (XXIII.55), or butter and honey (XXVIII.161).

5. Medicinal uses

Woodlice played quite a significant role in classical pharmacy, and the range of ailments against which they were said to be efficacious is not inconsiderable.

In Greek sources it is always pill-woodlice which are particularly specified for use, but presumably other varieties would have been employed as well. Once again such beliefs with regard to woodlice are found in the folklore of more recent times.[45] The creatures, then, are variously prescribed for jaundice (Dioscorides DMM II.35; Paulus Aegineta VII.3; Pliny XXX.93), δυσουρία (Dioscorides loc. cit., Eup. II.113; Cyranides p. 271; Paulus loc. cit.; Pliny XXX.68), asthma (Galen XIII.111,113 K; Dioscorides Eup. II.113; Pliny XXX.47), head complaints (Galen XII.565 K; Theodorus Eup. II.44), paralysis (Pliny XXX.86; Marcellus XXV.21), angina (Pliny XXX.35), lumbago (Pliny XXX.53), quartan fevers (as an amulet: Pliny XXX.101), various kinds of ulcers and skin conditions (Pliny XXIX.143, XXX.40, 75, 114; Marcellus XV.72, XXXII.43), complaints of the tonsils and throat (Pliny XXX.31), and ear disorders (in this case, to be boiled up with oil, etc., and instilled into the ear: Dioscorides DMM II.35, Eup. 1.54; Paulus loc. cit.; Galen XII. 366, 623, 634, 641; Cyranides p. 271; Cassius p. 44; Pliny XXIX.136; Marcellus IX.10.33). According to Galen (XII.366 K), woodlice were very popular with rural would-be medics as treatments for ear complaints, but the writer adds

---

[44] J. L. Cloudsley-Thompson, op. cit., p. 5.   [45] Ibid., p. 5.

that since these folk failed to exercise moderation in their prescriptions their activities were as likely to harm the patient as to cure him. Caelius (*Chron.* 119–20,129) condemns the use of woodlice altogether as one of the fanciful remedies of earlier writers rejected by his school of medicine, arguing forcibly that they are harmful to the patient and that it is impossible to comprehend how such things ever found their way into pharmacology. Marcellus provides a very elaborate account (VIII.128) of how to collect a certain secretion exuded by his *cutiones* for use as an ointment for treating certain eye conditions; he advises that if the creatures are discovered in numbers, their secretion should be collected and stored in a glass vessel because these animals are difficult to find.

6. Veterinary uses

Pelagonius (*Vet.* 49) prescribes the use of woodlice for treating ear complaints in horses.

7. Exotic species

Theophrastus (*HP* IV.3.6) gives a brief account of a form of *onos* which inhabits the deserts of Libya. The creature is described as being black in colour and similar to the *onoi* native to Europe, whose habit of rolling into a ball it shares. Being very common and naturally moist or juicy, it provides an important article in the diet of creatures such as snakes and lizards which have to live without taking water. The animal concerned here would seem to be one of the desert woodlice such as *Hemilepistus reaumeri* of N. Africa and the Middle East.[46]

[46] Ibid., p. 7.

II

SCORPIONS, SPIDERS, MITES AND TICKS

Arachnida: Scorpiones (Scorpions)

5. *Skorpios/Scorpio*

1. Identification: general

Scorpions were the most feared invertebrates in antiquity. Among the venomous animals that played such a considerable part in classical medicine, they ranked second in importance only to snakes, and were therefore prominent in the popular imagination to a degree shared by few other invertebrates.

Classical knowledge of scorpions extended over the whole of the known world. They are reported from Southern Europe, North Africa, Egypt, Ethiopia, the Middle East, Asia Minor, and India. With so wide an area in view, we are clearly dealing with a considerable variety of modern day species, but in much of the material at our disposal the group is viewed as a whole, and in such cases no more detailed identification is possible. It is impossible, for example, to ascribe the Aristotelian account to any particular species. There are in fact a number of species of scorpion which are entirely harmless, but although this was recognised by a number of authorities, one must presume that in the eyes of the general public no distinctions were made and the whole group considered as equally dangerous. The scorpion of the popular imagination had a universally sinister reputation.

Among the non-venomous species of scorpion, we have, in southern Europe, the widespread *Euscorpius flavicaudis* (de Geer), and also *E. italicus* (Herbst). *Buthus occitanus* (Amoreux) occurs in a moderately venomous form in Europe, but in more dangerous races in North Africa and the Middle East. In North Africa, the most significant venomous scorpion is *Androctonus australis* (Linn.). Such species produce very severe symptoms and quite often fatalities. Also regarded as dangerous in this region are *Androctonus aeneas* Koch, *A. amoreuxi* (Audouin & Savigny), and *Buthacus arenicola* (Simon). Africa also has harmless species such as those of the genus *Pandinus*. A severely venomous species from the Middle East, Egypt and part of Asia Minor is *Leiurus quinquestriatus* (Hemprich & Ehrenberg). *Androctonus crassicauda* (Olivier) has a similar distribution, except that it is absent from Egypt. Indian species include *Buthus tamulus* (Fab.).[1]

[1] Smith, *Insects*, pp. 418–23.

A number of classical authors, most notably in the medical tradition, attempt to distinguish particular varieties of scorpion, but it is not often possible to identify these with currently recognised species with any degree of certainty. A number of those reported from exotic regions are indeed plainly fabulous. Keller[2] has claimed that individual species can be recognised in the depictions of scorpions on a number of ancient gems, but this must be regarded as highly debatable. These gems, though not badly produced, can scarcely be seen as accurate zoological illustrations, and the slight differences in portrayal need be ascribed to nothing more than the individual styles of the craftsmen who produced them.[3]

A rare synonym of *scorpio, nepa*, is used by a number of Latin authors. According to the grammarian Festus (163 L), it is an African word; and it is mostly found as a proper name for the constellation known by the title of *Scorpio*. There are, however, two examples in literature of it being used for the creature itself (Cicero, *de Fin.* V. 42; Plautus *Cas.* 443).

1.2. Identification: species distinguished primarily by medical writers

In considering the various distinct varieties of scorpion recognised in antiquity we may conveniently distinguish between those characterised by their colour or general form and not located in any particular geographical region—these are solely the concern of the medical tradition—and the largely fabulous exotic forms which, although having a small place in medical writings, were of particular interest to geographers with their delight in fanciful descriptions of strange fauna from the fringes of the classical world.

The earliest writer known to have attempted to classify scorpions is Apollodorus of Alexandria, whose work *Peri Therion* provided the foundation for all subsequent treatments of venomous animals. According to Pliny (XI. 87–8), he recognised nine varieties in all, mainly distinguished by their colour but also including the exotic winged and double-stinged species. Since, however, Pliny regarded Apollodorus' enumeration as an *opus supervacuum* (on the grounds that he makes no mention of the relative toxicity of the various forms), he does not specify the other seven. Of the later authors dependent upon the *Peri Therion*, Nicander has a list consisting of nine varieties (which, since he omits the δίκεντρος, cannot have corresponded exactly with that of his predecessor), while Aelian (VI.20) has eleven. The extra species found here probably result from the further subdivision of certain of Apollodorus' categories. Philumenus (*de Ven. An.* 14), on the other hand, in contrast to his discussion of *phalangia*, makes no distinctions in his chapter on scorpions.

Physical descriptions of these reputed species are generally minimal and do not normally permit anything approaching conclusive identification. A number of suggestions have, however, been offered.[4] Certain manuscripts of Nicander and

---

[2] *Münzen und Gemmen*, p. 145 and pl. XXIV, *Ant. Tierwelt*, II pp. 471–2.

[3] That such depictions were not based on observation from life is evidenced by the fact that one gem shows its scorpion with an extra pair of legs and another with a pair missing (*Ant. Tierwelt* p. 472).

[4] Steier in PW 3A. 1804–6; Gossen, *Tiernamen*, Nos. 42–4; J. Scarborough, 'Nicander's Toxicology', *Pharmacy in History*, XXI, 1979, pp. 15–16.

Eutecnius' Latin paraphrase of his work contain illuminations which purport to illustrate these scorpions. These are discussed by Kadar,[5] who considers them to be of some diagnostic value, although in the opinion of the present author their excessive stylisation points to their being decorative embellishments of the text rather than attempts at accurate zoological illustration.

(i) Λευκός. The white scorpion, characterised as a harmless form, heads the lists of Nicander (*Ther.* 769–70) and Aelian. In the opinion of Gow and Scholfield[6] and Scarborough[7] it is to be identified not with any particular species of scorpion but with the newly born young of the whole group, which are of the appropriate colour. Alternatively we may be dealing here with the non venomous *Euscorpius* spp. of the Mediterranean, as Steier suggests,[8] although these are not white. Some of Ps. Callisthenes' giant Indian scorpions are described as white (III.17).

(ii) Πυροός (Nicander *Th.* 771 ff.) or πυρρός (Aelian). The reddish scorpion is identified by Steier[9] as *Buthus occitanicus*. Ps. Callisthenes describes giant red scorpions from India.

(iii) Ζοφόεις (Nicander *Th.* 775–6). The dusky scorpion, credited with causing insanity to its victims, has been identified[10] as a species of *Androctonus*, or[11] of *Euscorpius*. Nicander describes it as having a nine-segmented tail, which, if intended literally, would make it longer than the fabulous ἑπτασφόνδυλος: the scholia are aware of the problem and suggest, among other things, that the word is being used loosely to mean many-segmented. Aelian appears to subdivide this variety into two, since his list includes both a 'dusky', καπνώδης, and a 'black', μέλας, form.[12]

(iv) Χλοάων (Nicander *Th.* 777 ff.) or χλωρός (Aelian). The greenish scorpion is regarded as unidentifiable by Steier.[13] Scarborough[14] notes that species of *Androctonus* may be greenish in colour. According to the Cyranides (p. 46), those scorpions magically produced from the herb basil are green in colour.

(v) Ἐμπέλιος. The variety so named by Nicander (*Th.* 782 ff.), a livid or pale yellow form (Scholia ad loc.: πελιδνός . . . ὠχρὸς) is further described as being heavy bodied, which identifies with the γαστρώδης of Aelian's catalogue. In the opinion of Scarborough,[15] it is the yellow *Buthus occitanicus*.

(vi) Καρκινώδης. Aelian's catalogue includes one variety under this name, while Nicander (*Th.* 786 ff.) has two, distinguished by the particular form of crab they are each said to resemble. The first is described as being similar to the καρκίνος of the seashore, while the second is compared with the crooked-limbed rock crabs (ῥοικοῖσι ἰσήρεες παγούροις πετραίοισιν): the latter is said to further resemble

---

[5] *Zoological Illuminations*, pp. 41–2, 47.
[6] *Nicander*, p. 22.   [7] Art. cit., p. 16.   [8] Art. cit., 1804–5.
[9] Ibid., 1805.   [10] Steier, ibid.   [11] J. Scarborough, art. cit., p. 16.
[12] In the illuminated mss of Nicander the ζοφόεις is entitled μέλας.
[13] Art. cit., 1805.   [14] Art. cit., p. 16.   [15] Ibid.

παγούροι in its possession of heavy limbs and hard serrated claws. Nicander goes on to say that the similarity in appearance is the result of the scorpions being actually generated from dead crustaceans, a belief which is elsewhere applied to the whole group indiscriminately. They are named παγουροειδής in the illuminated mss. The likeliest proposed identification[16] is that of Scarborough[17] who suggests *Pandinus* spp.

(vii) Μελίχλωρος. Nicander's (*Th.* 797–8) honey coloured scorpion has been identified by its colour as *Androctonus australis*[18] or *Leiurus quinquestriatus*.[19] It is described as having a tail which is black at the tip, and in the scholia as κηροειδής . . . φλογοειδής. The use of the latter adjective suggests that we are to regard this variety as identical with the φλογώδης of Aelian's list. Aelian, however, appears to be confusing it with the following species appearing in Nicander, since he characterises it as the most venomous of all, a distinction which Nicander reserves for his πτερωτός, described by him as having flame coloured legs.

1.3. Identification: exotic species

(i) Πτερωτός. The winged scorpion is one of the most frequently mentioned of the horrific insects and reptiles which were popularly believed to abound in regions on and beyond the borders of the known world, an impression which was supported by the writings of the geographers who were well aware of the public interest in highly coloured reports of exotic flora and fauna. To credit with the power of flight a creature already feared for its deadly sting was to render it doubly fearsome, and as a result we find the winged scorpion a regular feature of the reputed fauna of such regions as Libya, Egypt and India. It is interesting to note that winged snakes are commonly reported from these same regions (cf. Herodotus II.75, III.107–9; Mela III.8.9; Pausanias IX.21.6; Strabo XV.1.37; Aelian XVI.41), and that some authors closely associate the two creatures and describe them in similar terms. We may also compare Nicander's airborne spider, the Egyptian κρανοκολάπτης.

According to Strabo (XVII.3.11), the region of Masaesylia on the Libyan coast was reportedly infested by 'a multitude of scorpions both winged and wingless'. A similar report concerning the desert region to the south is found in Lucian's essay on the *dipsas* serpent (*Dips.* 3), where we find that the creatures' wings are membranous like those of locusts, cicadas or bats. Pliny (XI.88) expresses some uncertainty as to whether these African forms really possess wings at all or whether they simply spread out their legs like oars and allow themselves to be carried along by the wind; but he is prepared to accept, on the authority of Apollodorus, that some at least may genuinely be regarded as winged.

---

[16] Gossen, art. cit., no. 44, and Scholfield, *Aelian*, II p. 37, suggest unconvincingly that they are not scorpions at all.

[17] Art. cit., p. 16.   [18] Steier, art. cit., 1805.   [19] Scarborough, art. cit., p. 16.

As regards Egypt, the location of Herodotus' winged serpents, we learn from Aelian (XVI.42) that Pammenes claimed to have personally observed the winged scorpion there.

Megasthenes, in his foundational work on the history and geography of India, reported (ap. Strabo XV.1.37 and Aelian XVI.41) that the area on the far side of the river Hypanis was inhabited by winged scorpions of immense size, and also by flying snakes two cubits in length, possessed of membranous wings like bats, which were nocturnal and dripped flesh-dissolving liquids onto their human victims.

Finally, from a less remote locality, we have Pausanias' very definite statement that an inhabitant of Phrygia brought to Ionia as a curiosity a scorpion with wings exactly resembling those of a locust.

Outside the geographical writers, we learn from Pliny (XI.88) that Apollodorus of Alexandria included the flying scorpion in his work on venomous animals. It thus appears in Aelian's dependent catalogue of scorpions (VI.20) and in the *Theriaca* (799 ff.) of Nicander. The latter characterises it as the most dangerous of the varieties with which he deals, capable of causing instantaneous death to children, and states that it has flame coloured legs and white wings like those of a locust (μάσταξ).

It has been concluded by some authors[20] that all these accounts are entirely fabulous, but in view of the constant features in the various descriptions and Pausanias' report of an actual specimen it does seem probable that some real creature lies behind the fanciful embellishments. Such a creature would need only a superficial resemblance to a scorpion for popular imagination to supply the remainder of the story. If, then, we are looking for a harmless but suitably large and dangerous looking winged insect, a number of possibilities present themselves. The oldest and in a sense most immediately obvious suggestion is that the creature concerned was some form of scorpion-fly (Panorpidae), these having the appropriate physical appearance,[21] but in view of their small size, the Panorpidae must in fact be ruled out. Bearing in mind the comparison with winged locusts and grasshoppers, one or other of the more bizarre Orthoptera might be possible candidates.[22] Bodenheimer[23] suggests quite plausibly the giant water-bugs of the family Belostomatidae, which fly to light and are suitably fearsome in appearance: such creatures are popularly known as water-scorpions. Alternatively, large dragonflies (Odonata), which have suitably elongated abdomens and are often regarded in folk belief as capable of stinging, may also be considered.

(ii) Δίκεντρος. The 'double-stinged'—and therefore doubly dangerous—scorpion is mentioned by a few authors in association with the above. According to Pliny (XI.87) it was recognised as a genuine variety by Apollodorus, whence it appears in the catalogue given by Aelian (VI.20). It is not, however, included by Nicander

---

[20] Keller, *Ant. Tierwelt*, II p. 478; Fernandez, *Nombres*, p. 77.
[21] Steier, art. cit., 1806; Gossen, art. cit., no. 63.
[22] Cf. Gossen, art. cit., no. 77.
[23] *Animal and Man*, p. 116.

(although it is referred to in the scholia on *Theriaca* 781). It was recorded by Pammenes (*ap.* Aelian XVI.42) as occurring in Egypt along with the winged variety. The δίκεντρος must be regarded as wholly fabulous.[24]

(iii) Ἑπτασφόνδυλος. A third reputed variety of scorpion, associated with the two preceding, was credited with a tail composed of seven segments (as opposed to the normal five). Pliny (XI.88) records that these were said to be more dangerous than their six (according to his enumeration) segmented relatives. Apollodorus appears to have recognised the species (scholia on Nicander *Ther.* 781), and it figures also in the list provided by Aelian (VI.20). Lucian (*Dips.* 3) reports that in the south of Libya there occur two forms of large scorpion, the winged variety and another which walks on the ground, of great size and having many segments (πολυσφόνδυλον).

While Strabo (XVII.3.11) in his description of the dangerous fauna of Masaesylia, on the coast of the same region, states that there occur there large numbers of wingless scorpions of remarkable size and having seven segments.[25]

2. Life history

Scorpions, unlike most invertebrates, do not lay eggs but give birth to fully developed young which immediately climb onto the female's back and remain clinging there until after their first moult.[26]

That scorpions are viviparous was certainly known to some ancient authorities, since we are informed of the fact by Aelian (VI.20), but it is uncertain whether this observation was familiar to Aristotle. The latter believed, of course, that all invertebrate eggs were strictly speaking undeveloped larvae, but even so his statement (*HA* 555a 23 ff.) that scorpions produce large numbers of egg-like σκωλήκια conveys the impression that he did not regard their life history as essentially different from that of related creatures such as spiders, whose eggs he describes in more or less the same terms. He goes on to assert that these σκωλήκια are incubated—the same expression as is used in the case of spiders—by the adult, which is plainly erroneous. Pliny simply follows Aristotle in stating (XI.86) that the offspring of scorpions take the form of *vermiculi ovorum specie*. He also reproduces the odd detail (XI.91) that the brood usually numbers about eleven. According to Aelian (loc. cit.), in a passage where something is missing from the text, scorpions do not mate, but this is denied by Galen (VI.640 K): the former author adds that they are more prolific under warm conditions.

According to popular belief, paralleled in connection with other venomous creatures such as *phalangia* and vipers, there was fierce antagonism between adult scorpions and their young. In the version of events given by Aristotle (loc. cit.) and followed by Antigonus of Carystus (*Hist.Mir.* 87), it is the latter who take the initiative by expelling and killing their parent. By Pliny (XI.91), however, we are told that the females are in the

---

[24] Although Gossen, art. cit., nos. 44, 63, attempts to identify it.

[25] It is not clear whether the latter adjectives apply to both winged and wingless forms, in which case we have a different picture from Lucian's.

[26] Cf. J. L. Cloudsley-Thompson, *Spiders, Scorpions, Centipedes and Mites*, p. 39.

habit of devouring their offspring, but that there is always one which is able to preserve itself by clinging onto its mother's body, and which subsequently kills her as well. We also read of newly born scorpions practising the use of their stings in preparation for adult life.

Alongside the 'scientific' view presented by Aristotle and his successors, there coexisted the older belief, originally applying universally to all invertebrate animals, that scorpions were produced not from others of their own kind but by spontaneous generation. The most popular view was that they arose from dead crabs buried in the ground, as described by Pliny (IX.99), Ovid (*Met.* XV.379–81) and Isidorus (*Or.* XI.4.3). According to the more detailed account given by Nicander (*Ther.* 788 ff.), the process is set in motion when fishermen draw the larger crabs out onto the land and allow them to crawl away into holes or crevices where they die to give rise to new scorpions. The superficial similarity, especially in their claws, between the two groups of creatures evidently gave credence to the story, as Nicander makes clear. Tertullian (*Scorp.* 1) says that scorpions are generated from the earth, a belief implied by Isidorus' (*Or.* XII.5.4) classification of them as a *vermis terrenus*.

It was also held that they could be produced from rotting wood (Schol. on Nicander *Ther.* 786), from human corpses (Augustine MPL 32.1372), or from the herbs mint (Antigonus Carystius 23) or basil (Pliny XX.119): the Cyranides (p. 46) explains that this supposed property of basil was utilised for magical purposes, and says that the scorpions so produced are green with seven-jointed tails. In Egypt, scorpions were said to arise from the dead bodies of crocodiles (Aelian II.33; Antigonus Carystius *Hist.Mir.* 19).

3. Feeding habits

Little attention was given in antiquity to the feeding habits of scorpions. They are in fact carnivorous, feeding upon other invertebrates and in some cases even small mammals, but all but one (Aelian VIII.13) of the few ancient references which we have on the subject credit them with a surprisingly innocuous diet in view of their sinister reputation. Pliny informs us that they feed simply upon earth (X.198; cf. Eusebius *Praep.Ev.* MPG 21 464) and that they also take considerable quantities of water (XI.88). Similarly, Nicander (*Ther.* 782 ff.) describes one of his scorpions as a voracious eater of grass and earth.

4. Popular beliefs and attitudes

As has been mentioned above, the scorpion had an extremely sinister reputation among the ancients, and was regarded as the most dangerous of all venomous animals with the exception of certain snakes which, however, it was held to surpass in its ubiquity and cunning. In view of this it is not particularly surprising that classical authors tended not to ask themselves under what circumstances the creature was likely to make use of its sting. Scorpions, of course, like other poisonous animals, employ their venom in hunting and as a means of self defence[27] but among the ancients it was widely held that they were in the

---

[27] Cf. Smith, *Insects*, p. 418.

habit of stinging out of sheer malignancy and viciousness at every possible opportunity. They are portrayed (Pliny XI.87) as constantly moving their tails and practising their strike in order not to miss an opportunity for attack, a pattern of behaviour to which Demosthenes (XXV.52) draws attention when he compares a certain individual to a scorpion darting to and fro with sting at the ready, always prepared to do some harm to someone. They were viewed, in anthropomorphic terms, as incorrigibly evil (*AP* XI.227; Porphyry *de Abst.* I.20; Eusebius loc. cit.; Arnobius *adv.Nat.* VII.23), being characterised by Philo (*de Spec.Leg.* III.103), as the type of creature to be killed on sight on account of its innate malevolence (ἐνυπάρχουσαν κακίαν). And accordingly they often appear as a point of comparison with evil, cunning or dangerous human beings (Demosthenes loc. cit.; Philo, *de Legat.* 205; Eupolis fr. 231; Cratinus fr. 77.; Diogenian VIII.8; Apuleius *Met.* IX.17; Suidas, Photius and Hesychius s.v. ὀκτώπουν; LXX *Ez.* 2.6, *Sir.* 26.7; Tertullian *Scorp.* passim; Augustine *Serm.* 105.9–10; Gregory Magnus MPL 76.879).

Not only were scorpions believed to miss no opportunity of exercising their stings on human victims, but it was held that they would go out of their way in order to do so. They were commonly visualised as lurking in crevices and under stones (Nicander *Ther.* 18,796), lying in wait for unwary passers-by; and a widely mentioned proverb counselling caution—'Under every stone there waits a scorpion'—gives expression to this idea (Sophocles fr. 138; Praxilla fr.4; Aristophanes *Thesm.* 528; Athenaeus 695d; Scolia fr. 22 B, 17 Edm.; Zenobius VI.20; Diogenian VIII.59). However, the general ascription to scorpions of a single-minded determination to sting and of considerable cunning in achieving their aim is illustrated most vividly by a story from Libya related by Aelian (VI.23). He describes how the inhabitants sleep on high beds with their legs set in vessels of water to provide security from scorpions, and how the creatures themselves, undeterred by these precautions, form a living chain which is begun by one of their number hanging sting downwards from the ceiling and which is extended until the lowest animal can reach the unsuspecting sleeper.

5. Predation
Despite their venomous qualities, it was believed that certain animals (Plutarch *Mor.* 87a), such as cocks (Plutarch *Mor.* 1049a, 87a; Augustine *Serm.* 105.9–10), hawks (Aelian X.14; Eusebius *Praep.Ev.* MPG 21.101; Ps. Eustathius *Hex.* MPG 18.478), ravens (*AP* IX.339), and perhaps deer (cf. Sextus Empiricus *Pyrrh.* I.57), were capable of eating them without suffering ill effects.

6. Scorpion stings and their treatment
Apart from the various species of snake, the scorpion has the most attention devoted to it of all the venomous creatures recognised in antiquity. It was considered a major threat to human beings, hiding itself under stones (cf.4 above, and also the myth of Orion: Hesiod fr. 18; Aratus *Phaen.* 637 ff.; Lucan *Phars.* 835–6; Nicander *Ther.* 13 ff.; Ovid *Fasti* V.541), on walls (Nicander *Ther.* 796) and in other inconspicuous places, and then emerging to sting the unwary. They were also known to enter houses (Fronto p. 79 N),

though this habit is mentioned more often in connection with exotic regions. Some scorpions, such as *Euscorpius* spp. of Southern Europe (Fronto's specimen which he found in his bed was probably one of these,[28] as is perhaps the κοινός σκορπίος of the Cyranides p. 47), are in fact not venomous at all, but although this fact was recognised by some (e.g. Pliny XI.89, who says that all those occurring in Italy are harmless) one would presume that the general public, with an understandable desire to take no risks, regarded all forms as being equally dangerous.

Although scorpion stings are often fatal, especially to children, they are by no means universally so.[29] In ancient authors there is a tendency to imply that a fatal result is inevitable if treatment is not forthcoming, but on the other hand one gathers that great confidence was placed in the numerous antidotes detailed by the medical writers, and that these were regarded as universally efficacious. According to Pliny (XI.86), the venom takes three days to have its effect, although other authors of a more popular nature (Lucan *Phars.* 834) regard death as occurring far more rapidly. In fact it is Pliny who is in error here, since the action of scorpion venom takes its effect within hours rather than days. He goes on to say that stings are universally fatal to young girls, almost always fatal to older women, but to men only in the morning, at which time the creatures will not have dissipated any of their venom. Pliny evidently believed that their supply was replenished during the course of the night.

It was held that male scorpions were more dangerous than their mates (Aelian VI.20; Pliny XI.87), and that the former could be differentiated by their more slender appearance (*gracilitate et longitudine*). Their toxicity was believed to vary according to the temperature, thus making them most dangerous at midday (Pliny XI.88). Those fortunate enough to survive an attack were credited with being thereafter immune from those of bees or wasps (Pliny XXVIII.32). According to Pliny (XI.90), their stings have no effect upon invertebrate creatures. Stories about whole populations being forced to evacuate from their homelands under pressure from animals of various kinds are not uncommon in ancient authors, and two of these, one concerning Rhoeteum on the Troad and another an area in Ethiopia (Agatharchides 59; Strabo XVI.4.12; Diodorus III.30; Aelian XVII.40; Pliny VIII.29,104), involve scorpions.

Certain foreign races, most notably the Psylli of Libya, were reputed to be immune to the stings of scorpions and other venomous creatures, and it was believed possible for an isolated individual in Europe to be born with this same capacity (Sextus Empiricus *Pyrrh.* I.82).

The symptoms of scorpion stings, as with those of other toxic creatures, are described by a number of medical authors, though without conspicuous signs of accurate observation. Nicander (*Ther.* 770 ff.) details symptoms for three of his species only, and without the lurid detail which characterises his treatment of the effects of other animals included in his poem. His source Apollodorus, we may recall, was criticised by Pliny for not stating which varieties were the more dangerous. Philumenus (*de Ven.Anim.* 14),

---

[28] Cf. Steier, art. cit., 1804–5; Leitner, *Zool. Terminologie*, p. 220.
[29] Smith, op. cit., p. 420.

Pseudo-Dioscorides (*Iob.* 6), Paulus of Aegina (V.8) an Aetius (XIII.21) give composite accounts for scorpions in general.

Since scorpions were regarded as ranking second only to snakes in medical importance, it is not surprising that ancient medical writers devote more space to the treatment of their stings than to those of any other invertebrate animal. The antidotes listed by the various sources consist, apart from those of a semi-magical nature, either of substances to be applied as an ointment to the affected spot or of those to be taken internally, but often it is left unspecified as to how a particular item is to be employed. They are generally similar to those prescribed for other poisonous creatures, but there are a number of unique items. It may be noted here that the medical writers specify a number of their antidotes as being effective in connection with all venomous animals without exception; these have not been considered in the present work.

The largest assemblage of supposed treatments, likely and unlikely, is to be found in Pliny's *Natural History*. Other lists, of varying extent, as well as isolated references in connection with individual *materia medica*, are to be found in the *Geoponica* (XIII.9) and in the works of Dioscorides (esp. *Eup.* II.126–7), Galen (esp. XIV.175–204 K), Philumenus (*de Ven.Anim.* 14), Paulus Aegineta (V.8), Aetius (XIII.9), Celsus (V.27.5), Q.Serenus (860–81), Marcellus Empiricus, Theodorus Priscianus (*Eup.* I.74), and Sextus Placitus. As is usual, the majority of antidotes are plants or extracts prepared from them: Pliny includes over seventy species (XX.8,25,32,41,46,50,62,68,117,121,125,129,133,145,155, 157,164,171,175,182,185,209,223,236,252,256,   XXI.118,131,141,149,152,162,163,170, 184,XXII.31,39,47,50,67,90,103,124, XXIII.6,111,112,118,123,128,138,160, XXIV.13, 16,22,29,44,45,108,115,117,136,175, XXV.119,121,122, XXVI.31, XXVII.5,42,50,68, 124, 127), many of which appear in the other authors cited above (in the references there noted plus the following: Dioscorides *DMM* 1.61,71,128, II.118,131–2, III.24,93,104,108, IV.190,192; Celsus V.27.6; *Geoponica* II.47; Ps. Apuleius *Herb.* 1.9, 63.1, 116.7). It was believed that scorpions themselves provided an antidote to their own stings, either immersed in oil (XI.90), or crushed and applied to the wound, or roasted and eaten, or burnt and its ashes taken in wine (XI.90, XXIX.91; Dioscorides *DMM.* II.11; Galen XII.366 K; Paulus VII.3; Celsus loc. cit; *Geoponica* XII.9.4): the burning of scorpions was thought to drive away others (*Geoponica* XIII.9.1). As for other non-herbal remedies, we find the following extensive array; bran in vinegar (Celsus loc. cit.); wine or vinegar (Pliny XXXI.99,127, XXIII.43,55; Q. Serenus loc. cit.; and frequently as a solution for other ingredients); honey (Celsus loc. cit.; Pliny XXXI.65); earthworms (XXIX.91); sea or river crabs (Pliny XXXII.53–5; Dioscorides *DMM* II.10, *Eup.* II.133; Cyranides p. 112) or shrimps or prawns (Cyranides p. 112); river snails (Pliny XXXII.56); red mullet (Pliny XXXII.44; Dioscorides *DMM.* II.22; Galen XII.365 K; Cyranides p. 272), *coracinus* (Pliny XXXII.56), or salted fish in general (Pliny XXXII.46); hen's brain or dung (Pliny XXIX.91; S. Placitus XXIX.21; Q. Serenus loc. cit.); geckoes (*Geoponica* XIII.97) in oil (Pliny XXVIII.155), or tortoise gall (Pliny XXXII.33), or the flesh of lizards (Pliny XXIX.91; Dioscorides DMM.II.64; Galen XII.334 K), or frogs (Pliny XXXII.48); *castoreum* from beavers (Pliny XXXII.30); goat's cheese (Q. Serenus loc.

cit.); hare's rennet (Pliny XXVIII.154; S. Placitus III.a15,b20); mouse flesh (Pliny XXIX.91; Dioscorides *DMM*. II.69; Galen XII.365 K); python's liver (Pliny XXIX.91); the dung of cows, calves, or goats (Pliny XXVIII.155,154); a pebble or earth-covered potsherd (Pliny XXIX.91); salt or seawater (Pliny XXXI.65,99; Dioscorides *DMM*. V.11,109; Celsus loc. cit.); sulphur (Pliny XXXV.177; Q. Serenus loc. cit.) or forms of agate (Pliny XXXVIII.139–42; Solinus V.26; *Orphic Lith*. 494–500, 622–4); the application of a silver ring (*Geoponica* XIII.9.2; cf. Cyranides p. 49); human ear wax (Pliny XXVIII.40), semen (Pliny XXVIII.52), or urine (Pliny XXVIII.67; S. Placitus XVII.16); and sexual intercourse (Pliny XXVIII.44). Nicander has a composite antidote list for venomous invertebrates in general, which is discussed below (7b,3).

In addition to prescribing remedies, medical writers deal also with a number of reputed preventive or deterrent measures. Certain herbs were believed to have insecticidal properties against scorpions (Pliny XXI.171, XX.50, XXII.60; Dioscorides *DMM* III.101, *Eup*. II.133; *Geoponica* XIII.9.1), and several writers record the curious belief that specimens immobilised instantaneously by the effects of aconite could be fully restored by applying white hellebore (Theophrastus *HP* IX.18.2; Apollonius *Hist.Mir*. 41; Aelian IX.27; Pliny XXVII.6; Dioscorides *DMM* IV.76, *Eup*. II.136). The gecko was not only regarded as an antidote when dead, but was supposed to immobilise or, according to some accounts, actually to kill scorpions when brought into proximity (Aelian VI.22; Pliny XXIX.90; Galen XIV.243 K): Aelius Promotus (p. 776) claims this can be tested by enclosing both animals in a glass vessel. By contrast, river crabs were said to attract scorpions to them (Pliny XXXII.55, XX.120). We read of fumigants involving herbs and animal substances credited with deterrent properties (Pliny XX.245; *Geoponica* XIII.9.3). A certain locality named Clupea in N. Africa and an island off the coast were reportedly free from scorpions, and accordingly the soil of these places was credited with being effective in destroying those in other regions (Pliny V.42, XXXV.202; Solinus XXIX.8). A number of plant species were believed to protect the wearer when carried (as were crabs, geckoes—Aelius p. 776—and coral—Orphic Lith. 510) or to be a source of protective ointments (Pliny XX.223,232, XXII.60, XXV.163, Dioscorides *DMM*. IV.8, *Eup*. II.133; *Geoponica* XIII.9.1).

Moving further into the realm of the semi-magical, we find it stated that if a person who has not eaten spits two or three times on a scorpion it will perish immediately (Galen VII.745, XII.289 K). Aelian's supposed explanation (IX.4) to the effect that spittle blocks the aperture through which venom is emitted from the sting is presumably a rationalisation of the same popular superstition. The use of magic against scorpions and other poisonous creatures is mentioned in passing by Plato (*Euthyd*. 290a); the magical papyri (*PMag* VII.193–6, cf. P3) describe the use of amulets bearing magical signs, and the Cyranides (pp. 19, 49, 112) gives instructions for engraved gems with the same purpose. We read of the employment of spells to prevent them from stinging or to cure stings (Pliny XXVIII.24; Cassius Felix p. 168; cf. Tertullian *Scorp*. 1) and also of the existence of a belief that, if stung, a person could transfer the ill effects to some unfortunate animal—the donkey is one specified—by sitting beneath it or simply by addressing it (Pliny XXVIII.155; *Geoponica* XIII.9.5–6; Cyranides p. 70; Aelius p. 776).

7. Scorpions and domestic animals

In veterinary medicine, as in human medicine, scorpions were perceived as a significant threat. The effects of their stings upon domestic animals, more specifically on cattle and horses, are described with slight variations by Vegetius (*Mul.* II.141,147) and the authors collected in the *Hippiatrica* (*Hipp.Ber.* 86.1,5–6 = Chiron 516–7), with the stated aim of enabling the observer to recognise the creature responsible. The accounts are expressed in restrained terms and do appear to display fairly accurate observation.[30] The prescribed ointments and potions given by the above mentioned sources (in the references cited, plus *Hipp.Ber.* 86.7. *Hipp.Can.*71.18), and *Geoponica* XVI.20 correspond with those employed for the treatment of humans, including as they do various herbal preparations, honey, human urine, the dung of certain animals, salt, and *aphronitron*. In addition, Vegetius (II.141) gives additional details concerning the care of affected animals. He prescribes that the location of the sting should first of all be fumigated with burning egg shells treated with vinegar and mixed with stag's horn or *galbanum*. This process is to be followed by blood letting or cauterisation, after which one or other of the listed ointments may be applied. In addition, the animal should be wrapped in blankets to induce sweating, and it should be given as fodder barley meal mixed with vine and ash foliage.

8. Medicinal uses

As well as being employed as an antidote to the stings of their own kind, dead scorpions are also prescribed for the effects of the reputedly venomous gecko (Pliny XXIX.72). We read of the ash (Pliny XXX.66) or the cooked flesh (Cyranides p. 47; Aelius p. 774) of scorpions as a treatment for bladder stones, and of the use of a scorpion tail amulet against quartan fevers (Pliny XXX.100). An ointment produced by placing scorpions in oil was believed to be effective against all kinds of fevers, as well as against epilepsy and demon possession, for which a scorpion tail amulet was also said to be effective (Cyranides pp. 46–7).

9. Exotic species

The three most widely mentioned and reputedly distinctive forms of scorpion, the winged, the two-stinged, and the seven-segmented, have been dealt with above (1.3). It may also be noted here that the catalogue of varieties found in the medical tradition and derived ultimately from the work of Apollodorus will have covered the fauna of N. Africa and Egypt as well as that of Europe.

  (i) *Caria*. According to the paradoxographers, the scorpions of Latmos in Caria were able to distinguish between the natives of their region and strangers, inflicting fatal stings upon the former but causing only a slight irritation to the latter (Aelian V.14; Antigonus *Hist.Mir.* 16; Apollonius *Hist.Mir.* 11; Pliny VIII.229). Carian scorpions are also mentioned in the inauthentic book IX of the *Historia Animalium*, but here

---

[30] Cf. Soulsby, *Helminths*, p. 522.

they are merely said to be of large size and especial toxicity (*HA* 607a 15 ff., reproduced in Pliny XI.90 where 'Scythia' is read for 'Caria').

(ii) *Persia.* We are informed by Aelian (XV.26) that along part of the route between Susa and Media the terrain is so infested by scorpions that when the king is to travel through he gives orders three days in advance for the local people to conduct an operation against them, with rewards for those who capture the most (Cf. Ptolemy *Geog.* VI.17).

(iii) *Libya.* The north coast of Africa was particularly noted for its large population of savage scorpions, not to mention other venomous animals. Pliny indeed characterises the scorpion as *hoc malum Africae* (XI.88; cf. Solinus XXVIII.33). Both winged and seven-segmented varieties are reported from the region, in addition to those of more conventional form; but even the latter were said to be of remarkable size (Strabo XVII.3.11; Lucian *Dips.* 3). Aelian has an entertaining description of the precautions taken by the inhabitants to preserve themselves from attack and the cunning exercised by the scorpions in circumventing these (see 4). A certain tribe from Libya known as the Psylli was reputedly immune to the bites and stings of all venomous animals including scorpions, and its members were believed to be capable of healing others not in possession of such immunity (Aelian 1.57, XVI.27–8; Pliny VII.14, XI.89): we also read of them operating in Italy, travelling the country selling protective amulets, etc. (Arnobius II.32); and Pliny (XI.89), who regards them with considerable suspicion as poison-vendors, accuses them of attempting to import foreign scorpions into Europe.

(iv) *Ethiopia.* Aelian (VIII.13) gives a brief account of an Ethiopian variety of scorpion known as σιβρίται, a tribal name.[31] These are said to feed upon a variety of invertebrates and reptiles and to produce a toxic excrement harmful to anyone treading on it. Agatharchides' report of a region evacuated on account of its being overrun by venomous animals has been mentioned above. Sextus Empiricus (*Pyrrh.* I.83) states that the Ethiopians living on the river Astapous include scorpions in their diet without suffering any harmful effects.

(v) *Egypt.* Both winged and two-stinged scorpions were reported from Egypt by Pammenes (Aelian XVI.42). Those at Pharos were reportedly harmless to human beings (Pseudo-Aristotle *HA* 607a 15 ff.; Pliny XI.89). By contrast, those from Koptos are said by Aelian (X.23) to be of large size and to possess unusually sharp stings causing instant death; however, he goes on to relate that despite this they never harm the women performing rituals in time of mourning at the sanctuary of Isis.

(vi) *Palestine.* Scorpions are mentioned by Philo (*de Vit.Mos.* I.192), following LXX *Deut.* 8.15, among the hazards of the Sinai desert. Cf. also NT *Lk.* XI.12.

(vii) *India.* From India we have reports both of the winged scorpion (Strabo XV.1.37;

[31] Cf. Fernandez, op. cit., pp. 236–7.

Aelian XVI.41) and of specimens remarkable only for their size (Aristobulus *ap.* Strabo XV.1.45). Giant scorpions also figure in Ps. Callisthenes' fanciful account of Alexander's adventures in this region: here we read of cubit long specimens near the Ganges (III.10), and of sand-burrowing (ἀμμοδύται) specimens of equal length, red or white in colour, heading an assortment of wild animals which emerge from the forest by moonlight to drink at a certain lake (III.17).

Arachnida: Pseudoscorpiones (False Scorpions)

6. *Skorpiodes*

Aristotle, in his account (*HA* 557b 9 ff.) of the very smallest spontaneously generated creatures, states—having described the clothes moth and the ἀκαρί—that 'other creatures are generated in books, some of them similar to those produced in clothes, others like scorpions without tails but very small'. Slightly earlier (532a 18) the same creature is mentioned in passing as a scorpion-like animal with claws (τὸ ἐν τοῖς βιβλίοις γιγνόμενον σκορπιῶδες). Pliny in his corresponding section (XI.117) does not refer to it explicitly. As has been recognised,[32] we have here a very clear description of the minute 'false-scorpions' (Order Pseudoscorpiones) which have exactly the appearance described.

Arachnida: Araneae (Spiders)

7. *Arachne/Araneus* and *Phalangion*

The terms ἀράχνη and φαλάγγιον represent two distinct groups of spiders recognised by the Greeks among both technical and non-technical writers. The distinction between them is essentially clear, though it cannot be drawn with absolute precision. Ἀράχνη may be used as a general word to cover both classes of spider, but in its more precise sense it comprehends those species which are known as being harmless to man, most particularly those which spin conspicuous webs. This is in keeping with the etymology of the word, which alludes to its spinning habits.[33] Φαλάγγιον, on the other hand, is applied to the much feared venomous spiders; plus those reputed to be venomous, since popular fear of poisonous creatures expanded this class of arachnid somewhat beyond its legitimate range. The primary distinction cannot be, as Sundevall[34] suggests, whether the spiders produce webs or not, since Aristotle recognises that some that he would class as *phalangia* have this habit. There are also certain hunting spiders, not constructing webs and usually known as λύκοι whose position in relation to the above two classes is not clearly defined. Aristotle, as one would logically expect, classes them as *arachnai*, but they are more commonly included as a sub-group of *phalangion*, presumably because their non-venomous character, admitted even by Nicander, was a matter of dispute. It should

---

[32] Sundevall, *Thierarten*, p. 231; Keller, op. cit. II. p. 478.   [33] Fernandez, *Nombres*, p. 25.

also be noted that certain creatures superficially resembling spiders, though not in fact arachnids at all, appear to have been recognised as *phalangia* through acquiring a reputation for toxicity.

The Romans had only a single native term, *araneus*, for spiders in general, and for the venomous species the Greek name was adopted in the form *phalangium*.

7a. *Arachne/Araneus* (sensu stricto)

1. Identification: general

As has been said, in their more precise sense these terms cover all spiders recognised as non-venomous, including many hundreds of present day species. In popular usage those types primarily in view would have been the orb-web spiders of the open country, much admired for the skill of their constructions, and the house spiders such as *Tegenaria* spp., well known as domestic pests and proverbial as emblems of decay and disuse.

In poetry the Greek term occasionally appears in the masculine forms ἀράχνης (for example, Hesiod *Op.* 777, Pindar fr. 296, Aeschylus fr. 121, Bacchylides fr. 3.7) and ἀραχνός (Aeschylus *Supp.* 887). There is also an alternative form ἀράχνηξ found in Hesychius and probably in the comedian Callias (fr. 2).[35] Fernandez[36] sees the Hesychian entry ἄρασιν (or ἄραριν) as etymologically related to ἀράχνη. A diminutive form ἀράχνιον occurs only in the *Historia Animalium* (555a27, 622b27) and the Cyranides (p. 62), simply as a variation on the normal word: ἀράχνιον normally means a spider's web. Corresponding Latin diminutives, *araneolus* (*Culex* 2) and *araneola* (Cicero *de Nat.D.* II.123), are also found.

The name φάλαγξ usually appears as a rare synonym of ἀράχνη, and never seems to have acquired the specialised usage given to its diminutive. The earliest examples of its use are found in Attic comedy, namely in Aristophanes (*Vesp.* 1509, *Ran.* 1314), who never uses ἀράχνη, and Plato (fr. 22); and in Plato and the *Frogs* it clearly refers to non-venomous house spiders. Similarly Xenophon uses it for web-spinning species (*Mem.* III.11.6), and it is significant that when, later in the same work, he wishes to speak about venomous ones he uses the diminutive. Aelian has two chapters devoted to spiders' webs, in one of which (I.26) he employs the word ἀράχνη, and in other (VI.57) φάλαγξ: he again uses the diminutive when venomous species are in view. There are, however, a few examples of φάλαγξ being used alongside the diminutive as a simple variation from the latter where poisonous spiders are concerned (Pseudo-Aristotle *HA* 609a 5, Nicander *Ther.* 654 and 715, fr. 31). The lexicographers are unhelpful with regard to this word, Hesychius defining it as an animal resembling a spider and Suidas synonymising it variously with *arachne* and *phalangion*.

The term θήραφος 'hunter', found only in the Cyranides (p. 62), is similar in meaning to ἀγρώστης and is thus synonymised with that name by Stromberg and Fernandez;[37]

---

[34] *Thierarten*, p. 234.   [35] Cf. Fernandez, op. cit., p. 25.   [36] Ibid.

[37] Strömberg, *Gr. Wortstudien*, p. 23; Fernandez, op. cit. p. 156.

but in the context, where it is glossed as 'φαλάγγιον ἢ ἀράχνιον, a small six-footed animal weaving cobwebs on walls and well known to all', it can only be a general synonym of *arachne*. In another ms of the Cyranides (p. 290), two further synonyms are given: καματερή (which also appears in Schol. Aristophanes *Ran*. 1349) and σαλαμίνθη.

1.2 Identification: the Aristotelian species

In *HA* 555a 26 ff. Aristotle, having considered the life history of spiders in general and having noted the considerable variety of methods employed by them for the protection of their eggs and young, goes on to name three distinct types, namely *phalangia* and two classes of *arachnai*.

The first of the latter he distinguishes by the name of αἱ λειμώνιαι ἀράχναι, the field or meadow spiders, which, he says, deposit their offspring in a web (ἀράχνιον) of which one half is attached to the parent and the other projects beyond; this the spider incubates until the young are able to fend for themselves. It has been generally recognised[38] that the author is here describing certain species of wolf spider (Lycosidae) whose females construct portable egg-sacks. Elsewhere Lycosid spiders seem to be covered by the name λύκος.

Aristotle's second class of *arachnai*, αἱ γλαφυραί, are mentioned only in passing as producing a relatively small number of offspring by comparison with the *phalangion*. The description is interpreted by some authors[39] as having the meaning of 'smooth spider'; but it seems more probable that these spiders are none other than the γένος σοφώτατον καὶ γλαφυρώτατον of *HA* 623a 7 ff., and that the adjective should be translated 'skilful' as in the latter passage. In this case the γλαφυραί would be the web-spinning *arachnai* as opposed to the non-sedentary λειμώνιαι.

1.3. Identification: the species in book IX of the Historia Animalium

In the non-Aristotelian ninth book of the *Historia* (622–3) we find a more elaborately defined threefold classification of spiders which may, however, be seen as more or less the same as that from book V discussed above. A number of authors[40] have erroneously assumed that the author is here subsuming three classes under the general heading of *phalangia*, but this is not in fact the case. Our author begins his account by stating that there are many types of '*arachnia*-and-*phalangia*', i.e. of ἀράχναι sensu lato, and then goes on to distinguish three of these types and to subdivide all three. Having considered *phalangia* and *lykoi*, he arrives at a group for which he has no specific name, but which he describes as τρίτον γένος σοφώτατον καὶ γλαφυρώτατον. These are the *arachnai* of popular parlance, the constructors of elaborate webs, most especially the orb-weavers and house spiders. He then moves immediately into a description of the process of construction followed by the orb web spiders, though without explicitly stating that this

---

[38] Aubert and Wimmer, *Thierkunde*, I. p. 161; Keller, op. cit. II. p. 463; Steier in PW 3A.1793.

[39] For example, D'Arcy W. Thompson, *Hist. An.*, V.27; Peck, *Hist. An.*, II. p. 203. This is also the definition given for this passage in LSJ.

[40] For example Aubert and Wimmer, op. cit., I. pp. 160–1; Fernandez, op. cit., p. 88.

does not apply to all members of the group. Orb-weavers belong to the families Argiopidae, including the well known *Araneus diadematus* Linn., and Tetragnathidae.[41]

Following the account of the orb web and its construction, a distinction is drawn between two types of '*arachnai* which are skilful and which weave thick webs'. Presumably this latter phrase is to be understood as distinguishing those *arachnai* which produce coarse and irregular webs (e.g. the house spiders) from the orb weavers previously considered with their much finer workmanship. Among these inferior weavers, then, we have firstly a large form with long legs which keeps watch suspended upside down from the underside of its web; and secondly a smaller variety, more well proportioned (συμμετρότερον), which keeps watch up above, hiding itself in a small hole. The first is probably *Pholcus phalangioides* (Fues.) while the second is perhaps the house spider *Tegenaria*.[42]

The same threefold definition of spiders is given by Pliny (XI.79–80), who renders the third type as *genus erudita operatione conspicuum*. He does not, however, deal with the two subgroups of web spinners considered in the preceding paragraph. Cicero (*de Nat.D.* II.123) has a twofold classification, distinguishing web spinners and the more mobile hunting species (e.g. Lycosidae).

1.4. Identification: species distinguished by medical writers

A number of varieties of spider (*s.str.*) are noted as distinct in medical sources.

 (i) A variety producing white webs (*telae candidae*), these latter being used for the treatment of mouth sores (Pliny XXX.27). This is perhaps to be equated with the 'kind of spider constructing a white web, small and thick', used, according to Dioscorides (*DMM.* II.63) in amulets for the prevention of quartan fevers.

 (ii) Certain small webs found on rafters (*quae in trabibus parvae texuntur*) are recommended by Pliny (loc. cit.) for the same purpose as (i). He also mentions a spider that constructs very coarse webs in this same habitat as a reputed cure for eye complaints when worn as an amulet (XXIX.132).

(iii) A variety termed by Pliny (XXIX.131) *araneus muscarius* whose web, and most especially its *spelunca* off at the edge of the web, was also said to be a cure for the eye condition *epiphora*: for this to be effective, particular procedures were required to be observed both by the collector of the web and by the patient. This spider is identified by Leitner[43] as a house spider (*Tegenaria domestica* (Cl.)), which is very probable. The same identification may be presumed for (ii) above.

(iv) A white spider with very long and thin legs employed as a treatment for eye ulcers (Pliny XXIX.132; Marcellus VIII.150). A white spider is also referred to in the

---

[41] *Araneus* spp. is the *Epeira* cited by Sundevall, op. cit., p. 234, and subsequent authors.

[42] Sundevall, op. cit., p. 235, Aubert and Wimmer, op. cit., I. pp. 160–1, and Keller, op. cit., II. p. 464, identify the two as species of *Tegenaria* and *Agelana*. However, none of these have the upside down habit typical of the long-legged *Pholcus*. *Agelana* is a non-domestic genus belonging to the same family as *Tegenaria*.

[43] *Zool. Terminologie*, p. 34.

Cyranides (p. 62) and recommended for the treatment of ophthalmia and hollow teeth. Judging from Pliny's description, this sounds like the long-legged domestic species *Pholcus phalangioides* (Fues.).[44]

(v) A black spider, larger than its white relative and making webs in trees, is mentioned in the Cyranides (p. 62) as a remedy for scrofula.

(vi) The 'spiders' webs (*araneorum fila*)' from vineyards mentioned by Marcellus (XXXI.20) are probably in reality the silken webs of foliage-feeding Lepidoptera which are termed ἀράχνιον by Theophrastus.[45]

2. Life history

In his account of spiders in *HA* 555a26 ff., Aristotle gives a brief general account of their development before going on to deal with particular types individually. After mating, he relates, the females produce what he describes as σκωλήκια μικρά which are round at first. These of course are in fact eggs, which subsequently hatch into miniature versions of their parents, but Aristotle's concept of the true nature of eggs forbids him to recognise them as such (cf. also *GA* 758b 9), although he rather inconsistently employs the verb ἐπῳάζω to describe the female's care of them. The offspring are laid in various kinds of ἀράχνια which differ according to species, some small and fine and some thick, some completely enclosed (ὅλως ἐν κύτει στρογγύλῳ) and some only partially so; and the female incubates them (Aristotle uses the verb normally used for nesting birds). Within three days the new-born spiders emerge, a process whch Aristotle has to describe in terms of the egg-like larvae taking on an articulated form. They do not hatch all at once, but once they have done so they are immediately able to jump about and to produce silk (cf. 623 a). They become fully grown in about four weeks (555b 16).

Pliny's account in XI.85 is a highly abbreviated version of the above, with the details about the variation in form of the egg-sack and the growth of the young spiders omitted. He renders Aristotle's σκωλήκια as *vermiculos ovis similes*, and describes the egg-laying process as follows: *pariunt autem omnia in tela, sed sparsa, quia saliunt atque ita emittunt*. The curious picture here presented by Pliny, as if to say that the females lay eggs at random all over their webs, would appear to arise from a misunderstanding of Aristotle's sentence about the unsimultaneous emergence of the newly formed spiders and their immediate commencement of activity. The Cyranides (p. 79) refers to the 'eggs' of spiders and *phalangia* being found along the roads at the beginning of spring: these would be the silken egg-sacks referred to by Aristotle and Pliny.

Isidorus, in his chapter *de Vermibus*, seems to reflect a popular belief that spiders were spontaneously generated (*Or.* XII.5.1–2). He defines *vermis* as any creature produced without sexual reproduction out of some inanimate material, and among possible materials he lists air; the spider is then defined as *vermis aeris, ab aeris nutrimento cognominata* (cf. also *Or.* XIX.27.4).

---

[44] Leitner, ibid., p. 10, erroneously identifies this as a harvestman (*Phalangium* spp.) and not a true spider.
[45] Cf. 25 below.

3. Feeding habits

Though it was well known that the main diet of web spinning and hunting spiders consisted of flies (Aristotle *HA* 488a 18, Philostratus *Im.* II.28, Nicander *Ther.* 735, Lucian *Musc.Enc.* 5, Pliny XXIX.87; Arnobius *adv.Nat.* VI.17; M.Aurelius X.10) and other winged insects (Aelian I.21; AP IX.372), there are occasional stories of their taking larger prey. Pseudo-Aristotle (*HA* 623b 1) and Pliny (XI.84) credit some spiders with the ability to capture small lizards and frogs, wrapping their mouths with silk to prevent them biting; and the latter author (X.206) even depicts the unlikely occurrence of a spider attacking and killing a snake, letting itself down onto its victim on the end of a silk thread.

4. Popular attitudes

Although some spiders were somewhat unpopular for making the interior of houses untidy with their webs, the attitude toward spiders in antiquity (except of course for the venomous *phalangia*) was generally a favourable one. They were noted for their industry and hard work displayed in their weaving, and were regarded as comparable in this respect with bees and ants (Aristotle *HA* 622b 23, *Phys.* 199a 22 ff.; Fronto p. 216.7 N; Isidorus *Or.* XII.5.2; Tertullian *adv.Marc.* I.14). The apparent skill and craftsmanship exercised in the construction of their webs, most especially the orb webs with their perfect symmetry—for it was recognised that some kinds were more skilful than others (*HA* 622b 23, 623a 7)—was greatly admired, very high expressions of praise being applied (Plutarch *Mor.* 966e–f; Seneca *Ep.* 121.22; Cicero *de Nat.D.* II.123; Pliny XI.82). As a result, they were credited with genuine intelligence (σοφία, *sollertia*: *HA* 623a 7; Aristotle *Phys.* loc. cit.; Cicero, loc. cit.); Aelian (VI.57) even ascribes to them an understanding of the principles of geometry. Plutarch (loc. cit.) describes spiders' webs as being the model for both women's looms and huntsmen's nets and the creatures themselves as combining the skills appropriate to the use of both: terminology from the activities of weaving and hunting is very commonly applied to the work of spiders, both in scientific and popular literature. Pliny (XI.84) says that to observe spiders handling large prey is a spectacle worthy of the amphitheatre.

It is interesting that amid all this eulogistic material there is no hint of spiders being considered as unattractive in appearance or in habits. There is, however, a description by Philostratus (*Im.* II.28) which refers to their savage nature, and Aeschylus (*Ag.* 1492) uses the spider's web as a sinister image, viewing it from the victim's point of view and emphasising the idea of cunning rather than that of skill.

The more elaborate spiders' webs were noted for the extreme fineness and lightness of texture of the silk threads used in their construction (Plutarch loc. cit.) and so often appear in literary similes as epitomising the idea of delicacy and fineness (*Odyssey* VIII.280; Lucretius III.383, IV.727; *Culex* 2, Ovid *Met.* IV.179, *Am.* I.14.7; Martial VIII.33). Fineness was the most sought-after quality in ancient fabrics, as the popularity of silk testifies, and it was remarked that spider silk surpassed in this respect anything that could be produced by human craftsmanship (Aelian I.21; Seneca *Ep.* 121.22). Associated with the idea of fineness is that of insubstantiality, and we find spiders' webs employed as a symbol of anything flimsy, easily broken or swept aside; for example, ineffective laws (Plato Com. fr. 22; Valerius Maximus VII.2.ext. 14), intellectual speculations (Basil *Hex.* MPG 29.8; Ambrose *Hex.* IV.17), futile schemes or precautions against disaster (LXX *Job.* 8.14, 27.18 and Gregory Magnus *Mor.* VIII.44; LXX *Is.* 59.6 and Cyril Alex. and Theodoret ad loc. MPG 70.1308, 81.461), and human life itself (LXX *Ps.* 38.12, 89.9 & Theodoret ad loc. MPG 80.1604).

As well as the much admired workmanship of the orb weavers and others, the more untidy spinnings of the domestic species also found their place in literature. Their webs frequently figure as symbols of age, decay, desertion, desolation and neglect, associated with disused and derelict buildings (Philostratus loc. cit.; Propertius II.6.35; Claudian *Rapt.Pros.* III.158) or with anything long forgotten or fallen into disuse (*Odyssey* XVI.35; Hesiod *Op.* 475; Cratinus fr. 190; Pherecrates fr. 142; Nicophon fr. 3 Ed.; Sophocles fr. 264; Lucian *Pseud.* 24, *Bis.Acc.* 3; Catullus 68.49–50; Propertius III.6.33; Jerome *Ep.* 107.1; Paulinus Nol.16.118). Associated with this imagery is the idea of cobwebs as a mark of poverty (Plautus *Aul.* 83–7; Afranius Com.411 Rb, 412 Dv; Catullus 13.7–8, 23.2). The picture of weapons or shields lying unused and covered with webs is employed on a number of occasions as a symbol of peace (Bacchylides 3.7 Edm., 13 Bk; Theocritus XVI.96; Nonnus *Dion.* XXXVIII.14).

5. Weather lore

It was an accepted fact of classical meteorology that the occurrence of threads of spiders' silk (spun in fact[46] by young spiders newly emerged from their egg sacks) drifting about in calm weather was a sign of windy or stormy conditions to come (Theophrastus *de Sign.* 29; Aratus *Phaen.* 1033; Aristotle *Prob.* 947a 33 ff.; Avienus *Arat.* 1771–3; Pliny XI.84). The cause of this phenomenon was, however, a matter of dispute. Aristotle, who devotes a section of his *Problems* to the question, suggests that it is something to do with the fact that spiders are particularly active in calm weather. Avienus and Pliny present the idea that spiders weave their webs in clear conditions and then take them apart again[47] when the weather is cloudy: it is these dismantled webs that float about and constitute the meteorological sign (cf. *Geop.* I.3.9). Hesiod (*Op.* 777) has the weaving of spiders as a seasonal sign, while Plutarch (*Mor.* 410e) states that a large number of webs was felt to be a mark of an unhealthy summer. Pliny (loc. cit.) reports the belief that the rising of rivers could be predicted by the observation of spiders setting their webs at a higher level; he also writes (VIII.103) that they can give forewarning of the imminent collapse of a building.

6. Construction and use of webs

The method of construction of spiders' webs was a subject of great fascination in antiquity, not only to writers on natural history, and is described in greater or lesser detail in a number of authors. The earliest account is that which is given in book IX of the

---

[46] Cf. J. L. Cloudsley-Thompson, op. cit., p. 155.

[47] *solvit*, Aveienus; *retexunt*, Pliny; not 'reweave' as Rackham translates.

Historia Animalium (623a 7 ff.). It is written in a somewhat compressed style and so is not luminously clear, but, as interpreted by D'Arcy Thompson,[48] it details the three basic stages involved in the construction of the orb web. Stage (i) covers the laying of the preliminary foundation of the web (ὑφαίνει γὰρ πρῶτον μὲν διατεῖναν πρὸς τὰ πέρατα πανταχόθεν), stage (ii) the laying down of the radii stretching from the centre to the circumference (εἶτα στημονίζεται ἀπὸ τοῦ μέσου), and stage (iii) the addition of the concentric or spiral threads and their attachment to the radii (ἐπὶ δὲ τούτοις ὥσπερ κρόκας ἐμβάλλει, εἶτα συνυφαίνει). Terminology drawn from the human craft of weaving is employed throughout.

Pliny's corresponding description (XI.80–1) is based upon that of the *Historia*, but with certain immaterial expansions. D'Arcy Thompson states that he has omitted stage (i) entirely, but the opening words of the section (*orditur telas*), after which the account is interrupted by some material on the origin of the silk, may be taken to correspond to this. Following the interruption, we have stage (ii) with the addition of some picturesque embellishments (*tam moderato ungue ... deducit stamina, ipso se pondere usus*) and an expanded version of (iii) (*texere a medio incipit, circinato orbe subtemina adnectens, maculasque paribus semper intervallis sed subinde crescentibus ex angusto dilatans, indissolubili nodo inplicat*). The idea of the spider employing its own body as a weight is not found elsewhere, but the picture of the creature working with its claws like the fingers of a human weaver recurs in a number of places (Aristophanes *Ran.* 1314; *AP* IX.372; Ovid *Am.* I.14.7).

The detailed description of the orb web given by Philostratus (*Im.* II.28) occurs in the context of a description of a painting, and depicts in static terms the same three component parts as are detailed in the *Historia*. We have (i) the foundation (τετράγωνος μήρινθος) which holds the whole structure together, (iii) the concentric threads (κύκλοι), and (ii) the radii (βρόχοι ἐκτενεῖς), in that order. Philostratus also mentions the spiders' habit of using silk threads to let themselves down to the ground and to climb back to their previous position. He is the only author to give anything approaching a physical description of the creatures concerned, referring to the artist's fidelity to nature in rendering their spotted appearance, their ἔριον ὑπομόχθηρον and their savage nature. The meaning of the middle phrase here is not entirely clear. It is interpreted in LSJ as a reference to the spiders' webs, but in the context and in view of the fact that a detailed portrait of the web is about to follow it would seem more likely to refer to an aspect of the creatures' external appearance.[49]

Among other descriptions of the orb web, Seneca's (*Ep.* 121.22) mentions only the concentric threads and the radii, while Aelian's (VI.57) merely emphasises its geometrical accuracy. Plutarch (*Mor.* 966e–f) draws attention to the web's thinness and evenness of texture, the fact that it holds together through having a viscous substance worked into it, and to the βαφὴ τῆς χρόας that gives it an airy or misty appearance.

There were two rival theories among classical zoologists as to the method by which

---

[48] *Hist. An.*, IX.39 note.

[49] This is how the phrase is understood by the Loeb translator, although his rendering 'repulsive fuzzy surface' gives rather the wrong impression: something like 'rather sparse coating of fur' would be better.

the silk was actually produced. The more accurate, ascribed to Democritus, was that it was formed internally as a kind of excretion (*HA* 623a; Pliny XI.80; Aelian I.21), and the alternative that it developed as some kind of external or surface product (φλοῖον in Pseudo-Aristotle; Pliny renders the idea by the phrase *lanigera fertilitas*).

Somewhere off the edge of the orb was what the *Historia* (loc. cit.) depicts as the maker's sleeping quarters and food storage space, a silk lined lair which Pliny describes as a *specus* carefully concealed at a distance from the centre and well insulated within as a protection against the cold (XI.82, cf. XXIX.131). Philostratus (loc. cit.) is under the rather picturesque misapprehension that spiders have two separate kinds of web: broad and open ones for use in the summer, and enclosed ones employed as winter quarters.[50] It was evidently a matter of popular belief that spiders spun not only to catch prey but also to keep themselves warm: Plautus (*Stich.* 347–50) has a little exchange in which a servant instructed to sweep down some spiders' webs replies that the poor creatures will feel cold (*Miseri algebunt postea*).

As far as the use of the web is concerned, it is noted that the spider keeps watch at the centre (*HA* loc. cit.; Aelian VI.57), waiting for insects to fly into it and become entangled, and sensing their arrival by vibrations transmitted to its legs through the threads (Chrysippus in *S.V.F.* II.879; Heraclitus fr. 67a Diels, 115 Marc.). As the trapped victim struggles to escape (Philostratus loc. cit.), the spider runs towards it, seizes it, and wraps it around with threads of silk (*HA* loc. cit.) before carrying it off to its lair to feed upon it by sucking its juices (*HA* loc. cit.; Pliny X.198; cf. also Aeschylus *Ag.* 1492, Xenophon *Mem.* III.11.6, Cicero loc. cit., Seneca loc. cit.). Plutarch (loc. cit.), who employs the analogy of a huntsman and his net, appears to be under the impression that when the spider senses that an item of prey has arrived it in some way manipulates the web and draws it together so as to enclose its victim; Pliny (XI.82), with a similar comparison in mind, depicts the spider's victims as colliding with the threads off the perimeter of the web and rebounding into the midst of it as into a hunting-net. Once it has dealt with its prey, the spider repairs any damage caused by it and returns to the watch (*HA* loc. cit.; Pliny XI.84). If a second victim should be caught while it is still handling or feeding upon a preceding one, it makes its way first to the centre and from there outward to despatch the new arrival (*HA* loc. cit.; Pliny XI.82): Pliny adds that this is so the latter will become further entangled by the shaking of the web. General repairs to the structure are made at sunrise or sunset (*HA* loc. cit.; Aelian VI.57), because, according to the *Historia*, these are the times when most insects are to be caught. According to the author of the *Historia* book IX (loc. cit.), it is the female alone who both weaves the web and captures prey, but the male has a share in what is taken. Pliny, however, as often, misreads the passage concerned and has a more even distribution of labour, with the male doing the hunting and the female the spinning (XI.84). As has been mentioned, it was noted that newly emerged spiders were able to spin silk straight away (*HA* loc. cit., 555b 5); Pliny (XI.83) depicts young ones practising their skill and learning how to weave.

[50] This reference to winter quarters could perhaps refer to the habit of many females of enclosing themselves with their eggs at the end of the summer.

7. Relations with man

House spiders such as *Tegenaria* spp. were well known inhabitants of human dwellings, causing a nuisance by spinning their untidy and unsightly webs on ceilings and walls (Plato Com. fr. 22; Pliny XXIX.87; Cyril Alex. MPG 70.1308) and especially among roofing timbers (Aristophanes *Ran.* 1313; Ovid *Met* IV.179, *Am.* I.14.8), both in homes and in public buildings (Arnobius *adv. Nat.* VI.17; Minucius Felix XXII.6). There are a number of references to the work of domestic slaves as including the sweeping away of accumulating cobwebs (Plautus *As.* 425, *Stich.* 355, 347–9; Titinius Com. fr. 36; Juvenal XIV.6.1) and the use of long rods or brushes for this purpose is mentioned (Ulpian *Dig.* XXXIII.7.12.22). Cf. Pliny XXX.27, XXIX.132. Certain creatures going by the name of 'spiders' are mentioned by writers on apiculture as being pests of beehives, but in all cases but one these references are evidently the product of confusion between the webs spun over the combs by the larvae of wax moths and the work of genuine spiders.[51] The sole mention of true spiders as pests of hive bees is given by Virgil (*G.* IV.247), who describes them as spinning their webs at the hive entrance (*laxos in foribus suspendit aranea casses*), presumably with the purpose of capturing the occupants as they make their way in and out.

8. Medicinal uses

Both spiders themselves and their webs figure in classical medicine and pharmacology. The creatures themselves, crushed either by themselves or with other ingredients, are prescribed for ear complaints (Dioscorides *DMM.* II.63, *Eup.* I.54; Cyranides p. 62; Cassius Felix p. 44; Pliny XXIX.138, XXX.26; Marcellus IX.11, 39, 97), eye ulcers (Pliny XXIX.132; Marcellus VIII.150), tumours (*condylomata*) (Pliny XXX.70), and boils (Pliny XXX.108); and in amulets for mouth complaints (Marcellus XIV.68), the eye condition *epiphora* (Pliny XXIX.132), fevers, scrofula, and insomnia (Cyranides p. 62), and to aid menstruation (Pliny XXX.129).

The most common use for spiders' webs was as a first aid application for wounds, to staunch the flow of blood (Pliny XXIX.114; Dioscorides *DMM.* II.63, *Eup.* I.198; Cyranides pp. 62, 258; Celsus V.2; Petronius 98): they were also said to prevent inflammation (Galen XII.343 K; Dioscorides loc. cit.). They are also recommended for application to fractured skulls (Pliny XXIX.114; Marcellus I.86; Q.Serenus 957), bruised joints (Pliny XXX.78; Marcellus XXXIV.65), for stopping bleeding from the nose (Pliny XXX.112; Marcellus X.24), for the removal of surplus flesh (Marcellus XXXI.20), and for the treatment of eye complaints (Cyranides p. 62; Pliny XXIX.131), mouth sores (Pliny XXX.27), breast conditions (Pliny XXX.131), fevers (Dioscorides *DMM.*II.63, *Eup.* II.20) and epilepsy (Theodorus Prisc. *Phys.* II.5). The egg-sacks of spiders are noted

---

[51] The spider-like beetle κλῆρος appears in Pliny and the agricultural writers as *araneus* or ἀράχνη. The φάλαγγες in Aelian's list of honey bee pests at the opening of I.58 could be taken to be these same creatures, but the correct reading here is probably φάλλαιναι, for which φάλαγγες appears as a false reading later in the same chapter.

in the Cyranides (pp. 79, 102) as effective against fevers. In the same work (p. 258) oil in which a spider has been drowned is said to provide an antidote to snakebite.

9. Spiders as omens

There are a number of recorded cases in both Greek and Roman history of spiders' webs in unexpected situations being noted as portents. The most famous of these took place at Thebes, where the sudden appearance of a distinctive web in the temple of Demeter is recorded as having been one of the signs of the impending capture of the city by the Macedonians in 335/4 BC. The most detailed account of the event is given by Diodorus (XVII.10.2.5), who describes the web as being extended to the size of a *himation* and as shining with iridescent colours. Aelian (*VH* XII.57) depicts the web as being spun over the face of the goddess' statue; while Pausanias (IX.6.6) has it over the door: the latter author says that it was black in colour, and contrasts it with a white web of good omen which appeared in the same position before the battle of Leuctra.

So far as Roman sources are concerned, we read of webs spun over legionary standards as portents of defeat and death for Pompey in 49 BC (Dio XLI.14.1) and Gaius Pansa in 43 BC (Julius Obsequens *Prod.* 69).

According to legend, the tree from which the victory garlands at Olympia were taken was indicated by being festooned with spiders' webs (Phlegon *Olympiades* fr. 1, *F.G.H.* 257 Jac.).

7b. Phalangion

1. Identification: general

The second of the main groups of spiders distinguished in antiquity consists of those species known as or reputed to be venomous. Much attention was devoted in antiquity to the study of venomous animals, which were regarded with a terror which can scarcely be said to be proportionate to the risks actually involved. Among these *phalangia* rank as one of the three main groups, after snakes and scorpions, and so figure prominently in that branch of medical literature which concerned itself with animal poisons.

From one point of view, the identification of the classical *phalangion* presents little difficulty, since there is only a single genus of truly venomous spiders native to the then known world. The most well known species is the so-called malmignatte, *Latrodectus mactans* (Fab.), a cosmopolitan spider which would have been found all over the classical world except for the north of Europe. It is very variable in colour (older authors tend to divide it into more than one species) and occurs in a number of geographical subspecies, the southern European one being *L. m. tredecimguttatus* (Rossi). Its bite results in very severe symptoms and may indeed prove fatal. Outside Europe there are two allied species, with at least the first of which writers such as Apollodorus, who was an Alexandrian, may well have been familiar. These are the middle eastern and north African *L. pallidus* Cambridge, and *L. dahli* Levi, recorded today from Iran. The bite of *L. pallidus* is said to be relatively mild in its effects.[52] However, identification is complicated

---

[52] Smith, *Insects*, pp. 426–8.

by the fact that the prevailing fear of venomous animals seems to have resulted in a number of species of harmless spider, plus certain spider-like insects, gaining a reputation for toxicity and being classed as types of *phalangia*. There are a number of spiders, for example *Chiracanthium punctorium* (Vill.), whose bites, though not venomous, can produce quite severe local symptoms. It is also probable that the harmless *Lycosa tarentula* (Rossi), much feared in mediæval Europe, would have had a similar reputation in antiquity.[53] The groups designated by the names λύκος and τετραγνάθον are clearly distinct from the rest of the so-called *phalangia* and can be identified with some degree of certainty, but in the case of some of the creatures on Nicander's list and the *phalangia* described by Aristotle conclusive identification is not always possible.[54]

As has been noted, the name φαλάγγιον is the diminutive of what appears to be a synonym of ἀράχνη: how it came to acquire its specialist meaning is unclear.[55] The word was adopted by Roman authors in the form *phalangium*, which in late sources (Marcellus, Cassius Felix, Vegetius, Chiron, Ps. Apuleius, *C.Gl.L.*) appears variously as *sfal-(sphal, fal)-angium(-ion, -ius, -io)*. Since φαλάγγια constitute a subgroup of *arachnai sensu lato*, a number of authors use *aranea* or ἀράχνη where one would expect to find φαλάγγιον or *phalangium* (*AP* IX.233; Paulus Aegineta V.7; Celsus V.26.7; Q. Serenus 860–81; Vegetius II.144; Sextus Placitus; Pliny, for example XX.117, who sometimes uses the vague phrase *aranei et phalangia*, e.g. XXIV.71, when in context it is strictly tautologous). A name τέκτων is given by Hesychius as referring to a 'type of *phalangion*': one suspects here that the definition is erroneous and that we are dealing with a popular name for the harmless γλαφυραὶ ἀράχναι.

1.2. Identification: the species in the Historia Animalium

In the authentic portion of the *Historia*, no mention is made of subdivisions under the general heading of *phalangia* with the exception of the reference to 'those *phalangia* which spin webs' (542a 12), implying that some do not (*Latrodectus*, incidentally, does produce webs). However, the author of book IX, categorising various types of spiders, gives brief descriptions of two varieties of what he describes as biting *phalangia* (622b 27 ff.).

The first of these is said to resemble the *lykos*, to be small, speckled in colour and fast moving[56] and to have the ability to jump. From the latter characteristic it possesses the specific name of ψύλλα or 'flea spider'. Pliny reproduces this description in XI.79, emphasising the creature's venomous nature and mistranslating ὀξύ as *acuminatum*.

---

[53] Ibid., pp. 425–6; Gow and Scholfield, *Nicander*, p. 23.

[54] O. Taschenberg, *Einige Bemerkungen zur Deutung gewisser Spinnentiere die in den Schriften des Altertums vorkommen, Zool. Annalen*, II, 1906–8, pp. 228–9, 245, disputes the assumption that Aristotle conceives of his *phalangia* as venomous, arguing that when he describes them as 'biting' spiders he is describing their voraciousness in relation to their prey, and not harmfulness to man; but in view of the phraseology used and the universal meaning of the word elsewhere it cannot but be concluded that he intends his readers to understand them as poisonous species. Taschenberg argues that the φαρμακοπῶλαι of the *Historia* would not have risked handling toxic spiders, but in fact dead specimens were used as antidotes to the bites of their own kind.

[55] Cf. Fernandez, op. cit., p. 89 on its etymology. In Cyranides p. 62 φαλάγγιον appears uniquely as a synonym of ἀράχνιον where non-venomous species are in view.

[56] ὀξύ: not 'tapering to a point' as D'Arcy Thompson translates, or 'aggressive' as Keller and others have it.

Judging simply by the physical description, the author here would seem to have in mind one or other of the jumping spiders (Salticidae), for which ὀξύ would be a very appropriate adjective to allude to their darting movements, and whose colouration could well be characterised by the term ποικίλος. This is the identification arrived at by Sundevall,[57] Aubert and Wimmer[58] and Taschenberg.[59] On the other hand, these jumping spiders are entirely harmless, and for this reason Keller[60] and Steier[61] prefer to identify the creature with *Latrodectus mactans*, even though this cannot jump. The simplest solution would seem to be that a jumping spider does lie behind the physical description, and that it has somehow been erroneously credited with toxicity.

The second variety is described as a large black spider which in its habits is in sharp contrast to its supposed relative, being sluggish in its movements, slow walking, not strong (οὐ κρατερόν) and never jumping; it also has long forelegs. Pliny mentions it briefly in XI.79. It is surprising that no reasonable proposals have been made as to this spider's identity.[62]

Our author concludes by noting that the above is not intended to constitute a comprehensive classification of *phalangia*, but that there are many other varieties known to pharmacologists, some of which give only a weak bite, and some of which do not bite at all.

1.3. Identification: the species known to the medical tradition

In classical medicine there is a tradition of writing on the subject of venomous animals which begins with the work of Apollodorus of Alexandria, upon whose lost treatise *Peri Therion* all later authors are dependent. Those authors whose work survives—Nicander's *Theriaca*, Philumenus' *de Venenatis Animalibus*, and parts of Pliny's *Natural History*—are all mere compilists with a tendency to distort rather than to improve upon the material they are reproducing. As we have it, the tradition is to a large extent highly fanciful, both as regards the details of the creatures described and the lurid portrayals of the reputed symptoms of their bites, so that Gow and Scholfield's comment[63] that 'the victim of snake-bite or poison who turned to Nicander for first-aid would be in a sorry plight' is entirely justified.

It is not known how many varieties of *phalangia* were detailed by Apollodorus. Nicander has a total of eight species, almost all of which occur in Pliny, while Philumenus lists seven, only four of which definitely correspond with those in the *Theriaca*. These lists clearly include spiders which were only venomous by repute, and also creatures which were not only non-poisonous but not even spiders at all. The illustrations of the various *phalangia* which appear in certain illuminated manuscripts of Nicander and Eutecnius'

---

[57] Op. cit., p. 234.   [58] Op. cit., I. p. 160.

[59] Art. cit., p. 230.   [60] Op. cit., II. p. 461.   [61] Art. cit. 1791.

[62] Sundevall's suggestion, op. cit., p. 234, of the water-spider *Argyroneta* is totally impossible, and that of Aubert and Wimmer, op. cit., I. p. 160, and Keller, op. cit., II. p. 463, of a member of the Solifugae is highly improbable. *Segestria* spp. could fit the description.

[63] Op. cit., p. 18.

Latin paraphrase are discussed by Kadar.[64] The depictions of all but two are wholly stylised, differing only in their colours. The ἀγρώστης or λύκος appears by some curious error as something resembling a deformed centipede, while the κρανοκολάπτης is portrayed as a yellowish four-winged creature with a thick tapering abdomen and a sort of two-pronged beak. They are therefore clearly of no help in identification. They were evidently intended not for diagnostic purposes, but as simple embellishments produced by an artist merely exercising his imagination on what was already before him in the text.

(i) Ῥώξ. The first variety on Nicander's list is described as being of a smoky or pitchy colour, as having legs which move in succession (ἐπασσυτέροις ποσὶν ἕρπων) and as having teeth in the middle of its stomach (Gow and Scholfield take this latter to be a simple anatomical misdescription). The scholiast on the passage explains that the spider acquired its name from its similarity to a grape (Th. 716 ff.). It appears in Philumenus (15.2 = Aetius XIII.20) under the alternative name ῥάγιον, where it is depicted as round and black, grape-like, with very small legs and a mouth in its stomach. Pliny's account (XXIX.86) corresponds to that of Philumenus, and he also uses the name rhagion. This is the only member of the group which appears outside the medical tradition; namely in Aelian (III.36), who names it ῥάξ, follows the details of its appearance as given by Philumenus, and credits it with causing instantaneous death: he also locates it as a native of Libya.

The description given, as regards colour and similarity to a grape, makes it fairly certain that we are here dealing with *Latrodectus mactans* itself; which is the conclusion arrived at by Scarborough and Steier.[65] One detail which is inappropriate to *Latrodectus* is Philumenus' reference to small limbs: this may perhaps be due to a misinterpretation of Nicander's ἐπασσυτέροις.[66]

Scarborough[67] suggests that the κόκκος defined by Hesychius as a type of *phalangion* is to be equated with the ῥώξ.

(ii) Ἀστέριον. The 'starry spider' is said by Nicander (725 ff.) to be characterised by gleaming coloured (λεγνωταί) stripes or bands (ῥάβδοι) on the upper surface of its abdomen. Pliny (XXIX.86) states that it is otherwise identical to the ῥώξ or rhagion. Particularly in view of this last statement, the most likely identification of this variety[68] is as a colour variant of *Latrodectus mactans* (such varieties were held to be separate species by authors of modern times).

(iii) Κυάνεον. This dark blue variety is described by Nicander (729 ff.) as woolly or downy and as darting about off the ground (πεδήορον ἀμφὶς ἀίσσει). Pliny

---

[64] *Zoological Illuminations*, pp. 43, 47.

[65] Scarborough, art. cit., p. 8; Steier, art. cit., 1790–1. Taschenberg, art. cit., pp. 236–9, proposes instead that the description fits some form of centipede, but this idea may safely be disregarded. Equally improbable, is the theory of Gossen, *Tiernamen*, no. 47, and art. in PW Supp. VIII.355, to the effect that Aelian's account refers to a disease-carrying tick: the ancients had no conception of the transmission of diseases by invertebrate animals.

[66] The ms reading in Aelian III.36 is μακρούς.

[67] Art. cit., p. 7.

[68] Steier, art. cit., 1792; Gow and Scholfield, op. cit., p. 184; Scarborough, art. cit., p. 8.

(XXIX.86) latinises its name to *caeruleus* and says that its fur is black. Steier[69] and Leitner[70] suggest that it is probably to be equated with the *araneus lanuginosus* with a large head described by Pliny in the preceding chapter [85] from a source other than Nicander. Pliny adds, on the authority of Caecilius, that when examples of this latter spider are cut open, they are found to contain a pair of *vermiculi* which, when wrapped in deer skin, are effective as a contraceptive amulet.[71]

(iv) Σφήκειον. According to Nicander (738 ff.), the σφήκειον is reddish and is so named because of its resemblance to a wasp. Pliny (XXIX.86) says that it differs from a wasp (*crabro*) only in its lack of wings. This is perhaps to be identified as one of the wingless hymenoptera of the family Mutillidae.[72]

(v) Μυρμήκειον. The *myrmekeion*, named from its supposed resemblance to an ant, has, according to Nicander (747 ff.), a fiery red neck, a sooty black head, a dust coloured body, and an abdomen spotted or starred all over its upper surface. It appears in Philumenus (15.3 = Aetius XIII.20) as resembling a large ant: his description abbreviates that of Nicander. The scholia on Nicander preserve some material from a lost work of Sostratus Περὶ βλητῶν καὶ δακέτων (fr. 4 Wellman) in which the creature, said to be also known by the names of μυρμηκοειδές and μύρμηξ ἡρακλεωτικός, is described as possessing a black head, a fiery red neck, and a black abdomen dotted with small white spots like a star.

Pliny has two more or less identical descriptions of this *phalangion* (XXIX.84,87), the repetition being due to the fact that he is working from two separate sources. The first of these, which does not use the specific name, characterises it as like an ant but much larger, with a red head, and with the remainder of its body black with white spots; and adds that its bite is more painful than that of a wasp (*vespa*) and that it especially lives *circa furnos et molas*. The second description, which is the one in sequence with the rest of Nicander's varieties, omits the habitat reference, declares the sting to be 'as painful' as a wasp's rather than more so, and particularises the head as the especial point of resemblance with ants while making no mention of its colour.

Fernandez[73] suggests that the name σίφων defined by Hesychius as εἶδος θηρίου μυρμηκοειδές is to be seen as a further synonym. Keller[74] proposes as an identification a small harmless jumping spider, *Myrmarachne formicaria* (Deg.) which is somewhat ant-like with a white-dotted abdomen. We may be more or less certain, however, on the basis of the fairly detailed and clear description, that the correct identification is that put forward by Taschenberg;[75] namely that the

[69] Art. cit., 1801.   [70] Op. cit., p. 34.

[71] Most authors declare this to be unidentifiable, though Taschenberg, art. cit., p. 267, offers a suggestion.

[72] Taschenberg, art. cit., p. 243, suggests a member of the genus *Scolia*, but these are not wingless. Scarborough, art. cit., p. 12, suggests a jumping spider of the genus *Sarinda*.

[73] Op. cit., pp. 37–8.   [74] Op. cit., II. p. 467, followed by Leitner, op. cit., p. 37.

[75] Art. cit., p. 243, supported by Gow and Scholfield, op. cit., p. 185, and Scarborough, art. cit., p. 13.

creature concerned is a hymenopteron, a member of the Mutillidae or 'velvet ants' such as *Mutilla europea* Linn., whose wingless female does superficially resemble a large ant, has a painful sting, and possesses a red thorax, black head and black white-marked abdomen.

(vi) Τετραγνάθον. The 'four-jawed' spider is absent from Nicander's list but is described by Philumenus (35.1–2) and Aetius (XIII.19) in a different section from the rest of their *phalangia*, though they say it is a 'type of *phalangion*', and by Pliny (XXIX.87). It is also reported as an exotic species by a number of geographers. The name occurs in the alternative forms *tetragnathion* (Pliny) and τετραγνάθος (Agatharchides). According to Philumenus, there are two varieties: one rather flat and whitish, rough-legged, with two excrescences on its head at right angles to one another, conveying an impression of two mouths and four jaws: and the other having a line dividing its mouth into two equal parts, also presenting a four-jawed appearance. One suspects that we have here what were originally not descriptions of two separate varieties, but rather alternative explanations of how the same creature acquired its name. Aetius combines the two into a single composite description. Pliny also distinguishes two forms of this spider, but his account does not correspond with that of Philumenus. He states that the more dangerous of the pair is that which is marked by two white lines crossing each other in the centre of its head, while its less venomous relative (*lentior*) is ashen in colour becoming white towards its hind end.

In a number of geographical works and authors dependent on them (Agatharchides 59; Strabo XVI.4.12; Diodorus III.30; Aelian XVII.40; Pliny VIII.29), we find the account of a certain vacant tract of land—located by the first three authors in Africa near the territory of the Akridophagoi, by Pliny in Africa near the Cynamolgi, and by Aelian in India near the Rhizophagoi—which was formally inhabited, but whose population was compelled to permanently evacuate the area when it was invaded by vast swarms of scorpions and τετραγνάθα (φαλάγγια in Diodorus, *solipugae* in Pliny) which caused numerous fatalities and could not be resisted. There is a further reference in Strabo (XVII.3.11) to *phalangia* of unusual size occurring in large numbers at Masaesylia on the coast of Libya.

Taschenberg,[76] Steier[77] and subsequent authors[78] agree in identifying the τετραγνάθον with species of the order Solifugae, arguing that the name is appropriate to the formidable mouthparts of these creatures which consist of a pair of pincer-like chelicerae. Gossen[79] cites the common North African species *Galeodes arabs* Koch, as does Scarborough.[80] Although they are not in fact venomous, these solifuges are large and sinister-looking, and are able to inflict a painful bite.

---

[76] Art. cit., p. 234.   [77] Art. cit., 1798–9.

[78] Leitner, op. cit., p. 37; Scarborough, art. cit., pp. 10–11; J. L. Cloudsley-Thompson, op. cit., p. 89; F. S. Bodenheimer, *Animal and Man*, p. 75.

[79] Art. cit., no. 45.   [80] Art. cit., pp. 10–11. Another large species is *G. araneoides* Koch.

As has been mentioned, Pliny, in his version of the geographers' story, uses the name *solipuga* for the offending creatures. Elsewhere he says (XXIX.92) that this is the name given by Cicero to a kind of venomous ant rare in Italy and called in Spain (Baetica) *salpuga*. This latter spelling is used by Lucan (IX.837), who was himself a native of Spain, to refer to one of the venomous creatures of the Libyan desert, dangerous when its lair is trodden upon. Solinus (IV.3), followed by Isidorus (*Or.* XII.3.4, XIV.6.40), spells the word *solifuga*, explaining that the animal is so named because it flees the daylight. He describes it as the one venomous creature native to Sardinia, an *animal perexiguum aranei forma* which inhabits the silver mines, creeping about unnoticed until it is inadvertently sat upon, whereupon it delivers its deadly bite: the local spring water provides an antidote (IV.6). According to the grammarian Festus (389.4.L), the name is *solipugna* and refers to a *genus bestiolae maleficae* which derives its name from the fact that it becomes more active under the heat of the sun. In the Latin glossaries the word is variously spelt *solipaga*, *salpinta*, and *salpiga*, and variously defined as 'a kind of fly' (II.185.57), a many legged creature (III.433.7), and a *serpens quae non videtur* (V.97.19).

Taschenberg,[81] on the basis of Pliny's usage, simply equates the Latin *salpuga* with the Greek τετραγνάθον as referring to members of the Solifugae. He points out that, although there are no members of this order in Sardinia, there is a species native to Spain, *Gluvia dorsalis*: this latter is said to be a familiar sight in the regions where it occurs.[82] These conclusions are supported by Steier[83] and Scarborough.[84] Keller[85] suggests that the Latin term had a wider application and was in effect a native synonym of the Greek φαλάγγιον. This view finds some support from Pliny (XXII.163), where *solipuga* apparently translates Theophrastus' (*HP* VIII.10.1) φαλάγγιον, and from Arnobius (*adv.Nat.* II.23, VII.16), who twice uses *solifuga* in passing in lists of venomous animals where a general rather than a specialised word would be expected. Alessio,[86] supported by Fernandez, puts forward the view, based upon an ingenious but unconvincing identification of the creatures concerned, that the Latin term derives from a hypothetical Greek original. It is, however, most likely that *salpuga* represents a native Spanish word from which the various alternative spellings derived by popular etymology.[87]

An interesting name ἡλιοκεντρίς known only from a single glossary entry

[81] Art. cit., pp. 248–52.   [82] Cloudsley-Thompson, op. cit., p. 89.

[83] Art. cit., 1798–9.   [84] Art. cit., pp. 9–10.   [85] Op. cit., II. p. 462.

[86] G. Alessio, 'Zoonymata', *RFIC*, 1938, pp. 152–6; Fernandez op. cit., pp. 150–1. The jumping spiders referred to by Alessio are too small and inoffensive to have acquired a sinister reputation, and the courtship behaviour supposed to have given rise to their name is not something which would draw the attention of a casual observer.

[87] Ernout and Meillet, *Dict. Ét.*, p. 591. The habits of various Solifugae would fit well the contrasting popular etymologies: though most species are nocturnal (hence *solifuga*)—for example *Galeodes*, others—e.g. *Gluvia*—are active in daylight and in some parts of the world are called 'sun spiders' (hence *solipugna*); cf. Cloudsley-Thompson, op. cit., p. 89.

(II.185.57), where it is defined as μυίας εἶδος: *solipaga*, is taken by Alessio and Fernandez[88] as being a Greek rendering of the Latin *solipuga*, interpreting the latter as meaning 'stinging under the influence of the sun' (which is the etymology given by Festus). Fernandez takes the μυίας εἶδος literally and identifies the creature as a kind of biting fly, but this is unnecessary in view of the fact that *phalangium* itself is in four glosses defined as *musca venenosa*.

(vii) Λύκος. The so-called 'wolf-spiders' were recognised in antiquity either as a distinct sub-group under the general heading of *phalangia* (as they are by the medical writers), or as a class in their own right standing alongside *arachnai* and *phalangia*. The feature that was supposed to distinguish them from other spiders is never in fact defined: the name certainly covers the active hunting spiders (Lycosidae) which do not construct webs, but it also comprehends some varieties that do spin webs, though where webs are mentioned it is emphasised that these are distinctly poor efforts. Book IX of the *Historia Animalium* makes mention of three distinct varieties of λύκοι: (1) a small form with no web; (2) a larger variety which spins a web of poor quality over holes on the ground or on dry stone walls and keeps a watch inside its hole, holding onto the end threads, ready to leap out and seize upon any creature that becomes entangled; and (3) a speckled form spinning a small and inferior web under trees. Pliny's version of this account (XI.80) includes only the first two varieties and describes the larger of them as spreading out small forecourts in front of their holes in the ground (*maiores in terra et cavernis exigua vestibula praepandunt*): he translates the Greek name as *lupus*. Species (1) is identified by Sundevall[89] as a small Lycosid; (2) appears to be, judging from the very clear description of its web-surrounded tunnel, a member of the genus *Segestria* or the common Mediterranean species *Filistata insidiatrix* (Forsk.);[90] while (3) has been generally identified[91] as a member of the genus *Theridium*.

There was evidently some difference of opinion among medical authorities as to whether the wolf-spider could be considered dangerous or not. Nicander (734 ff.) in fact states that its bite is harmless, but the inclusion of it in his work indicates that it had acquired a reputation as venomous. Philumenus (15.1–2 = Aetius XIII.20), who along with Nicander and Pliny (XXIX.85) counts it as a form of *phalangion*, makes no such disclaimer, so one must assume that he for one regarded it as venomous. Both writers characterise the spider as a web spinner which captures and devours flies and any other creature that is unfortunate enough to find its way into its trap. Nicander does not describe its physical appearance, but Philumenus

---

[88] G. Alessio, art. cit.; Fernandez, op. cit., p. 236.

[89] Op. cit., p. 234. Aubert and Wimmer, op. cit., I. p. 160, Keller, op. cit., II. p. 463, and Steier, art. cit. 1794, suggest a harvestman (Opiliones) rather than a true spider, but this is unlikely.

[90] Cf. Cloudsley-Thompson, op. cit., p. 166. Sundevall, op. cit., p. 234, Aubert and Wimmer, op. cit., I. p. 160, and subsequent authors suggest a member of the genus *Agelana*, but these do not build the distinctive hole-based webs described by Ps. Aristotle and Pliny.

[91] Sundevall, op. cit., p. 234, Aubert and Wimmer, op. cit., I. pp. 160–1.

says that its body is broad and rounded, that its neck region is furrowed or incised, and that it has three smooth processes arising from its mouth.

Nicander in dealing with this species uses an alternative name ἀγρώστης which has the meaning of 'huntsman',[92] appropriately enough, and is not found elsewhere.[93]

The λύκος was the only form of *phalangion* which was itself used for medicinal purposes other than as an antidote against the poison of its own kind. Dioscorides (*DMM*. II.63) includes it as a treatment for tertian fevers and mentions that it was also known by the name of ὁλκός.[94] It is also prescribed by Pliny for complaints of the spleen (XXX.52) and for fevers (XXX.104), either made into a plaster along with its web or used in the form of an amulet: in these two passages he latinises the Greek name to *lycos*.

The wolf-spider of Nicander and Philumenus has been variously identified, but the majority of authors[95] regard it as a member of the Lycosidae. It may be noted that the famous tarantula (*Lycosa tarentula*), popularly regarded as dangerous in mediæval Europe, belongs to this family.

To be equated with *lykos* are the λειμώνιαι ἀράχναι of Aristotle and the second class of spiders distinguished by Cicero (*de Nat.D*. II.123), comprising those species which rather than spinning webs to capture prey simply lie in wait and seize upon anything which is passing.

(viii) *Πιθήκη*. According to Aelian (VI.26) the πιθήκη or 'monkey' is a venomous spider also known as ὀρειβάτης, ὑλοδρόμος or ψύλλα. It is produced among trees, is hairy, and has a slight incision in its abdomen so that it looks as if it has been cut into by a thread. Its bite, the symptoms of which are described, is very dangerous, but river-crabs serve as an antidote. Fernandez[96] assumes this creature to be identical with the ψύλλα described in the *Historia Animalium* as a type of *phalangion*, though there is nothing in the descriptions themselves which corresponds, while Gossen[97] identifies it as *Latrodectus mactans*.

(ix) *Κανθαροειδής*. Only Nicander (752 ff.) gives a detailed account of this variety, for which he gives no specific name (the above is the title given to it in illuminated manuscripts). He describes it as a small *phalangion* resembling a blister-beetle (εἴκελα κανθαρίδεσσι φαλάγγια τυτθά), fiery in colour, which darts about in swarms in fields of pulses and other legumes and bites those harvesting the crop by hand. It is not unlikely that it is to be equated with the *phalangia* said by Theophrastus (*HP* II.18.1) and Aelian (IX.39) to be produced among bitter vetch, and with Pliny's *solipugae* found among legumes (XXII.163; cf. XVIII.156).

---

[92] Cf. Fernandez, op. cit., pp. 176–7.

[93] Strictly speaking he speaks of it as the 'ἀγρώστης ... which resembles the λύκος in form', but this is simply a misleading mode of expression.

[94] Cf. Fernandez, op. cit., pp. 155–6.

[95] Taschenberg, art. cit., p. 242; Steier, art. cit., 1793; Scarborough, art. cit., pp. 11–12.

[96] Op. cit., p. 51.   [97] Art. cit. no. 46.

There are a number of theories as to the identity of these creatures. Taschenberg[98] takes the view that, judging from the description, they must in fact be true blister-beetles, e.g. *Mylabris* spp., even though these do not bite or sting. He adds, however, that the symptoms of blistering described and the habitat are reminiscent of the harvest-mite, and that, even though the mite itself is too small to be the creature depicted by Nicander, its bite could have been erroneously ascribed to something more conspicuous such as a brightly coloured beetle. Gow and Scholfield[99] point out, quite correctly, that *Latrodectus mactans* itself, which lives in fields, constitutes a considerable hazard to those harvesting crops. This identification, however, would leave the comparison with the κανθαρίς unexplained; Scarborough,[100] therefore, returns to the theory that the offending creature is a blister beetle.

(x) *Κρανοκολάπτης*. At the end of Nicander's list of *phalangia* (759 ff.) there appears what is arguably the most extraordinary creature in his entire poem. Nicander himself does not put a name to it, but the scholia inform us that it was known as κρανοκολάπτης or κεφαλοκρούστης. It is described as native to Egypt and as developing among the leaves of the persea tree. In appearance it resembles a nocturnal moth with wings which are dense or felted (στεγνά) and downy, like the hands of someone who has handled dry dust or ash. It has a fearsome head which νεύει αἰὲν ὑποδράξ and is hard, and a heavy body. But its most important feature is its deadly sting, with which it attacks the head and neck of its victims causing instantaneous death. We learn from the scholia on this passage that the creature was considered by Sostratus in his lost work Περὶ βλητῶν καὶ δακέτων: they quote this author (fr. 2 Wellmann) as saying that an antidote to its sting is a member of its own kind drowned in oil.

Philumenus' statement (15.5 = Aetius XIII.20) that the creature is greenish (ἔγχλωρος) in colour would appear to be the result of a misreading of Nicander or some other source (some manuscripts and the scholia to Nicander read ἔγχλοα instead of ἔγχνοα): this would not be the only such error evident in this author's work. He also describes it as elongated (ὑπόμηκες), and as having a sting under its neck with which it attacks the head of its victims: the curiously situated sting is probably another misunderstanding of Nicander's source.

Under the name of κρανοκόλαπτα φαλάγγια these creatures are also referred to by Dioscorides (*DMM*. I.187), who mentions that they are found on the persea tree and are especially prevalent in the Thebaid. It is very probable that Pliny's venomous moth (XXVIII.162) is based upon some account of the κρανοκολάπτης (*papilio quoque lucernarum luminibus advolans inter mala medicamenta numeratur. Huic contrarium est iocur caprinum*).

It is clear from Nicander's description that lying behind the highly coloured

---

[98] Art. cit., pp. 244, 258.   [99] Op. cit., p. 185; cf. Smith, op. cit., p. 427.
[100] Art. cit., p. 13. He suggests *Lytta*, but this is the wrong colour, unlike *Mylabris* spp.

accounts of the κρανοκολάπτης is some form of lepidopteron, a large furry moth which because of its dramatic appearance was erroneously credited with being highly dangerous and then classified among the *phalangia*. The scale-covered wings of lepidoptera are perfectly described. Taschenberg[101] suggests one of the hawk moths (Sphingidae), with their long probosces, shining eyes and bulky abdomen; and some of these are indeed sufficiently impressive to have produced such fear in the minds of the uninformed public. Fernandez[102] considers the creature to be fabulous, but the ascription to it of a particular food plant and the reference to its antidote strongly support the view that we are dealing with a real animal, however its description may have been embellished.[103]

(xi) Σκληροκέφαλον. The 'hard-headed' spider appears only in the catalogues given by Philumenus (15.4) and Aetius (XIII.20) who describe it as having a hard stony head, and as having all over its body markings like those of moths that are attracted to light. It is, however, quite evident as Taschenberg[104] points out, that its occurrence as a separate variety is purely the result of a ludicrous misunderstanding on the part of Philumenus of the account of the κρανοκολάπτης given by Nicander or some other source. The latter creature is described in the *Theriaca* as resembling the moths that fly to light and as possessing a hard head.

(xii) Σκωλήκιον. Also mentioned only by Philumenus (15.4) and Aetius (XIII.20) is the 'worm-like' spider, described as elongated and somewhat spotted (ὑποσπίλον), especially in the region of its head. Taschenberg[105] is of the opinion that it is to be equated with the κανθαροειδής of Nicander, but there is no evidence for this.

2. Life history

Aristotle describes the courtship of certain *phalangia* which spin webs in *HA* 542a 12 ff. He states that the females and males encounter one another on their webs, and that the former pull upon the web from the centre, whereupon their partners pull in response from the other direction: after they have gone through this process several times they come together for mating. The next stage is related in 555b 10 ff., where we are told that φαλάγγια in general lay numerous eggs (or, in Aristotle's view, immobile larvae) in a thick 'basket' (γυργαθὸν παχύν) spun from their silk. The males assist the females in the incubation of their offspring. When the young have grown in size, they surround the female and kill her, and often the other parent as well if he can be captured. The latter piece of erroneous information is repeated by Antigonus (*Hist.Mir.* 87) and the scholiast on *Theriaca*. 11.[106] Pliny (XI.85) reproduces the details about egg-laying and subsequent

---

[101] Art. cit., p. 245.   [102] Op. cit., p. 112.

[103] Scarborough's theory, art. cit., pp. 14–15, involving confusion between some form of large wasp and burnet moths (*Zygaena* spp.) producing a toxic secretion, is unlikely since the latter are butterfly-like rather than moth-like, are diurnal, and are in no way sinister in appearance.

[104] Art. cit., p. 257.   [105] Ibid., p. 258.

[106] This story could be due to observation of female spiders which enclose themselves with their eggs at end of summer and die so enclosed.

developments, interpreting Aristotle as saying that the eggs are laid *in ipso specu* and that the young spiders actually eat the females.

3. Phalangion bites: their symptoms and treatment

Alongside snakes and scorpions, *phalangia* constituted one of three main classes of venomous animals feared in antiquity. In the case of the *Latrodectus* spiders, whose bites do cause severe symptoms and occasional fatalities, these fears were justified, but the activities of this genus resulted in many harmless species acquiring an equally sinister reputation. Spiders, of course, only bite in self defence when disturbed, but suspicion of them in antiquity was such that, as with scorpions, they were popularly believed to attack humans out of sheer malice and indeed to go out of their way in order to do so (Plutarch *Mor.* 525 f.; Philo *de Somn.* II.88; Ps. Demosthenes XXV.96). Among non-technical authors, we find references to their bites as having fatal consequences (Philo loc. cit. Diogenes L.VI.44; Solinus XI.13), and as causing suppuration (Aristotle *Top.* 140a 4) and necessitating the amputation of a limb (*AP* IX.233); according to Xenophon (*Mem.*I.3.12) their victims are driven insane, and there is an account by Strabo (XI.4.6) of a certain Albanian variety which causes its victims to die of laughter.

In his *Theriaca* Nicander purports to give the distinctive symptoms for each of the *phalangia* in his catalogue, and these he depicts in his characteristically lurid and highly coloured manner, sparing his readers none of the gruesome details. The symptoms detailed for the ῥώξ (719 ff.) do bear some relation to those recorded for *Latrodectus mactans*,[107] but the remainder are simply the product of over vivid imagination. The details given by Nicander on this subject are reproduced in highly abbreviated form by Pliny (XXIX.86–7) for each of the varieties he takes from his source's list. Composite lists of symptoms for *phalangia* in general are given by Pseudo-Dioscorides (*Iob.* 5), Paulus Aegineta (V.6–7), Philumenus (15.6–9, 35.2–3) and Aetius (XIII.18–20), the latter two giving separate accounts for τετραγνάθα, and for *arachnai* erroneously viewed as distinct from *phalangia*.

In view of the reputation of *phalangia*, as of other poisonous invertebrates, it is not surprising that classical medical writers devote considerable attention to the detailing of antidotes against their bites. Nicander (839–956) has an enormous general catalogue of treatments for invertebrate bites in general, the bulk of which is composed of over a hundred species of plant, a very high proportion (probably well over three-quarters) of all the varieties recognised in antiquity. In addition to these, he prescribes rabbit curd, sheep dung, river crabs, salt, red ochre, and nitre, as well as treatment by cautery and by immersion of the patient's body in a skin of wine, and blood-letting by the use of cupping glasses or leeches. His list is discussed and tabulated by Scarborough.[108] In Pliny's *Natural History*, we find thirty-two references to herbal remedies (XX.117,133,175,182, XXI.119,141,149,170,   XXII.64,   XXIII.135,155,   XXIV.61,71,120;   XX.62,106, XXI.162,   XXII.60,   XXIII.43,63,160,   XXIV.16,62,79,154,167,   XXV.119,163,

---

[107] Scarborough, art. cit., p. 8.   [108] Ibid., pp. 80–7.

XXVII.30,32,124,134) along with the prescription of tortoise flesh (XXXII.33,45), *castoreum* from beavers (XXXII.30), the young of weasels (XXIX.84), red mullet (XXXII.44), cock's brains (XXIX.88), and the ash of sheep's dung (XXIX.88), ants taken in drink (XXIX.88), sea water (XXXI.65), wine or vinegar (XXIII.43,63). Other similar, but briefer, antidote lists are given by Galen (XIV.175–204, 248K), Dioscorides (*Eup.* II.126–7, cf. *DMM.* II.10,22,64,69, V.11), Philumenus (15.10–16), Aetius (XIII.20), Paulus Aegineta (loc. cit.), Celsus (V.27.6), Marcellus (XVII.15), Q. Serenus (860–61), and Cassius Felix (pp. 166–8), who gives something approaching a programme of treatment (cf. also Cyranides p. 55; Sextus Placitus 3a.10, 28.1; Ps. Apuleius *Herb.* 3.8, 99.4).

It was believed that actual specimens of *phalangia* could serve as effective antidotes against the bites of their own kind, and dead ones were kept in stock by medics for this purpose (Pliny XXIX.84; Galen XIV.248K). It would seem that the reference by Pseudo-Aristotle (*HA* 622b) to the supplying of *phalangia* by φαρμακοπῶλαι refers to this practice: it is interesting that the author states that even non-venomous varieties were included. Pliny writes that simply to show the patient one of the spiders was regarded as effective. He also notes that the creatures' *cortices*, by which he perhaps means their shed skins, may, when discovered, be pounded up and taken internally. The use of magic is mentioned by Plato (*Euth.* 290a) and Philo (*de Somn.* II.88).

Phalangia were regarded as a serious threat not only to man, but also to his domestic animals (*Hippiatrica Ber.* 86, *Cant.* 71; Vegetius *Mul.* II.141,144–5; Chiron 512–14). As well as biting animals out in the open (86.3), they might also breed among the hay used as fodder (86.9,10) with the consequent threat of their young being eaten as well as of attack by the adults. The adults themselves might also be accidentally swallowed (86.15; *Mul.* II.144). The symptoms of *phalangion* poisoning in animals are described (86.1,9,10,12,14; 71.13; *Mul.* II.144–5; Chiron 512,514), and various remedies, some preparations to be taken internally, some to be applied externally to the injured spot, are prescribed. These remedies, as with those in human medicine, are mainly herbal, but also include such items as salt, wine, honey, earth from ant nests (86.3), the nestlings or nests of swallows (86.3), stag's rennet (86.9) and the offending spiders themselves (Chiron 512). Cutting round or cauterising the affected spot (86.9,10,12), letting of blood (86.14; *Mul.* II.144), immersion in warm water (86.12), and the use of religious invocations (86.3) are also mentioned. According to Pliny (VIII.97) stags are able, if bitten by *phalangia*, to cure themselves by eating crabs.

Arachnida: Acari (Mites and Ticks)

8. *Kroton/Ricinus*

1. Identification

The creatures known by the ancients under the above names may be identified as being primarily those arachnids which we know today as ticks (Ixodoidea).[109] Being well known external parasites, in the first place of domestic animals, but also of man, they were closely associated with lice (φθεῖρες); but, since they are quite distinctive in appearance and habits, the possibility of confusion between the two groups may for the most part be disregarded. Of the two chief families of ticks, the Argasidae comprises in Europe only pests of birds, and these are not explicitly mentioned in ancient sources. We are therefore dealing essentially with the Ixodidae. Ticks of this family are blood-sucking parasites whose habits differ from those of lice in that they do not remain upon the one host for the whole of their existence but are more mobile. The most important species do not restrict themselves to a particular host species, but tend to occur upon a range of domestic and wild animals. From domestic animals they may be transferred to human beings, but they do not infest man in the way that lice do. The most well known and ubiquitous species is *Ixodes ricinus* (Linn.), which may be found upon sheep, cattle, dogs, etc., and upon man. Other significant species include *Haemaphysalis punctata* Can. & Fanz., *Dermacentor reticulatus* (Fab.), and *D. marginatus* Schulze.[110] A number of authors[111] have suggested that the parasitic Diptera of the family Hippoboscidae were included by the ancients under the names κροτών and *ricinus*. These creatures would certainly have brought themselves to the attention of farmers: the only question is what in fact they would be most likely to have been called. It is quite probable that the sheep-ked, *Melophagus ovinus* (Linn.), being wingless, sedentary, and therefore superficially similar to a tick, would have gone under this name, but *Hippobosca equina* Linn. is winged and quite active, and would therefore probably have been classed as a fly.[112]

2. Life history

Unlike lice, female ticks lay their eggs in one batch away from their hosts. The larvae which emerge from these climb onto grass blades and shrubs and cluster there until a suitable host passes, whereupon they attach themselves to it.[113] It would seem that Aristotle's (*HA* 552a 15) statement that ticks are spontaneously generated out of the grass known as ἄγρωστις is based upon observation of the larvae awaiting passing animals in this way.

3. Description and habits

There is a considerable difference in size between gorged and ungorged specimens of ticks, since the abdomen is highly extensible and swells as it fills with blood. Observation of this fact gave rise to the erroneous belief, recorded by Pliny (XI.116, XXX.82), that the unfortunate creature had no outlet for its food but simply continued to swell up until

---

[109] This identification has been generally accepted by previous authors, for example Sundevall, *Thierarten*, p. 227; Keller, op. cit., II. pp. 398–9; Steier in PW 3A.1810; Gossen in PW Supp.VIII.354; W. Richter in PW Supp. XIV.981–4.

[110] Cf. Smith, *Insects*, pp. 440–6; Soulsby, *Helminths*, pp. 472–3, 483, 485, 487.

[111] Sundevall, op. cit., p. 227; Aubert and Wimmer, op. cit., I. p. 166.

[112] Cf. Soulsby, op. cit., pp. 450–2.

[113] Ibid., pp. 470–1.

it burst. It was also held that it remained in one spot for as long as it survived (Pliny XI.116).

4. Proverbial expression

The phrase ὑγιέστερος κροτῶνος, 'healthier than a tick', was evidently a popular and well known one. It is found twice in the surviving remains of Menander (fr. 305A, 318), in the Greek Anthology (AP IX.503), and in Aelian (Ep. 10). Though there is no doubt that in the examples given above the proverb was understood as an allusion to the creature we are discussing, there was some dispute even in antiquity[114] as to whether this was the original meaning of the saying. According to some (Strabo VI.1.12; Eustathius Comm. in Dionysium Periegetem 369) the original and correct version was Κρότωνος ὑγιέστερος referring to the city of that name as having a reputation for a healthy environment, and only by way of a joke was it altered to give an apparent allusion to ticks. Zenobius, on the other hand, states that the expression does indeed originate from the animal, and with good reason; for, he says, λεῖόν ἐστιν ὅλον καὶ χωρὶς ἀμυχῆς καὶ μηδὲν ἔχων σίνος. A similar explanation is given by the lexicographers (Suidas and Photius) τὸ γὰρ εἶναι πάντοθεν ὅμοιον καὶ μηδεμίαν ἔχειν διακοπήν, ἀλλ' εἶναι λίαν ὁμαλῶς. In other words, the tick always appears as if it is healthy and well fed, i.e. when it is engorged.

5. Relations with man

Although ticks are known as pests of human beings, they do not cause the constant infestations characteristic of lice. Accordingly direct references to their association with man are almost non-existent. Pliny (XXX.82) describes the habits of ticks and characterises them as foedissimum animalium, but does not actually mention that they suck the blood of man as well as animals. Petronius (57), though, does introduce them as a human parasite in association with lice.

6. Relations with domestic animals

The main significance of ticks to man lies in their attacks on his domestic animals, and descriptions of their activities and of means of combating them are found in a number of agricultural contexts.

(i) Asses. According to Aristotle (HA 557a 15 ff.), followed by Pliny (XI.116), asses are free from ticks, but this is in fact incorrect.

(ii) Cattle. Cattle ticks are noted by Aristotle (loc. cit.) and Pliny (loc. cit., XXII.47). These would be especially I. ricinus and H. punctatus.[115] Columella (RR VI.2.6) describes them as fastening onto the animal's thighs, and says that they should be removed by hand. Later, however (VII.13.1), he advises against this practice on the grounds that sores may result, but says that the creatures will drop off if touched

---

[114] As among modern authors: cf. Gossen, art. cit., 354; Richter, art. cit., 983; Fernandez, Nombres, p. 162.

[115] Boophilus, mentioned by Gossen, art. cit., 355, is a non-European genus.

with a preparation of pitch and lard or bitter almonds. According to Pliny (XXII.47) and Dioscorides (*DMM*. III.9) the herb *chamaileon* is fatal to ticks as is cedar oil (*DMM*. I.77).

(iii) *Sheep and goats*. Both of these animals figure in the lists of potential hosts of ticks given by Aristotle and Pliny (loc. cit.). With regard specifically to ticks on sheep (especially *I. ricinus* and probably also *Melophagus ovinus*), Cato (*de Ag*. 96) recommends that after shearing the animals should, as a preventive measure, be smeared with a mixture of *amurca*, wine dregs and water in which lupin has been boiled. Galen (XI.19 K) mentions the use of cedar oil for the same purpose, and the *Geoponica* (XVIII.16) prescribes the use of the same as well as a number of other herbal preparations.

(iv) *Pigs*. Ticks from pigs are mentioned only in the context of a magical prescription ascribed to Nigidius (Pliny XXX.84).

(v) Ticks such as *I. ricinus* and *D. reticulatus* are common pests of dogs, and their presence on these animals is noted by Aristotle (loc. cit.) and Pliny (XI.116, XXII.47, XXX.83,134). By Aristotle they are given the special name of κυνοραιστής 'dog destroyer'[116] which is regarded as an earlier synonym of *kroton* by Hesychius and Eustathius (*in Od*. 1821.44). This word occurs as early as the *Odyssey* (XVII.300) where the hound Argos is depicted in old age lying neglected on a dung heap ἐνίπλειος κυνοραιστέων. In this context, though, it does not have so specific a meaning but simply implies 'vermin' in general. As Steier[117] points out, such an animal would not be 'full of ticks' only.

As well as including dogs in his list of the hosts of ticks which he reproduces from Aristotle, Pliny (XI.116) states that they have a particular pest of their own which he describes as *volucre . . . aures maxime lancinans*. Despite the literal meaning of *volucre*, it seems most likely that this is no more than a second reference to ticks, since the ears are one of the areas upon which they particularly congregate. The fact that ticks attach themselves especially to the ears of dogs is also noted by Plutarch (*Mor*. 55e), Varro (*RR* II.9.14), Isidorus (*Or*. XII.5.15), and the *Geoponica* (XIX.2.10). The two agricultural sources mentioned recommend the animals' ears—and also the space between their toes—to be rubbed with a preparation of bitter almonds in water to prevent attack. The *Geoponica* (XIX.3.3) also recommends that to combat ticks on the rest of the body the same methods as prescribed for sheep should be employed.

It may be noted that there are no references to ticks on horses. That ticks infest not only domestic but also wild animals was also observed: they are mentioned in connection with foxes in an Aesopic fable (Aristotle *Rhet*. 1393b; Plutarch *Mor*. 790c). Aristotle uses the Homeric term κυνοραιστής in this context, while the latter employs the more usual name.

---

[116] Cf. Fernandez, op. cit., p. 137.
[117] Art. cit., 1810.

7. Medicinal uses

The 'blood' of ticks (this is everywhere specified)—that is, presumably, the blood that they have taken from their hosts—is prescribed by Latin medical writers for the treatment of *ignis sacer* (Pliny XXX.106) and ulcers (Q.Serenus 692), and for the removal of prevention of unwanted hair (Pliny XXX.134; Q.Serenus 665). There is a tendency to specify the particular kind of host from which the ticks should be collected. In addition to these more 'conventional' medicinal uses, Pliny records that a dog tick worn as an amulet was held to relieve all pains (XXX.83), that according to Nigidius those who have collected one from a pig are avoided by dogs for a day (XXX.84), and that according to the Magi a person carrying a tick could discover if a patient would recover by asking appropriate questions (XXX.83).

9. Akari

The ἀκαρί is a minute creature mentioned under this name only by Aristotle (*HA* 557b 7 ff.) whose identification is somewhat unclear.[118] According to the text of the *Historia* as we have it, it is spontaneously generated in ageing wax (κηρῷ παλαιουμένῳ) and in wood, is white in colour, and is considered to be the smallest of all living creatures. Previous authors have more or less ignored the reference to wood but have tended to consider that κηρῷ cannot be the true reading. Aubert & Wimmer read τυρῷ and Peck πικερίῳ and accordingly identify the creature[119] as the common cheese mite *Acarus siro* Linn. Certainly this creature would fit the description, though most other mites would as well.[120] However, the manuscript reading of Aristotle is supported by Pliny (XI.115), who without naming the creatures renders the details from the *Historia* in the words *etiam cerae id gignant quod animalium minimum existimatur*, and by Aristophanes Byzantinus' epitome of the *Historia* (36), where the alternative name κηροδύτης is employed (ἐκ δὲ τοῦ κηροῦ οἱ κηροδύται λεγόμενοι). Steier suggests[121] that the reference to wax may imply that we are dealing here with some pest of beehives such as the minute wingless fly *Braula coeca* Nitzsch.[122] This creature, known as the bee-louse (and probably to be identified also with the φθεῖρες or 'lice' found as bee pests in the *Geoponica* (XV.2.13), which recommends fumigation as a means of combating them), is sufficiently small to be a plausible candidate, but there are also true mites which are known as infesting hives. One might speculate that *akari*, though not found in any other author, was a general name for all those small arachnids which we know as mites and perhaps for other creatures of similar size such as the bee-louse.

---

[118] On the etymology of the word cf. Fernandez, *Nombres*, pp. 94–5, and Chantraine, *Dict. Ét.*, p. 46.

[119] Aubert and Wimmer, op. cit., I. p. 159; Peck, *Hist. An.* p. 213. The same identification is given by Sundevall, *Thierarten*, p. 230, and Gossen in PW Supp. VIII.355.

[120] A number of mites besides *A. siro*, for example *Glycyphagus domesticus* (DeG.) referred to by Gossen, art. cit., 355, are found in stored food products.

[121] *Ap.* Leitner, *Zool. Terminologie*, p. 20.

[122] Cf. Imms, *Gen. Textbook*, pp. 1021–2.

III

GRASSHOPPERS, COCKROACHES
MANTIDS, MAYFLIES etc.

Insecta: Collembola (Springtails)

10. *Petaurista*

At the close of his discussion, which does not claim to be comprehensive, of the smaller spontaneously generated animals (*HA* 557b 11 ff.), Aristotle states that in general creatures are found to be produced 'in almost anything, both in dry things that are turning moist and moist ones which are becoming dry, anything which contains life'. This immediately follows his description of the *ses* or clothes-moth, the *akari* discussed above, and the book-scorpion. It is in this same context that Pliny (XI.115), reproducing Aristotle's account, speaks in rather more specific terms of certain jumping insects generated from dirt by the sun's rays and others produced from damp dust in caves, saying *Alia rursus generantur sordibus a radio solis, posteriorum lascivia crurum petauristae; alia pulvere umido in cavernis volucria*. The term *petaurista* is as Fernandez[1] points out a latinised form of the Greek πεταυριστής, though there are no examples of the latter being used as an insect name. Fernandez and Leitner[2] suggest that we have here another name for the flea and that Pliny is reproducing Aristotle's comments (from 556b 25) about the generation of that insect. However, since fleas are considered elsewhere in book XI of the *Natural History*, it seems more probable that Pliny is here independent of Aristotle, as he clearly is with regard to his insects in caves. These latter, though termed *volucria*, are not necessarily to be looked for among winged insects, since *volucre* as a noun in Pliny seems to mean no more than 'insect'.[3] Steier[4] identifies them as springtails, and it is probable that the *petaurista* should be regarded as such also. The Collembola are small soil-dwelling insects which can jump very effectively by means of an organ situated beneath their abdomen. Due to their ubiquity and their occurrence in large numbers they are not inconspicuous.

---

[1] *Nombres*, p. 177.
[2] *Zool. Terminologie*, p. 196.
[3] Cf. 8.6.v above.
[4] *Aristoteles & Plinius*, p. 52.

Insecta: Orthoptera (Grasshoppers, Locusts, Crickets)

11. *Akris/Locusta*

1. Identification: general

The corresponding term ἀκρίς and *locusta* have essentially the same scope as the modern order Orthoptera, but are mostly used to refer either to locusts or to those grasshoppers and crickets conspicuous by reason of their song. In both Greek and Latin there exist distinct names for what we would consider as crickets (Grylloidea), but these are rarely found in literature, and it would seem that the more general terms were often employed for these insects. We have no record in Latin of any specific term for locusts, but in Greek there are a number of words which are sometimes used instead of *akris* to designate these pests, in distinction from more innocuous species. Most of these locust names would have been simply local synonyms, though there is evidence that some of them may have referred to distinct stages in the insects' life history. Because of their song, the harmless species of *akris* are often closely associated with cicadas, and it is probable that there was some degree of confusion between the two groups of insects. As regards the harmless grasshoppers (Acrididae), bush-crickets (Tettigoniidae) and crickets (cf. 12.1 below), there is little purpose in enumerating particular species; the swarming locusts of antiquity, however, belong to three species: *Locusta migratoria* Linn., the most universal in distribution, the Mediterranean and N. African *Dociostaurus maroccanus* (Thunb.), and the African and Middle Eastern *Schistocerca gregaria* (Forsk.).

1.2. Identification: individual names

As is the case with the cicada, we have record of a large number of names referable to grasshoppers and locusts whose precise import it is rarely possible to determine. The majority are preserved only by the lexicographers, so that we do not in many cases have examples of their actual usage. It is likely that most of these words are simply local synonyms of *akris*, or perhaps names distinguishing the swarming locusts from their harmless relatives. It is not impossible, however, that in a varied group like the Orthoptera some of the lexicographers' ill-defined terms may have been applied to distinct types within these two major divisions.[5]

(i) 'Αττέλαβος (other forms: ἀττέλεβος, ἀττάλαβος).[6] Aristotle uses this term for one of two species of Orthoptera which he distinguishes by their life history. To the other he applies the general name *akris* following his practice of using popular terminology in a new more precise sense. Though Aristotle fails to give any details by which we can conclusively determine which species he understands

---

[5] Gossen in PW Supp.VIII.179–81 has, unfortunately departing from his more modest conclusions in PW VIII.1382–6, proposed a large number of specific identifications for these names: these are mostly pure guesswork, and it is not necessary to consider them all individually.

[6] Cf. Fernandez, *Nombres*, pp. 237–8.

under each name, it is perhaps correct to presume, as previous writers[7] have done, bearing in mind usage elsewhere, that by *attelabos* he is trying to distinguish locusts from the rest of the Orthoptera, although in fact his life history of *akris* seems to be based mainly on observation of the former.

In the *Historia* as we have it, the *attelabos* is said (556a 9 ff.) to lay its eggs in the same way as the *akris* previously described. Severe autumn rains will destroy the eggs, but if there is a drought adults will be produced in large numbers. This last remark is the element from the *attelabos* section that Pliny incorporates into his single account of locusts and grasshoppers given under the synonym of *akris locusta*, though he manages to turn the autumn rains into spring ones. A little later in the *Historia* (556b 1 ff.) we find the statement that *attelaboi* lay eggs in uncultivated ground by means of a pointed organ behind, which is why they are plentiful around Cyrene.[8]

Theophrastus (fr. 174) also draws a distinction between *akrides* and *attelaboi*, stating that the latter are more dangerous than the former. A number of non-specific writers use *attelabos* to refer to locusts (Herodotus IV.172; Plutarch *Mor.* 380f & 637b; Antisthenes *ap.* Diog. L.VI.1.1; *AP* I.265), while others simply employ the general term *akris* for these insects. The scholiast on Aristophanes *Ach.* 150 gives it as a synonym of πάρνοψ, another locust name. Hesychius defines the word as 'ἀκρίς μικρά', which Gossen[9] links with Pliny's one use of the name. As has been noted, in his zoological section Pliny ignores Aristotle's distinction between the two species of Orthoptera, but in his treatment of *materia medica* (XXIX.92) he mentions as a remedy for or charm against insect stings 'the smallest of the *locustae*, without wings, which are called *attelebi*'. The LXX translators, whose choice of Greek words to translate the considerable number of Hebrew locust names appears to be more or less arbitrary, use ἀττέλεβος once (*Nah.* III.17; Vulgate *attelabus*) as a rendering of *'arbeh*, the most frequent of these names, elsewhere rendered by ἀκρίς (most frequently) and βροῦχος. On the basis of Jerome's comments on this passage (MPL 25.1264–9), where he describes the creature in question as *parva locusta inter locustam et bruchum, et modicis pennis reptans potius quam volans*, Gow[10] suggests that ἀττέλαβος and βροῦχος are the more specific names for the immature 'hoppers' of locusts. It may be noted that Pliny (XI.101) describes newly emerged locust nymphs as *pinnis . . . reptans*. It appears that in the original Hebrew of *Nah.* 3, and of *Jl.* 1, the various locust words do refer to successive stages of the insects' life history,[11] but it must be

---

[7] Sundevall, *Thierarten*, p. 198, identifies Aristotle's *attelabos* as the locust *L. migratoria* and his *akris* as a bush-cricket, and is followed by Keller, *Ant. Tierwelt*, II p. 458. Aubert and Wimmer, *Thierkunde*, I pp. 159, 161, suggest the former may be locusts and the latter grasshoppers.

[8] Aubert and Wimmer, op. cit., I pp. 161, 532–3, delete this second passage as inauthentic.

[9] *Zool. Glossen*, p. 11.

[10] A. S. F. Gow,. 'Notes on the 5th Idyll of Theocritus', *CQ*, XXIX, 1935, p. 67.

[11] I. Aharoni, 'On some Animals mentioned in the Bible', *Osiris*, V, 1938, pp. 475–6. This is the authority followed by Kohler and Baumgartner, *Lexicon in Veteris Testamenti Libros*, Leiden, 1953.

doubted whether, as Aharoni suggests, the Greek terms were ever so used in normal parlance. Jerome, therefore, when he says that the *bruchus* is the wingless first stage which can only crawl, *attelabus* the succeeding stage with partially developed wings and capable of jumping, and *locusta* the fully winged adult, is simply trying to explain the passage in question and not speaking of the meaning of the words in ordinary Greek or Latin usage (his definition would not even hold for other parts of the LXX).

(ii) *Πάρνοψ.* Also found in the dialectical forms πόρνοψ and κόρνοψ,[12] this is, apart from *attelabos*, of which it appears to be a synonym, the most frequent of the classical names referring specifically to locusts in distinction from the harmless Orthoptera. Fernandez discusses a number of suggested etymologies for the name, one of which[13] would link it with ἀκορνός, ὀκορνός and κάρον as being based on a root with the meaning of damage or loss—hence an allusion to the insect's destructive activity—but concludes that none of them can be said to be entirely convincing.

The Attic form πάρνοψ occurs in a number of contexts alongside *akris*, though never alongside *attelabos*, indicating that it stands in the same relation to the former as does the latter. Aelian (VI.19) distinguishes three kinds of Orthoptera; namely *akris* (grasshoppers and ?bush crickets), *parnops* (locusts) and *troxallis* (crickets).[14] Nicopho (fr. 1) has *akrides* and *parnopes* in conjunction, as does Galen (II.177 K). Aristophanes (*Ach.* 150; *Av.* 185,588; *Vesp.* 1311) uses *parnops* as his standard word for 'locust', and the word is also used in this sense by Pausanias (I.24.8), and in an unspecified sense by Sextus Empiricus (*Pyrrh.* I.49). Artemidorus Daldianus (II.22), evidently rather ill informed with regard to zoological nomenclature, lists *akrides*, *parnopes* and *mastakes* as if they were distinct pest species rather than synonyms.

The name appears as a synonym of *akris*, in its wider sense as covering locusts, in Strabo (XIII.1.64), who also records the alternative forms πόρνοψ and κόρνοψ from Boeotia and the Aeolians in Asia Minor, and Thessaly respectively. It is similarly synonymised by Hesychius and Photius. The latter also identifies it with *attelabos*, as does the scholiast on Aristophanes *Ach.* 150. This is the scholiast's second definition, the preceding being εἶδος ἀκρίδος, found also in Suidas, Photius (as a separate gloss from the one noted above) and Schol. Aristophanes *Av.* 185.

Suidas gives two curious alternative definitions, namely ἀγρία μέλισσα and κώνωπες. This last is also given by Schol. Aristophanes *Av.* 185. A diminutive of κόρνοψ, κορνώπιδες,[15] is similarly rendered by Hesychius. Also listed by Hesychius alone is πρανώ, a form of πάρνοψ,[16] defined as εἶδος ἀκρίδος.

---

[12] Cf. Fernandez, op. cit., pp. 239–42.   [13] Strömberg, *Gr. Wortstudien*, pp. 16–17.
[14] Gossen, *Tiernamen*, no. 49, for some reason makes the first two distinct kinds of locust.
[15] Cf. Fernandez, op. cit., p. 65.   [16] Cf. ibid. p. 241.

(iii) *Βροῦχος*. This name is found in a considerable number of different forms,[17] most of which are preserved only by Hesychius. Apart from its use in the LXX and dependent sources, it occurs outside the lexicographical tradition in Theophrastus fr. 174, who lists it—in the form *βροῦκος*—as the most harmful of three kinds of Orthoptera, the others being *attelabos* and *akris*, and in the astrological writers Ptolemy (*Tetr.* II.8.84) and Hephaestio (*Apot.* I.20.29), who clearly are using it simply as a synonym of *akris* in the sense of 'locust'. *Βροῦκος* is said by Hesychius to be the Ionian term for 'a kind of *akris*' and the Tarentine name for *attelabos*, though 'others say it is the praying mantis'.

The other forms of the word are as follows: *βρύκος*, synonym of *attelabos* in Hesychius; *βρούκα*, Cypriot name for 'the green *akris*', in Hesychius s.v.; *βρέκος*, Cretan for the 'small *akris*', corrected to *βρεῦκος* by editors of Hesychius including Latte, though Fernandez does not accept this as valid; *βραῦκος*, also Cretan, in *Anec. Bekk.* 233; *βραύκη*, synonym of *akris* in Hesychius; *βρόκος*, equated with both *akris* and *attelabos* by Hesychius; *βερκνίς*, defined as *akris* by Hesychius; and *βροῦχος*, the form given in *EM* 216, where it is synonymised with *μάσταξ* as *εἶδος ἀκρίδος*, and employed by the LXX translators.

From the above definitions it may be seen that there is considerable evidence for regarding *βροῦχος* as a synonym of *attelabos* and *parnops*, with the distinctive meaning of 'locust'. It may be noted that the phrase 'small *akris*' is one that is also used to describe *attelabos*. There is, however, an indication of a more extended usage. The Cypriot 'green grasshopper' would be unlikely to be a locust, but rather something resembling the large bush-cricket *Tettigonia viridissima* Linn., as Gossen[18] suggests. A name applicable to such Orthoptera as this could easily have been applied to the praying mantis, which is often defined as a species of *akris*.[19]

Βροῦχος is the second most frequently used term in the LXX to translate the various Hebrew locust names, though there is no consistency in usage. It is used to render the main locust name *'arbeh* (adult locust) as well as *chasil* and *yeleq*, which are terms for the early stages of the insects' life history.[20] Its Latinised form is *bruchus*, found e.g. in the Vulgate, patristic authors, and Prudentius (*Ham.* 228); it is defined in glossaries as *locusta* (IV.594.28, V.493.11) or *genus locustae quod volat* (V.348.20, 403.73). Among patristic commentators, some agree with Jerome (loc. cit.) in making it a stage in the locust's life history (Augustine *in Ps.* 104.34 ... *locustae et bruchi*, ... *altera est parens altera fetus*; Prosper *in Ps.* 104.34; Cassiodorus *in Ps.* 104 MPL 70.750; Eucherius *Instr.* II.12), while others make it a distinct form of locust (Theodoret MPG 81.1808; Gregory Magnus *Mor.* XXXIII.39) or synonymous with *akris* (Cyril Alex. MPG.71.838).

(iv) *Μάσταξ*. This apparent synonym of *attelabos* is derived from *μασάομαι*,

---

[17] Ibid., pp. 148–50.   [18] *Zool. Glossen*, p. 18.

[19] Gossen in PW Supp.VIII.179–80 parcels out the various forms of the name in a most curious fashion among the Orthoptera on his list, giving no indication of the process by which he has arrived at his conclusions.

[20] I. Aharoni, art. cit., pp. 475–6.

μαστάζω, which is appropriate enough for a locust name.[21] Though there are few surviving references from antiquity, it survives in some modern Greek dialects[22] and so was evidently more commonly used than our sources might at first seem to indicate. Nicander (*Ther.* 802–4) compares a reputed winged scorpion with the 'corn-eating *mastax* . . . which flitting over the top of the corn feed on husked grain, and haunt Pedesa (in Caria) and the vales of Cissus'. Artemidorus Daldianus (II.22) lists it as a damaging pest species along with *parnops* and *akris*, being evidently unclear that the three words refer to the same insect. In the *Etymologicum Magnum* (216), the name is given as an equivalent of *brouchos*, and the comment is added that 'Cleitarchus says that among the Ambraciots the *akris* is called *mastax*'. The equation with *akris* is also given by Hesychius, Photius, and Eustathius (*Comm. in Od.* 1496.53 ff.), and we are told that the word was used in this sense by Sophocles in his Phineus (fr. 716). Fernandez[23] supports the view that the μάσταξ being fed by mother birds to their young in *Iliad* IX.323–4 and Theocritus XIV.39 is not, as commonly interpreted,[24] equivalent to μάσημα, i.e. a mouthful of chewed up food, but specifically a locust. This is very likely in view of the frequent mentions in literature of the important role played by various species of bird in preying upon these insects.

(v) Ὀκορνός. This word is listed as an equivalent of *parnops* (Photius) and *attelabos* (Hesychius and Photius, the latter citing it as an Ionian name), and as meaning τὰ ἀκριδώδη (Hesychius). Photius says that it was used by Aeschylus in his Philoctetes (fr. 256). A variant form, ἀκορνός, is found only in Hesychius, who again equates it with *attelabos*. It has been suggested that the word in its two forms is actually related to *parnops/kornops*, though Fernandez regards this as doubtful.[25]

(vi) Ἀλίβας. According to Hesychius, an equivalent of *brouchos*, though its primary meaning is that of 'corpse'. Gossen,[26] evidently with this latter idea in mind, identifies it as a specific name for the cave-dwelling cricket *Troglophilus*. LSJ do not include the zoological meaning of the word, though it is noted briefly by Fernandez,[27] who sees an allusion to the external appearance of the insect involved.

(vii) Ἄρκυμα. Hesychius: ἀκρίς ὑπὸ Περγαίων.[28]

(viii) Βρέττανα. Cretan name for 'the small *akris*' (Hesychius).

(ix) Κάρον. Defined as 'large *akris*' by Hesychius. It has been suggested[29] that this word is related to *parnops/kornops*, but this is disputed by Fernandez[30] who sees it

---

[21] Fernandez, *Nombres*, pp. 107–8.
[22] Fernandez, ibid.; G. P. Shipp, *Modern Greek Evidence for the Ancient Greek Vocabulary*, Sydney, 1979, pp. 382–3.
[23] Ibid., supported by G. P. Shipp, ibid.   [24] For example in LSJ.
[25] Op. cit., pp. 166–7.   [26] *Zool. Glossen*, p. 6.   [27] Op. cit., p. 202.
[28] Cf. Fernandez, op. cit., p. 242.   [29] Strömberg, op. cit., p. 16.
[30] Op. cit., p. 148.

as derived from a root with the meaning of 'to jump'. This would imply that it is a non-specific appellation.

(x) *Κέκρα*. Equivalent to *akris*, according to Hesychius. In the view of Fernandez[31] the word alludes to the insect's song.

(xi) *Κωριδάμνας*. *Akris* in Hesychius. Gossen[32] sees a relationship with *χώρα* and therefore suggests a reference to the earth-burrowing habits of the mole-cricket *Gryllotalpa*. Fernandez, however,[33] does not regard the word as etymologically explicable.

(xii) *'Ολίγιας*. A type of *akris* (Hesychius). This name is presumably in allusion to the small size of the insect to which it was applied[34] and so may be a genuine species name rather than a general term.

(xiii) *"Ορπας*. A term for the younger stages of grasshoppers and/or locusts, mentioned by Hesychius.[35]

(xiv) *Τριοπίς*. Hesychius writes; 'Three-eyed. Some say it is an animal similar to the *akris*'. We are no doubt dealing here with a distinct kind of Orthopteron, or possibly with some insect, such as the mantis, thought to be related.[36]

(xv) *Φαρμακίς*. Alongside its primary definition of 'witch', Hesychius notes this word as meaning also *akris*. Nicander uses the word as an adjective meaning 'venomous'. Gossen[37] proposes two alternative identifications, both Orthoptera which emit noxious fluids. Considerably more likely, however, is the proposal of Fernandez[38] that *φαρμακίς* is a name for the praying mantis, which was popularly thought of as having a sinister influence on any creature that might encounter it.

(xvi) *Πετηλίς*. Name derived from an adjective meaning 'winged', and defined as *akris* by Hesychius.[39] Fernandez notes the existence of related modern Greek and dialect words referring to Lepidoptera.

(xvii) *Τετραπτερυλλίς*. A Boeotian word, 'four-winged', found in Aristophanes *Ach.* 817, and explained as a name for a grasshopper or locust by the scholiast ad. loc. and Suidas.

(xviii) *Στιτθόν*. A type of *akris* in Hesychius. The word is considered by Fernandez[40] to be of onomatopoeic origin, relating to the insect's song.

(xix) *Καλαμῖτις*. A name used to denote a pet grasshopper or cricket kept for the sake of its song in *AP* VII.198. It is derived from *καλαμή* and presumably relates to the insect's habitat, as does the adjective *ἀκανθοβατίς* applied to it in the epigram.

---

[31] Ibid., p. 128. Gossen, *Zool. Glossen*, p. 50, seems to base his identification on a supposed connection with *κέρκος*, 'tail'.

[32] Art. cit., p. 66.   [33] Op. cit., p. 243.   [34] Ibid., p. 95.   [35] Cf. ibid., p. 242.

[36] Ibid., p. 92. Gossen in PW VIII.257 suggests that it is to be identified as a dragonfly, a group of insects not explicitly mentioned in any surviving classical work, but this is very unlikely in view of the comparison drawn by Hesychius.

[37] In PW Supp.VIII.180–1.   [38] Op. cit., p. 196.   [39] Ibid., p. 77.   [40] Ibid., p. 131.

(xx) Ἀσίρακος, ἀσείρακος (Galen). Dioscorides (*DMM*. II.52) describes a kind of *akris* so named, also called ὄνος, 'wingless and large-limbed', which 'when dried and taken in wine is of great assistance to those who have been bitten by scorpions, and is used a very great deal by the Libyans around Leptis'.[41] Similar details are given by Galen (12.366 K), though he gives Egypt as the region in which the insect was so employed.[42] Fernandez[43] suggests that the name *onos* is connected with the antennae of the insect, but it is more likely to relate to its general external appearance. There are a number of large apterous desert-living crickets in the region which would fit the description.[44]

(xxi) Ἀττάκης. A name, probably of non-Greek origin[45] found only in the LXX and in dependent authors such as Aristeas (145) and Philo (I.85). In these latter it is given as ἄττακος. It is used by the LXX translators to render the Hebrew *sol'am*, identified by Aharoni[46] on the basis of their description in the Talmud as belonging to the genus *Tryxalis*, which occurs only in the list of edible Orthoptera in *Lev*. XI.22. There is an alternative form ἀττακύς (Al.*Lev*. XI.22), and a latinised *attacus* found in the Vulgate and in *C.Gl.L*. V.562.5. Isidorus' statement (*Or*. XII.8.9) that the Greek ἀστακός, like the Latin *locusta*, can be used as an insect name as well as that of a marine crustacean, is perhaps the result of a confusion with the above word.[47]

(xxii) Σκορπιομάχος. The Pseudo-Aristotelian *De Mirabilibus* (844b 24 ff.) contains an account of a form of *akris* so named because it preys upon scorpions. We are told that it circles around its intended victim, stridulating all the time, before it finally moves in to kill and devour it. The writer adds that the creature may be eaten as an antidote to scorpion stings.[48]

(xxiii) Ὀφιομάχης, ὀφιομάχος (in Hesychius). A species of grasshopper reputed to have the same mode of life as the above. The first mention of it, though it is not actually named, appears in the inauthentic book IX of the *Historia Animalium* (612a 34), where we read that the *akris* has been observed in combat with a snake, gripping its victim around the neck.[49] This story is reproduced by Pliny (XI.102), who applies it to *locustae* in general. The specific names are noted by Suidas and

---

[41] Wellmann in his edition of Dioscorides substitutes τρωξαλλίς without mss support, but there is no justification for this correction.

[42] Gossen in PW Supp.VIII.181 identifies it as a form of stick insect, but this is very unlikely.

[43] Op. cit., p. 50.

[44] For example members of the subfamilies Ephippigerinae and Bradyporinae. Bodenheimer, *Geschichte*, p. 232, suggests the family Pamphagidae.

[45] Fernandez, op. cit., p. 238.

[46] Art. cit., pp. 477–8.

[47] Fernandez, op. cit., pp. 51–2.

[48] In the view of Gossen, in PW VIII.1384, this is a hunting wasp rather than an orthopteron, but one of the larger carnivorous bush-crickets could well have acquired such a reputation.

[49] Aubert and Wimmer amend ἀκρίδα to ἰκτίδα, but the parallel story of the σκορπιομάχος is evidence in favour of the former reading.

Hesychius, who both give the definition of 'a type of *akris* without wings'. Ὀφιομάχης is employed in the LXX, in the list of edible Orthoptera in *Lev.* XI.22, to render the Hebrew *chargol*, a word found only in this context and which has been identified by Aharoni[50] as one of the large green bush-crickets of the region such as *Tettigonia viridissima* or the carnivorous *Saga viridis*. Hence it appears in Philo (*de Op. Mund.* 163–4), who draws a moral lesson from its habits. Fernandez[51] suggests that the translator responsible for this rendering of *Leviticus* XI, having exhausted all the locust names at his disposal and with one Hebrew term left to translate, simply invented the word ὀφιομάχης for the occasion, having in mind the unnamed predatory orthopteron described in the *Historia Animalium*. However, though such an extraordinary expedient would not be out of keeping with the LXX translators' general disregard for zoological accuracy, this theory is placed in doubt by the fact that there are still several well-known locust names available for use in this context, one of which (*attelebos*) is indeed found elsewhere in the LXX. It is probable, therefore, that the term was already in existence; though whether its selection was due to some popular tradition concerning the Hebrew *chargol* cannot of course be determined. The word appears in the Vulgate latinised as *ophiomachus* and in *C.Gl.L.* V.574.27 as *opinacus*.

(xxiv) Ἐρυσίβη, which in normal Greek refers to the fungal disease of crops known as rust, is used by the LXX translators to render the Hebrew locust names *chasil* (in four places) and *selasal* (in Deut. 28.42). The latter has been identified as the mole-cricket *Gryllotalpa gryllotalpa* (Linn.).[52]

(xxv) Ὀξύγη. This term is recorded only in *C.Gl.L.* II.31.47, where it is defined as εἶδος ἀκρίδος and equated with *bufo et cufo*. *Bufo* elsewhere in Latin refers to the toad, but if it was an insect name also it may perhaps be equated with the *bubo* found in Polemius Silvius' list of invertebrates.[53]

2. Life history

Aristotle (*HA* 556b 18 ff.) provides a quite detailed account, fairly accurate in its details, of the life history of his *akris*, adding much briefer details for his *attelabos*. Pliny (XI.101–7), disregarding the division into two species, gives a composite account under the name *locusta*, incorporating details ascribed to *attelabos* in the *Historia*: his account omits some of the details given by Aristotle. After mating, we are told, the females produce their

[50] Art. cit., p. 478. Aharoni states that the equivalent word in Arabic refers to *Saga*, so that despite the reasons he gives for thinking otherwise this may well be the species referred to by the Hebrew. Being predacious and the largest orthopteron of the region, it, unlike the inoffensive *T. viridissima*, might have given rise to erroneous beliefs about its prowess.

[51] Op. cit., pp. 138–9.

[52] I. Aharoni, art. cit., p. 478.

[53] Ἀκρίδιον and κάλαμος, included by Gossen, in PW Supp.VIII.181, are not genuine insect names.

offspring by fixing in the ground the tube (καύλον) at their hind end.[54] They lay them all together in one spot, so that the effect is like that of a honeycomb. (This is evidently based on accurate observation of Acridid egg-pods, though, according to Aristotle's theories on insect development, they could not strictly speaking be labelled as eggs.) Pliny, disregarding Aristotle's technical terminology, speaks simply of *ova condensa*. Aristotle, in a passage passed over by Pliny, goes on to say, rather confusingly, that after this initial 'oviposition' there arise (γίγνονται) egg-like larvae (σκώληκες), enveloped by a kind of light earth as by a membrane, in which they develop to maturity. These κυήματα, situated a little beneath the ground surface, become so soft that they are compressible when touched. When they have come to their full development, small black *akrides* emerge from their earth-like envelopes; their outer skin splits, and immediately they become larger. Pliny's version of events is here partly independent of the *Historia*; he describes the nymphs as small black creatures without legs, crawling by means of their wings (*pinnisque reptans*), but does not deal with their transformation into the adult. The females, Aristotle continues, lay at the end of summer and then die.

There follows in the *Historia* a disjointed and in one case internally contradictory passage which Aubert and Wimmer[55] rightly reject as inauthentic. We read first that the nymphs emerge in spring, and then two sentences later that the eggs[56] overwinter in the ground and that the insects emerge from them when summer arrives. Pliny resolves the contradiction by specifying the time of emergence as the end of spring. In this same passage of the *Historia*, the curious statement is made that the females die at the end of summer as a result of the action of certain *skolekes* which develop around their neck (*qui eas strangulat* adds Pliny). D'Arcy Thompson[57] believes that this statement is based on actual observation of some parasitic larvae, but it is far more likely to be simply an unfounded paradoxographic tale. As well as being reproduced by Pliny, the story is also repeated in a slightly different form in a fragment of Theophrastus (fr. 174).

In a statement not drawn from Aristotle, Pliny notes the opinion of some that there are two generations of *locustae* each year, one laying its eggs at the rising of the Pleiades and dying at that of Sirius, and the second emerging at the setting of Arcturus.

Alongside the scientific account given by Aristotle, there coexisted the earlier popular belief that locusts and grasshoppers, like other insects, were spontaneously generated. We discover from Antisthenes fr. 129 (*ap.* Diog.L.VI.1.1) that they were thought of, like cicadas, as being earth-born; and Plutarch (*Mor.* 637b) relates that in historical times a swarm of locusts was generated from unburied corpses in Sicily. Similarly, the Cyranides (pp. 33, 241) relates that storms in Libya cast up large fish on the shore, which generate

---

[54] D'Arcy Thompson, *Hist.An.*, V.28 note, thinks Aristotle is referring to an ovipositor such as is possessed by bush-crickets rather than grasshoppers and locusts. But female locusts do deposit their eggs in the ground by means of their greatly extensible abdomen, and it is most probable that this is what Aristotle is referring to. The life history as depicted here fits Acrididae in general, but particular details, such as the black nymphs, point to locusts as being the origin of these observations.

[55] Op. cit., I. p. 531.

[56] The reference to 'eggs' represents non-Aristotelian terminology.

[57] Op. cit., V.28 note.

σκώληκες; these turn successively into μυῖαι and then into *akrides* which make their way across the sea as a swarm.[58] Isidorus (*Or.* XI.4.3) states that locusts are generated from dead mules as bees were believed to be from oxen.

3. Habitat

In the apparently interpolated passage of the *Historia* mentioned above, we read that the preferred habitat of *akrides* is level and broken ground, since they require cracks in the earth in which to lay their eggs. Pliny (XI.102) abbreviates to *non nascuntur nisi rimosis locis*. In certain epigrams in the Greek Anthology, singing *akrides* of the kind kept as pets are described as living in fields, in distinction from tree-loving cicadas (κατ' ἄρουραν ἀηδόνι, VII.190; ἀρουραίη, VII.195; singing in the furrows, VII.192).

4. Sound production and habits

Among the Orthoptera, grasshoppers and locusts (Acrididae) stridulate by friction of their hind legs against their forewings (tegmina), while bush-crickets and true crickets (Tettigoniidae, Gryllidae) do so by friction of the tegmina alone. The first of these modes of sound production, being readily observable, is the one of which we have clear mention in antiquity, and would presumably have been ascribed to all of the above three families. Aristotle (535b 12) states that the *akris* sings by rubbing of its paddle-shaped hind-legs. This of course omits mention of the part played by the wings, whether by inadvertence or ignorance being unclear; though the corresponding passage in Pliny (XI.266) mentions both (*locustas pinnarum et feminum attritu sonare*). Although he is correct in this place, elsewhere in the *Natural History* Pliny provides a contradictory version (XI.107). Here he writes that the sound appears to emanate from the back of the insect's head, where at the point where the shoulders join (*commissura scalparum*) they have as it were teeth which are rubbed together to produce a sound. Ernout and Pepin[59] suggest that this statement is by some confusion based upon the Aristotelian account of cicada song production, but the latter is reproduced by Pliny elsewhere. It is more probable that these details are based on observation of the common diurnal bush-cricket *Ephippiger ephippiger* (Fieb.) or its relatives which stridulate with very short wings set as it were 'at the back of their neck'. Meleager (*AP* VII.195) speaks of the *akris* beating its wings with its feet, while Mnasalces (*AP* VII.192, 194) and Phaennus (*AP* VII.197) say only that the song is produced from the wings. Aelian (VI.19) in a list of animal songs uses the verb βομβοῦσαν to describe that of the *akris* (here presumably in its meaning of grasshopper/bush-cricket rather than, as Scholfield suggests, true cricket[60]).

In Greek literature, the song of the *akris* is closely associated with that of the cicada, and, though mentioned less frequently, seems to have been thought of as equally pleasant. Poets like Theocritus and those of the Greek Anthology seem to use *tettix* and *akris* more

---

[58] There is some confusion in the text with the migratory habits of quails, one version making the locusts metamorphose into these birds.

[59] A. Ernout and R. Pepin, *Pline l'Ancien, Histoire Naturelle XI*, Paris 1947, p. 151.

[60] Aelian, II. p. 35.

or less indiscriminately where mention of a singing insect is required, demonstrating a probable confusion in popular parlance between the two groups of insects. The cicada is often credited with the same mode of stridulation as the *akris*.[61] In Latin poetry, however, the cicada is the only singing insect named. It is likely that the poetic *akris* includes true crickets as well as grasshoppers and bush-crickets, although the former did possess a distinct name (*troxallis*).

The terms in which the song of the *akris* is described are generally the same as those used to refer to the cicada's. The same verb ἀείδειν is employed, and the sound is similarly described as clear or shrill (λιγύς, λιγυρός etc., *AP*. VII.189,192,195,197; ξουθός, VII.192). Like the cicada, the *akris* is compared to the nightingale (*AP*. VII.190 τᾷ κατ᾽ ἄρουραν ἀηδόνι), while in Aristotle (*de Aud.* 804a 23 ff.) the songs of these two creatures, along with that of the cicada, are together categorised as λεπταὶ καὶ πυκναί and λιγυραί.

Theocritus introduces *akris* and *tettix* in identical contexts (cf. VII.41 & V.29, V.34 & VII.159), though his awareness of the distinction between them is clear from the fact that although, in the passages cited, he sets their songs on an equal level of merit, he does mention the former also as a pest species (V.108–9). Where pet *akrides* are described in the Anthology, their singing is on three occasions either stated (VII.194, πανέσπερον ὕμνον) or implied (VII.195,197) as taking place in the evening, a clear distinction from the habits of cicadas, but which is appropriate to those of some crickets and bush-crickets.

Pliny (XI.107) mentions that whereas the song of the cicada is mainly heard around midsummer, that of the *locusta* is a feature of the two equinoxes.

5. Predation

In antiquity it was considered that the most important predators of locusts and grasshoppers were various species of bird. Because of the immense damage that could be done to agriculture by the irruption of locust swarms, these birds were regarded as being of great service to mankind. We read of particular species being especially honoured in this regard in different geographical regions.

General references to the importance of birds in combating locusts are found, not unexpectedly, in Aristophanes' play of that name (185,588). We are informed by Aelian (III.12) that the people of Thessaly, Illyria and Lemnos honour jackdaws for destroying the eggs and young of locusts, and have decreed that they be fed at public expense. Of these three localities, only the practice in Lemnos is mentioned elsewhere, by Pliny (XI.106) and Plutarch (*Mor.* 380 f.). The former uses the word *colunt*[62] for the respect paid to the jackdaws there, and somewhat overdramatises the account of their activity, implying that they fly out like an army to meet invading locust swarms head on.[63] Plutarch simply notes that the Lemnians honour the birds for consuming the eggs of

---

[61] Cf. 16.6.1 below.

[62] Not 'keep', as in Rackham's translation, which implies captivity.

[63] Cf. the ibises flying out to intercept the winged serpents on their way into Egypt from Arabia, in Herodotus II.75. These 'serpents' have themselves been identified as locusts (cf. Davies and Kathirithamby, *Greek Insects*, pp. 148–9).

locusts: he also has a different species involved, larks instead of jackdaws. Eusebius (*Praep.Ev*. MPG 21.101) states that the Egyptians honour the ibis for its destruction of snakes, locusts and caterpillars.

A somewhat similar story is told of certain birds known as σελευκίδες and identified as the Rose-coloured Pastor (*Pastor roseus* Linn.), which does migrate into locust-infested areas in the way described,[64] from Asia Minor. Pliny relates (X.75) that these arrive in response to prayers offered up to Jupiter by the people in the region of Mt. Cadmus (in Caria) when locusts are devastating their crops; they are not, he goes on, native to the region, but only arrive when their help is needed, and it is not known where they come from or where they go once their mission has been fulfilled. The same species is mentioned in passing by Galen (VIII.397 K), who locates them more vaguely 'in Asia' and states merely that they devour locusts insatiably all through the day. They are given prominence in patristic accounts of the natural world as examples of divine beneficence (Basil *Hex*. MPG 29.181; Ps. Eustathius *Hex*. MPG 18.736; Ambrose *Hex*. V.82). It has been generally assumed[65] that these *seleukides* are identical with the unnamed birds referred to in an account from Eudoxus of Cnidus preserved by Aelian (XVIII.19), though they are located in a different area, namely East Galatia. Here the sacrifices seem to be offered as direct invocations to the birds themselves, which respond by arriving en masse. The writer adds that to capture one is an offence punishable by death, since if such a penalty were not carried out the birds would take offence and refuse to respond to the next occasion their help was required. There is a single curious reference (*de Mir. Ausc.* 847b 4) to moles as predators of locusts in Aetolia.

6. Locusts: their importance as pests

Locusts, not surprisingly, were the most feared of all agricultural pests in antiquity. Most of the references we have to this subject relate to the irregular invasions of locust swarms, though there are some which imply a more general activity as pests. They are spoken of by Aristophanes (*Av*. 588) and Theocritus (V.108–9)[66] as attacking vines.

It was the unpredictable and devastating nature of the locust's invasions (Pliny XI.103– 6, Quintus Smyrnaeus II.196 ff., Aristophanes *Ach*. 150), so different from the more controllable activity of other noxious insects, standing corn being described as the main subject for their attack (Nicander *Ther*. 803–4; Quintus), that made such an impression upon the people of antiquity. Incoming swarms are compared with an invading army (Quintus; LXX *Jl*. II.1 ff.; Basil *Hex*. MPG 29.181; Josephus *BJ* IV.536), moving inexorably onward in strict formation, allowing nothing to obstruct their progress (LXX

---

[64] F. S. Bodenheimer, *Animal and Man*, pp. 120, 136; D'Arcy W. Thompson, *A Glossary of Greek Birds*, London, 1936, pp. 258–9; F. Capponi, *Ornithologia Latina*, Genova, 1979, pp. 457–8.

[65] For example by Scholfield, *Aelian*, III p. 347.

[66] A. S. F. Gow, art. cit., p. 67, and *Theocritus*, 1952, II pp. 110–11, takes these to be immature locusts, since they are described as hopping rather than flying. But these vine pests may not necessarily be locusts at all, but other Orthoptera, for example, *E. ephippiger* (cf. Bodenheimer, *Geschichte*, p. 156) or *Anacridium aegyptium* (Linn.) (cf. I. Aharoni, art. cit., p. 478).

JL. 2.1 ff. and Cyril Alex. ad. loc.), often so vast in number as to obscure the sun (Pliny), like a large raincloud overshadowing the earth (Quintus). It was observed that they were capable of great feats of endurance, flying continuously day and night without food (Pliny 103) and traversing large tracts of land and sea (Pliny 103–4). It was known that they were able to cross the Mediterranean, since Pliny notes that the swarms invading Italy emanate mainly from Africa (105; cf. Livy XLII.10.7).[67] The noise of their wings as they fly is described as ־eing like that of a flock of birds (Pliny 104; LXX *Jl.* II.5) or a grassland fire (Pliny 104; LXX *Jl.* 2.5 and Cyril Alex. ad. loc.). Wherever they finally land, the result is total devastation, all available plants being consumed (Pliny 104; Quintus; Tacitus *Ann.* XV.5; LXX *Ex.* X.15; Philo *de Vit.Mos.* I.121, *Praem.et.Poen.* 128; Josephus loc. cit., *Ant.* II.306). Pliny says that they scorch up much of the foliage by their touch (*multa contactu adurentes*) and that they gnaw at everything, even at the doors of houses.

Pliny states that swarming locusts finally perish when they are carried away en masse by the wind and deposited in the sea or in some marsh (103; he adds that this occurs simply by chance, and not, as older writers had said, as a result of their wings being made wet by the dampness of the night). This type of occurrence is described by Julius Obsequens (*Prod.* 30), taking his account from Livy, and in LXX *Ex.* X.19 (cf. Philo *Vit.Mos.* I.122). Pausanias (I.24.8) notes three possible ways in which swarms can be destroyed; namely, by being carried off by a strong wind, or by the effect of high temperature after rain, or intense cold. That locusts could be disabled from flight by low temperatures or by their wings becoming wet is commonly stated by patristic commentators on *Nah.* 3.17 (Cyril Alex. MPG 71.838; Theodoret MPG 81.1808; Theodore of Mopsuestria MPG 66.424; Jerome MPL 25.1264).

It is not difficult to imagine the helpless dismay that was occasioned by the advent of such swarms. Pliny (104) vividly pictures folk looking anxiously up at the invading insects, unable to do anything except hope that they will pass over their land and not alight on the spot. Quintus of Smyrna (II.198) speaks of them as bringing famine to men by their total destruction of crops, and they figure in Artemidorus' *Onirokritikon* (II.22) as representing disaster to farmers. A number of incursions were of sufficient severity to be recorded in secular history as major disasters (Plutarch *Mor.* 637b; Livy XXX.2.10, XLII.2.5, 10.7; Julius Obsequens *Prod.* 30 = Augustine *Civ.* III.31). Concerning the last of these, in the area around Cyrene, it is reported that after the swarm had been blown into the sea and cast up on the beach later, the decaying insects caused a plague among livestock and an enormous number of human fatalities.[68]

As Pliny (104) tells us, the arrival of locust swarms was often interpreted as being an occurrence of supernatural origin, a mark of divine displeasure (Arnobius *adv.Nat.* I.3,16). In Livy and Julius Obsequens (loc. cit.) their appearances come under the heading

---

[67] Bodenheimer, *Animal and Man*, p. 138, identifies these locusts emanating from Africa as *S. gregaria.*

[68] J. L. Cloudsley-Thompson, *Insects and History*, London, 1976, p. 190, suggests that this account can be explained by famine and epidemic resultant upon loss of crops to the locusts.

of portents. According to Pliny (105), the Sibylline books were on occasions consulted and appropriate action, unspecified, taken on the basis of their instructions. Pausanius (I.24.8), in his account of the antiquities on the Acropolis, refers to a statue of Apollo Parnopios, said to commemorate an occasion when the god drove away a swarm that was causing devastation in Attica. Likewise Strabo (XIII.1.64), giving examples of divine epithets connected with animals, tells us that the Aeolians in Asia honour Apollo Pornopios, and that the Oetoeans honour Herakles under the title Kornopion ἀπαλλαγῆς ἀκρίδων χάριν. Cf. also the practices connected with locust-eating birds noted above. Writers on astrology and omens include locust swarms among the major natural disasters that befall mankind and which are therefore foreshadowed in various ways (Ptolemy *Tetr.* II.8.83–4; Hephaestio *Apotel.* I.20.27,29, 21.10,11,33; Lydus *de Ost.* 70.20, 72.26).

The only effective measure known in antiquity for the control of locusts was the physical collection of the insects themselves, preferably in their younger stages. In areas where the danger from swarming was particularly great, such systematic extermination was actually regulated by law, something unknown with regard to any other agricultural pest and a measure of the seriousness with which the threat from locusts was treated. Thus in Cyrene (cf. *HA* 556b 1 ff.), so we are told by Pliny (105), there was a law prescribing three locust collections per year, involving on respective occasions the eggs, young and adults, with penalties for those failing to participate. And in Lemnos (106) a similar piece of legislation prescribed a certain quantity of dead locusts that each man was required to present to the magistrates. In Syria (106) the people were compelled *militari imperio* to assist in the insects' extermination. Livy (XLII.10.7) describes how in 173 BC, when a severe infestation occurred in Apulia, a special official was sent out to organise the men in collecting and destroying them.

The use of fire as a means of driving off locust swarms seems to be described in the *Iliad* (XXI.12–14), the earliest classical reference to these insects. In his comments on this passage Eusthathius (1220.38 ff.) states that this was a method employed in Cyprus and Libya.

Apart from the above, the classical countermeasures of which we read are generally of an unlikely or magical nature. The Magi, according to Pliny (XXXVII.124), prescribed certain gemstones, amethysts and engraved emeralds, to be used with appropriate incantations as charms to avert locust swarms. Other suggestions are listed in the *Geoponica* (XIII.1): the writer here states that there were many others he could have mentioned, but that he has given the most straightforward. From this we discover that it was believed that if everyone ran and hid indoors when a locust swarm was sighted the insects would pass over (also in Palladius *RR* 1.35.12); and that they could be deterred by sprinkling with preparations of bitter lupine or wild cucumber (also in Palladius), by the tying of bats onto the tops of trees (also in Cyranides p. 279), by the burning of captured locusts (also in Palladius) or by the use of herbs *apsinthion*, *prason* and *kentaurion*. It is also recommended that locusts be boiled up into a *garos* and the resulting liquid poured into holes in the ground, in which the living insects will collect and thus be easily captured.

7. Grasshoppers and locusts as food

Although most references to the use of Orthoptera as food are from outside Europe, we read in some Greek sources of these insects being collected and eaten (*Aesopica* 387 and *Vit.Aes.* 99; Galen 17B.484 K; Athenaeus 63c): they were considered a cheap food especially characteristic of the poor (Schol. Aristophanes *Ach.* 1116; Scholiast *ap.* Th.L.L.VII.1605.65) or of soldiers (Aristophanes *Ach.* 1116; Porphyry *de Abst.* I.25).

Both Columella (*RR* VIII.11.14) and Palladius (I.28.5, 29.2) recommend *locustae* for feeding to young birds (cf. *Geoponica* XIV.23.3).

8. Grasshoppers and crickets in captivity

It was a popular custom in antiquity, among the Greeks at least, for singing insects to be kept as pets in homes. Although there is some evidence, not undisputed,[69] for the keeping of cicadas, it is *akrides* (Latin *locustae*) which are usually mentioned in this connection. These insects were evidently kept in some form of cage or ἀκριδοθήκη.[70] This term is only found in Theocritus (I.52–3) and Longus (I.10), where the items are described as being made out of plant stems or reeds. Here they are just playthings, so that we lack a description of whatever more permanent form of cage may have been used for such domestic pets as those in the Anthology (Borthwick[71] takes the εὐερκὴς δόμος of *AP* VII.193 to be such a cage).

The epitaphs for these pets found in the Anthology (VII.189,190,194,197–8) express great affection for them. In two of them (190,198) a small 'tomb' is described as being made. Their song was very highly regarded (VII.189–193): Mnasalces (VII.194) describes one as filling the house with music as it begins its evening song (πανέσπερον ὕμνον ἀείδειν) and Phaenus (VII.197; cf. 195) writes of one bringing sleep to its owner. In one epitaph a lifespan of two years in captivity is specified (VII.198). There is only one reference to what these insects were fed upon: this is from Meleager (VII.195), who mentions vegetable food in the form of a leek, and drops of water.[72]

As to the precise identity of these insects, they are likely, in view of the quality ascribed to their songs, to have been crickets or bush–crickets rather than grasshoppers. Gow[73] argues that the references to their singing in the evening are strong evidence in favour of this conclusion. The field cricket *Gryllus campestris* Linn., the chief species that has been kept in captivity in modern times, sings through the night as well as by day; while the house cricket *Acheta domesticus* Linn. and *Oecanthus* spp. are typically active by night. Similarly, the large bush–cricket *T. viridissima* sings chiefly from early evening onwards.

[69] Cf. 16.9 below.

[70] N. Douglas, *Birds and Beasts of the Greek Anthology*, London, 1928, pp. 186–8, and A. S. F. Gow, *Theocritus*, II pp. 12–13, argue that, whatever the correct reading here (v.l. ἀκριδοθήρα), a cage must be meant rather than a trap.

[71] E. K. Borthwick, 'A Grasshopper's Diet—Notes on an Epigram of Meleager and a Fragment of Eubulus', *CQ*, XVI, 1966, p. 105.

[72] E. K. Borthwick, ibid., pp. 103–6, suggests that this reference to the supply of liquid results from the transfer to the *akris* of the dew drinking motif associated with the cicada (Cf. 16.4 below).

[73] 'Mnasalces', *CR*, NS VI, 1956, pp. 92–3. N. Douglas, op. cit., pp. 186–8, 192–3, similarly identifies them as crickets.

9. Medicinal uses

The only recorded medicinal uses for *akrides/locustae*, apart from the employment of the distinct species ἀσείρακος as an antidote for scorpion stings, are of whole specimens for *dysouria* (Dioscorides *DMM.* II.52; Galen XII.366; Paulus VII.3; Pliny XXX.123) and of limbs only for skin disorders and *scabritia* of the nails (Pliny XXX.30, 111).

10. Grasshoppers and locusts outside Europe

(i) *Libya.* On locusts as plentiful around Cyrene, cf. Aristotle *HA* 556b 1 ff., Pliny XI.105 (and 6 above), Eustathius *Comm. in Il.* 1220.38 ff., and Jerome (*adv.Jov.* MPL 23.308), who speaks of them as a popular source of food in the region (cf. Porphyry *de Abst.* I.25). Pliny (VIII.104) writes of an African tribe being driven from their territory by locusts. Herodotus (IV.172) mentions a Libyan tribe, the Nasamones, as drinking sun-dried and ground locusts mixed with milk.

(ii) *Egypt.* The swarming locusts described as occurring in Egypt (LXX *Ex.* 10.12–19; cf. 6 above) are identified by Bodenheimer[74] as *S. gregaria*.

(iii) *Palestine.* There are numerous references throughout the LXX to swarming locusts in the Middle East (cf. 1.2.i and iii, 6 above). These would be *D. maroccanus* and *S. gregaria*.[75] A number of different species of Orthoptera, including the latter as chief, but also certain harmless species, were used as food in the region (LXX *Lev.*11.22, cf. I.2.xxi and xxiii above; NT *Mt.* 3.4, *Mk.* 1.6).

(iv) *Parthia.* Pliny (XI.106) records the use of locusts as food in this area, and Tacitus (*A.* XV.5) mentions a destructive swarm.

(v) *Syria.* Cf. Pliny XI.106, cited under 6 above.

(vi) *Ethiopia.* Strabo (XVI.4.12), Diodorus (III.29), Agatharchides (58) and Pliny (VI.195) make mention of a certain tribe in Ethiopia known as the *Akridophagoi*, owing to the fact that *akrides* make up a large part of their diet (or the whole of it, according to Pliny). From Diodorus, who gives the most details, we learn that a swarm of locusts—of large size, with wings of a dingy and unpleasant colour—arrives in their territory each spring, driven by the wind. On their entrance they have to cross a certain ravine, and at the bottom of this the *Akridophagoi* light fires so that the insects are overcome by the smoke as they pass over and so fall to the ground. After several days, during which piles of dead locusts have accumulated, these are mixed with salt and made into cakes (Jerome MPL 24.364 may refer to this). These locusts, and those of Libya, would be *S. gregaria* and *L. migratoria*, used for food in modern times.[76]

(vii) *India.* Pliny (VII.29) notes on the authority of Agatharchides that certain Indian tribes have locusts as an article of their diet. In his zoological section, he records

[74] *Animal and Man*, p. 77.
[75] Ibid.
[76] F. S. Bodenheimer, *Insects as Human Food*, The Hague, 1951, pp. 139, 160, 202–6.

(XI.103) the reputed occurrence in India of *locustae* three feet in length, whose dried limbs could be used as saws.

(viii) *Arabia*. Aelian (X.13), declaring that the variety of form and colour among the fauna of Arabia would be a challenge to any artist, however skilled, gives as an example the *akrides* there, which are beautifully patterned with markings that resemble gold.

12. *Troxallis/Gryllus*

1. Identification and habits

Τρωξαλλίς and *gryllus* represent the most clearly distinguished of those insects coming under the general heading of *akris* or *locusta*.[77] They refer generally to the true crickets (Gryllidae), such as the field cricket *Gryllus campestris* Linn, and the house cricket *Acheta domesticus* Linn., insects clearly distinct in their appearance from the rest of the Orthoptera, and also the closely related burrowing mole-crickets (Gryllotalpidae). Though in technical literature they are clearly distinguished from the remainder of the Orthoptera, it is clear from, for example, the Greek Anthology that *akris* in its most generalised sense could be used to include them. Certainly some of the terms used to describe pet *akrides* in the Anthology do indicate that crickets are in view.[78]

In Greek sources we find considerable confusion between *troxallides* and cockroaches. In Hesychius, the μυλακρίς is defined as a cockroach found in meal, or the *troxallis* from the same environment, or a 'σιτοφάγος ἀκρίς'. And in Photius the parallel term μυλαβρίς is rendered as a cockroach or *troxallis*. Similarly, the Latin translation of Dioscorides' *DMM* translates the chapter heading Περὶ σιλφῆς (II.38) by *De grilis* (II.19). This confusion clearly results from the fact that the domestic cricket *Acheta domesticus* has the same habitat as cockroaches, similar light-avoiding habits, and a superficial similarity in appearance.[79]

Aelius Promotus (p. 775) makes reference to 'the *troxallides* found in baths and other places where they sing (κράζουσιν)', a clear reference to *A. domesticus*. Aelian (VI.19) includes the *troxallis* as a singing insect alongside two other kinds of Orthoptera, *akris* (grasshopper) and *parnops* (locust), supporting its identification as a cricket. In fragment 15 of Alexis, however, it figures as an insect pest devouring vegetables, and Gow[80] suggests that therefore it may represent the locust. Fernandez,[81] using this second passage,

---

[77] Keller, op. cit., II. p. 460, and Leitner, *Zool. Terminologie*, p. 132, correctly identify the insect with the true crickets *Gryllus* and *Acheta*. Gossen in PW Supp. VIII.180 maintains that the Greeks did not distinguish crickets from cicadas and included them all under the name *tettix*; according to him *troxallis* is the bush-cricket *Ephippiger*, but this is not supported by the evidence.

[78] Cf. 11.8 above.

[79] Gossen, art. cit., 180, due to his misidentification of *troxallis*, considers the association to be due to similarity of appearance only.

[80] A. S. F. Gow, 'Mnasalces', *CR*, NS VI, 1956, p. 92. However, *Gryllotalpa* can constitute a pest of crops by devouring the roots.

[81] *Nombres*, pp. 104, 124.

sees an etymological distinction between the terms τρωξαλλίς and τριξαλλίς (along with its variant τριξέλλας), viewing them as names of different origin which later came to be used interchangeably due to their similarity. He regards the former as a derivative of τρώξ with diminutive suffix and thus as referring to the voracity of the insect originally denoted by it (presumably the locust, though some horticultural pest might be considered to be in view); and the latter as an onomatopoeic name related to the verb τρίζω, used of insect sounds, and thus originally the authentic term for crickets.

That the Latin *gryllus* (or *grillus*, as it appears in the glossaries) was considered to be the term corresponding to *troxallis/trixallis* is clear both from Pliny (XXX.49) and from a number of glossary entries (in these varying spellings are employed: τρωξαλλίς in C.Gl.L. II.460.58, τριξέλλας in II.459.25, and τοξαλλίς in III.188.44 and 258.25). In the passage cited Pliny states that the *trixallis* is a creature resembling a wingless locust, which has no Latin name, although many authorities maintain that it is identical with the insect called in Latin *gryllus*. Elsewhere in his medical section he used *trixallis* twice (XXX.117,129) and *gryllus* on five occasions. In XXIX.138 he states that the *gryllus* lives in a hole (*caverna*) which it excavates in the earth, that it walks backward, and that it sings (*stridat*) by night. These details correspond very well with the habits of both *Gryllus campestris* and the mole-cricket *Gryllotalpa gryllotalpa* (Linn.).[82] Isidorus' description (*Or.* XII.3.8) is identical to that of Pliny. Elsewhere in Latin literature the *gryllus* or *grillus* appears as a meadow insect alongside *scarabaei* and *locustae* (Phaedrus *App.* 32) and as an inhabitant of cornfields that sings and lives in a hole (*Dirae* 72–4). [83]

Leitner,[84] following Ernout and Pepin, identifies the *scarabaei . . . focos et prata crebris foraminibus excavant nocturno stridore vocales*, included by Pliny (XI.98) in his section on beetles, as crickets rather than, as in Rackham's[85] translation (following the conjecture *focos et parietes*), as wood-boring beetles like the death-watch. In this case, the *focos excavans* would be *A. domesticus* and the *prata excavans G. campestris* or *Gryllotalpa*.

Hesychius includes in his lexicon the term άδιγορ, said to be a Scythian name for the *troxallis*.

In addition to the glossary definitions noted above, there are a number of curious items which require comment.[86] The equation of *gryllus* with *attelabos* (C.Gl.L. II.250.31; Gl.Lat. II.159.GA16) is a readily explicable inaccuracy, while the definition *vermis in igne manens similis muscae* (II.581.43) evidently belongs rightly to the fabulous *pyrallis* or *pyrigonos*.[87] C.Gl.L. II.511.23 seems to preserve an otherwise unknown Greek name κυνακρίς, a compound of κύων and άκρίς.[88] The gloss (V.422.40) *Chantari* (?κάνθαροι)

---

[82] Both live in excavated burrows, both stridulate by night (though *Gryllus* sings by day also), both may emerge backwards from their burrows (though running backwards is more characteristic of *Gryllotalpa*). Cf. D. R. Ragge, *Grasshoppers, Crickets and Cockroaches of the British Isles*, London, 1965, pp. 138–43, 150–4.

[83] Cassiodorus' *talpinum animal*, referred to by Gossen, art. cit., 180, is simply the mammal of that name.

[84] A. Ernout and R. Pepin, op. cit. p. 149; Leitner, op. cit., p. 132.

[85] *Nat. Hist.*, III. p. 493.

[86] *Gryllus* is onomatopoeic and is considered by Ernout and Meillet, *Dict. Ét.*, p. 283, and Fernandez, op. cit., pp. 125–6, to be native Latin rather than a borrowing of the Greek γρύλλος.

[87] Cf. 60 below.

[88] Strömberg, *Gr. Wortstudien*, p. 20; Fernandez, op. cit., p. 54.

vermes qui cantant nocte sicut lucusta is clearly, as Gossen has noted,[89] a reference to crickets, though there is no reason to suppose, as he does, that some distinct species is in view.

2. Crickets in captivity
On this subject, cf. 11.8 above.

3. Medicinal uses
Pliny (XXIX.138) notes that medicinal properties were ascribed to the *gryllus* by Nigidius and by the Magi. The latter, he goes on, capture them by introducing into their holes an ant tied onto a hair, so that *formicae complexu* the cricket may be drawn out (= Isidorus XII.3.8). The use of *grylli* is recorded for ear complaints (XXIX.138,143), for sore throats and colds (*destillatio*) (XXX.32 = Marcellus XV.21), for *tolles* (Q. Serenus 287–8), scrofula (XXX.38) and *ignis sacer* (XXX.106). In the latter two cases, it is recommended that the insect be used together with earth from its own excavations, having been dug out with an iron implement. Under its Greek name *trixallis* it is prescribed for ulcers (XXX.117) and to aid menstruation (XXX.129; Aelius Promotus 755.9)—their ash being used in these two cases—and also, roasted and taken in wine twenty at a time, for asthma.

Insecta: Dictyoptera (Cockroaches and Mantids)

13. Silphe/Blatta

1. Identification: general
Σίλφη and *blatta* are the standard Greek and Latin terms for the various species of cockroach (Dictyoptera: Blattaria) well known as domestic pests in antiquity. This identification has been generally accepted by previous authors.[90] Also included under these names would have been the tenebrionid beetles *Blaps* spp. which have a similar habitat and appearance.

The cockroach is one of the few significant insect species which is not dealt with in any detail by Aristotle. In the *Historia Animalium* its sole appearance is in a list of creatures which shed their skin (601a 3): this is in book VIII, some portions of which are inauthentic. We understand from Pliny (XX.140) that the older medical writers which he employed as sources classified cockroaches into a number of distinct varieties, of which he cites three, but this classification is not preserved by any surviving work of Greek medicine. Dioscorides, Galen and others simply deal with these insects under one general heading. It is therefore not clear to what extent Pliny's supposed species correspond with the various distinct Greek names for cockroaches which figure in lexicographical sources.

---

[89] Art. cit., 180.

[90] Sundevall, *Thierarten*, p. 199; Aubert and Wimmer, op. cit., I. p. 70, Leitner, *Zool. Glossen*, p. 57. Unlike *Blaps*, the two chief domestic species of cockroach, *Blatta orientalis* Linn. and *Blattella germanica* (Linn.), are neither of them native to Europe, having been introduced from N. Africa. So that, as Leitner points out, we do not know for certain if either had reached there in antiquity. Probably they had, at least by Roman times, when other pest species were certainly being introduced to areas where they were not native (cf. 36.1 below).

Only one of Pliny's varieties is given a Greek name, the other two being denoted by some feature of their physical appearance.

The evident confusion, at least as regards nomenclature, between cockroaches and crickets, due to the common domestic habitat of the former and the house-cricket *Acheta*, has been previously discussed.[91] A number of entomological terms preserved mainly by the lexicographers (viz. μυλακρίς in Hesychius, μυλαβρίς in Photius, and μολουρίς, μολυρίς, μελουρίς in Suidas, *EM*, Sch. on Nicander and *Etym. Gud.*) are defined as being referrable either to cockroaches or to crickets. Evidence for a similar confusion in Latin is provided by the translation of Dioscorides' *DMM*, where Περὶ σιλφῆς is rendered by *De grilis*. (The apparent ascription of sound production to the *silphe* in Cyranides p. 42 is probably due to this confusion.[92])

The cockroach is classed by Pliny (XI.99) as a form of beetle (*scarabaeus*), which is in keeping with its external appearance.[93]

1.2. *Identification: varieties and alternative names*

(i) *Τίφη* or *τίλφη*. In normal Greek usage τίφη appears as a simple synonym of *silphe*. It seems to have been employed interchangeably with the more common name. Aelian (I.37, VIII.13) uses both, as does Lucian (*Gall.* 31, *Adv.Ind.* 17). Pollux (VII.19) uses it instead of *silphe*, and the scholiast on Aristophanes *Ach.* 920 says it is the Attic equivalent of the latter. The form τίλφη is found only in Lucian *Adv.Ind.* 17. However, there is strong evidence[94] that, as Fernandez[95] suggests, the name under consideration originally referred to an entirely different insect and only later became confused with *silphe* due to its similarity in sound.[96]

(ii) *Scarabaeus rutilus*. After describing the common cockroach—*blatta*, considered as a species of beetle, Pliny (XI.99) goes on to mention certain *ex eodem genere rutili atque praegrandes scarabaei* which excavate dry earth and construct combs (*favos*) which resemble a small porous sponge (*fistulosae modo spongiae*) and contain poisonous— or medicinal (*medicatus*)—honey. No explanation has been offered for this curious application of hymenopterous behaviour to so unlikely an insect as a species of cockroach or beetle.[97]

[91] Cf. 12.1 above.,

[92] Fernandez, *Nombres*, pp. 169–70, suggests that the association of at least the term μυλακρίς with crickets was due merely to popular etymology, and that there was no genuine connection between the insect so named and any of the Orthoptera. This is because he identifies the μυλ- complex of names not as referring to cockroaches, but rather to the mealworm *Tenebrio molitor*, which, however, must have been classed as *kis/curculio* in antiquity (cf. 36 below).

[93] Cf. the scholion on Aristophanes *Ach.*920 whch defines σίλφη as ζῷον κανθαρῶδες. *Blaps*, of course, is a true beetle.

[94] Cf. 18 below.   [95] Op. cit., p. 239.

[96] The form τίλφη could, as Fernandez says, be either the result of a fusion of the two words, or a false atticism for σίλφη.

[97] A. Ernout and R. Pepin, op. cit., p. 150, suggest that *scarabaeus rutilus* is a rendering of the Greek χρυσοκάνθαρος, a name for the rose chafer, but there is nothing in Pliny's description to suggest this, Steier, *Aristoteles*, p. 50, and Leitner, op. cit., p. 216, suggest a soldier beetle of the genus *Cantharis*.

(iii) *Blatta mollis*. The 'soft' species is one of the three kinds of cockroach which Pliny (XXIX.140) mentions as being distinguished by earlier medical authorities: it has its unique medicinal use, as a remedy for warts when boiled in oil. The name is immediately reminiscent of the definition μαλακή σίλφη given by Hesychius and Photius for the words μυλακρίς and μυλαβρίς respectively. But the obstacle to the identification of the two is that *mylakris/-abris* is clearly synonymous with Pliny's second species, which he names *myloecos*. It is not improbable, however, that he may have had before him information from separate sources under the names μυλακρίς=μαλακή σίλφη and μύλοικος and been unaware that the same insect was being spoken of in both cases. Steier's suggestion[98] that the *blatta mollis* is simply the cockroach nymph, and the other two kinds the adult males and females respectively is inherently unlikely—the visible differences would not be sufficient to give the impression of distinct species—and takes no account of the Greek evidence.

(iv) *Blatta odoris taedio invisa, exacuta clune*. Pliny's third type of cockroach (XXIX.141) is described as having an unpleasant smell and of being pointed at the end of its body, and as with the other two is credited with distinct medicinal properties, namely as a treatment for bruises and various skin conditions. The description fits very well the tenebrionid beetle *Blaps*,[99] which is similar in appearance and domestic habit to true cockroaches and would be most likely to have been taken to be one in antiquity.

(v) Μύλοικος (preserved only in the latinised form *myloecos*). Pliny's second species, so named because of its preferred habitat (*circa molas fere nascens*). It is recommended in medicinal use for skin ailments. The word is not found in any surviving Greek source, but evidently represents the same insect as does *mylakris* and allied terms.

(vi) Μυλακρίς and μυλαβρίς. The μυλακρίς is defined by Hesychius as ἡ μαλακή σίλφη, white in colour and found in mills and among meal, along with the alternative renderings '*troxallis* found among meal' and ἀκρίς σιτοφάγος. In Pollux VII.19 it appears as the τίφη from mills, and likewise in II.189, where an example of its use by Aristophanes (fr. 583) is cited (here it is in the diminutive form μυλακρίδια as an insect pest in association with σκώληκες). The word appears to be a compound of μύλος and ἀκρίς, though Fernandez[100] does not accept this. The alternative form μυλαβρίς is given by Photius, who renders the first two definitions of Hesychius in abbreviated form. Fernandez[101] takes it to be

---

[98] Steier, op. cit., pp. 45–6; Leitner, op. cit., p. 57. Three obvious types would be the house cricket, the cockroaches, and *Blaps*, but the evidence does not seem to fit. The term 'soft', however, could well distinguish crickets and cockroaches from the hard-shelled *Blaps*.

[99] Leitner, op. cit., p. 57.

[100] Op. cit., p. 169, against Strömberg, *Gr. Wortstudien*, p. 20. The curious statement in Hesychius that the *mylakris* is white in colour could perhaps be taken to mean 'pale', as is the house cricket by comparison with the two domestic cockroaches and *Blaps*.

[101] Op. cit., p. 170.

merely an erroneous spelling, but it is accepted as a genuine word in LSJ:[102] the latter view is supported by the fact that both occur together as synonyms in Pollux (VII.180).

(vii) Μυληθρίς. Related to μυλωθρός, 'miller', this term is found only in Pollux (VII.19), as a synonym of mylakris.[103]

(viii) Μολουρίς, μολυρίς and μελουρίς. In Nicander (Ther. 416) a certain species of snake is depicted as preying upon frogs and μολουρίδες in meadows. The latter is a diminutive of μόλουρος, which figures later in the same poem (491) as a reptile. The scholiast on the passage, however, gives a definition which would make the word synonymous with mylakris, describing it in identical fashion as a creature compared by some with the akris and by others with the silphe. The Etym. Gen. is in agreement with this. Definitions of similar import are given for the alternative forms μελουρίς (τὴν σιτοφάγον ἀκρίδα; cf. Hesychius s.v. Mylakris) and μολυρίς (τὰς ἀκρίδας) by the Etym. Magnum and Suidas respectively.[104]

(ix) Κυνόλφη. Corrected to κυνόσφη in LSJ, an emendation not accepted by Fernandez, this name appears only in Hesychius as a synonym of silphe. Fernandez associates it with the verb κυνέλφει, 'to hide', appropriate for a cockroach.[105]

2. Life history and habitat

The cockroach was not among those insects whose life history was recorded by Aristotle. The sole reference to the insect in the Historia Animalium (601a 3) simply consists of its inclusion among a list of creatures which shed their skin during the course of their development. Augustine (c. Faustum MPL 42.363) speaks of cockroaches being spontaneously generated in obscuris cubiculis; and Arnobius (adv.Nat. VI.16) suggests that they make nests like mice.

Most of the classical references to these insects depict them as living in association with man (see 4 below). They were notorious for lurking in dark places and shunning the light (Pliny XI.99; Isidorus Or. XII.8.7; Augustine loc. cit.; Arnobius loc. cit.): Virgil (G. IV.243) describes them as lucifugae blattae.

3. Habits

There is a curious account by Aelian (I.37) of silphai which attack the eggs of swallows,

---

[102] And with reservations by Chantraine, Dict. Ét., p. 721.

[103] Cf. Fernandez, op. cit., p. 170.

[104] Gow and Scholfield, Nicander, p. 178, consider that Nicander intends a reptile to be understood here. Perhaps the word had two meanings. Gossen's suggestion in PW Supp. VIII.257 of a dragonfly is very improbable.

[105] Op. cit., p. 156. C.Gl.L. V.492.65 synonymises blatta with tinea, which is either simply a mistake, based on both insects being associated in literature as book pests, or an example of the general extension of the range of meaning of tinea in late Latin (cf. 26.1 below). C.Gl.L. II.478.58 interestingly glosses blatta by the Greek χρυσαλλίς, which implies a confusion with the cockchafer or rose chafer (cf. 32.1.2. below).

and against which the birds protect their offpsring with celery foliage, the latter being presumed to have a deterrent effect.[106]

4. Relations with man

The cockroach was considered as one of the more unpleasant insect pests of antiquity, well known by reason of its domestic habitat. Pliny (XXIX.140) feels it necessary therefore to apologise for introducing so disgusting a creature (*animal inter pudenda*) into the medical section of his *Natural History*. Galen describes it (XII.632 K) as a creature infesting houses (κατοικίδιος), and general references to it as a pest species are found in Lucian (*Gall.* 31), Laberius (*Mim.* 94) and Lactantius (*de Ira* MPL 7.118), and as infesting public buildings in Arnobius (loc. cit.). Though its main diet in homes would have been human foodstuffs, there is no actual reference to this in surviving sources. Instead the majority of allusions portray it as a creature destroying books and papers (Lucian *Adv.Ind.* 17; Martial VI.61—in association with *tineae*, XIII.1, XIV.37) and making its home among them: there is an entire epigram (*AP* IX.251) which addresses the cockroach as Ἐχθίστη Μούσαις σελιδήφαγε, λωβήτειρα, φωλάς . . . κελαινόχρως and laments its depredations. It is also noted, in association with the clothes moth (*tinea*), as feeding upon clothes in store (Horace *Sat.* II.3.119).

Cockroaches were also known to infest bath-houses, according to Pliny (XI.99) encouraged by the warm vapour (*umido vapore prognatae*). Dioscorides (*DMM.* II.38), Galen (XII.123 K) and Paulus of Aeginas (VII.3) characterise the insect as being found in bakehouses (ἀρτοκόπεια). Those cockroaches which were of importance as pests in mills appear in Greek sources under the special names μυλακρίς, μυλαβρίς, μυληθρίς and μύλοικος, synonymised with σίλφη by the lexicographers as well as by Pliny (cf. 1.2 v–vii above).

No deterrent or insecticidal preparations against cockroaches are mentioned in classical sources. Pliny, however, does make mention of two herbs, one named after the insect (*blattaria*), which when thrown down on the floor were said to draw together all the cockroaches in the house so that they could be conveniently disposed of (XX.171, XXV.108). The cockroach was also considered important by beekeepers, being listed among those pests infesting hives (Virgil *G.* IV.241–3; Columella *RR* IX.7.5; Palladius *RR* I.37.4): yearly fumigation was recommended to combat such pests.

5. Medicinal uses

Despite their reputation as noxious pests, cockroaches figure prominently among the animals employed in Greek medicine. Where a mode of preparation is specified, they are most commonly described as being boiled with oil or rose-oil (Pliny XXIX.139; Dioscorides *DMM.* II.36; Galen XII.366,632,861,123 K; Paulus Aegineta VII.3). Use of their ash is noted once by Pliny (XXIX.142), who says it was recommended to be kept ready for use in horn boxes. There is also mention of a kind of fat (στέαρ) as a product of

---

[106] Scholfield, *Aelian*, I pp. 56–7, suggests that the parasitic fly *Stenopteryx hirundinis* lies behind this story; but this insect would seem too small to be thought of as a *silphe*.

these insects, though there is confusion in our sources as to whether this is some exudation from the living insect (Aetius VIII.35 'σίλφης κατοικιδίου τῆς βδεούσης τὸ στέαρ'; Galen XIX.726, 731 K 'σίλφαι βδέουσαι', Cf. Isidorus *Or.* XII.8.7 *siquidem et comprehensae manum tingunt*) or a substance extracted from dead specimens (Galen, XII.861 K σίλφης τῆς κεφαλῆς δεούσης τὸ στῆρ, which Pliny XXIX.139 interprets as *quaedam pinguitudo blattae, si caput avellatur*). Cockroaches or preparations from them are most commonly prescribed for ear ailments, it being recommended that the liquid resulting from boiling them with oil be dropped into the ear (Pliny XXIX.139; Dioscorides *DMM.* loc. cit. and *Eup.* I.54; Galen XII.123,366,632,641,861; Paulus Aegineta loc. cit.) or that the preparation be inserted on wool or linen (Pliny XXIX.139). Pliny's three distinct varieties are all recommended for the treatment of warts, boils, various skin conditions, and for bruises and contusions. In addition, cockroaches in general are prescribed for the removal of objects embedded in the flesh (Pliny XXIX.142; Marcellus XXXIV.40), for the treatment of the uterus (Pliny XXX.131), for toothache (Galen XII.861 K), for swollen feet and ankles (Marcellus XXXIV.40), and for asthma, catarrh, and jaundice (Pliny XXIX.142).

14. *Mantis*

1. *Identification*

Although it is mentioned in only one surviving literary work from antiquity, our lexicographical sources indicate that this insect must have been widely known to the general public in the ancient world. This is indeed what one would have expected to be the case with so large and distinctive a creature. Despite the scantiness of the information available to us, there is fortunately no doubt as to the identity of the classical insect with that which is known today as the praying mantis. Keller, giving this identification,[107] cites the most familiar European species, *Mantis religiosa* Linn., but it should be noted that there are a number of related types which would not have been distinguished in antiquity. The mantids as a group (suborder Mantodea) are conspicuous insects which sit motionless upon foliage with their pincer like forelegs held in a distinctive posture in front of them, ready to seize upon other insects that come within range. It was this habit which attracted the curiosity of the ancients, though there is no evidence of any observation that its purpose was predation.

Curiously, the mantis seems to have been of more interest to the general public than to the zoological tradition: the creature does not figure in the works of either Aristotle or Pliny. It was commonly thought of as a form of grasshopper or cricket (*akris*). The surviving example of the word's use is found in Theocritus (X.18), where the author employs it in a metaphorical sense to describe a woman (μάντις τοι τὰν νύκτα χροιξεῖθ' ἀ καλαμαία). This passage is accompanied by a series of nine scholia, which seek firstly to elucidate the import of the uncomplimentary comparison, and secondly to explain how

---

[107] *Ant. Tierwelt*, II. p. 460.

the insect acquired its name. Much of this material, along with a few additional details, appears also in the lexica of Hesychius and Suidas, and in the anthology of proverbs compiled by Zenobius.

The corresponding Latin term to *mantis* is not known, since the insect is only mentioned in Greek sources. In Greek, though *mantis* was the commonest name, going back at least to the time of Aeschylus (fr. 40 N) according to the scholia, a number of alternative terms were also in use:

(i) Γραῦς σέριφος or σερίφη. This term is given as a synonym of *mantis* by scholion (a) on Theocritus. According to Zenobius (II.94) the name was a Sicilian one and was also used as a proverbial expression applied to elderly spinsters. Hesychius and Suidas give similar details. Zenobius and the two lexicographers all give σερίφη as an alternative spelling. The manuscripts of Hesychius and some of those of Zenobius read ἔριφος and ἐρίφη but in view of the word's probable connection with σέρφος the form with the consonant would seem to be the correct one. Zenobius cites the opinion of Apollodorus (*FGH* 244.301 Jac.) to the effect that the proverbial expression he is dealing with derived from an actual woman of Seriphos rather than from the insect, but this evidently is erroneous. In the view of Fernandez,[108] the element σέριφος is the actual insect name, with γραῦς—'old woman'—in apposition to it: σέριφος would be related to the terms σέρφος and σύρφος.[109] In this case, the 'old woman' component might refer either to the insect's appearance or to its reputation of being endowed with the 'evil eye'.[110] It has also been suggested, however, that the first was originally an entomological term as well, having been formed by popular etymology from an original κραῦξ, this theory being based on the entry in Hesychius s.v. κραυγή.[111]

(ii) Καλαμαία or ἀρουραία. Two terms deriving from the fact that the mantis was noted as an inhabitant of cornfields. They are used either as adjectives qualifying *mantis*, as in Theocritus (cf. Zenobius II.94, ἀρουραίαν ἀκρίδα=Hesychius s.v. *Graus eriphos*; Suidas s.v. ditto ἀρουραίας ἀκρίδος, and s.v. *Arouraia mantis*) as cited above, or in an absolute sense (καλαμαία Hesychius s.v., defined as εἶδος ἀκρίδος ἦν καὶ μάντιν καλοῦσι, ἀρουραία; Scholion(a) on Theocritus—γραῦν σέριφον, οἱ δὲ ἀρουραίαν ὀνομάζουσιν).

(iii) Βασκανία. This word appears to be given as a synonym of γραῦς σερίφια in Hesychius' third definition under κραυγή (in Latte's ed., καὶ γὰρ ἡ γραῦς Σεριφία ἀκρίς ἐστιν ἡ λεγομένη βασκανία). The text, however, has not been transmitted clearly at this point, and this meaning of the word is not included by LSJ. Fernandez[112] recognises it as a genuine insect name. As he points out, a word with a meaning like 'witchcraft' would fit in with the insect's other name of *mantis* and with its reputation for having a malign influence on other living creatures.

---

[108] *Nombres*, pp. 191–2.   [109] On which cf. 62 below.
[110] Cf. Davies and Kathirithamby, *Greek Insects*, p. 177.
[111] Fernandez, op. cit., p. 192 fn.   [112] Ibid., p. 196.

(iv) Φαρμακίς. Defined as *akris* by Hesychius, it seems likely in view of its primary meaning of 'witch' that this word is in fact a popular synonym of *mantis* rather than a name referring to a grasshopper and cricket.[113] As has been mentioned, the term *akris* was sufficiently wide to include the praying mantis within its compass.[114]

(v) Καταχήνη. This is recorded as a name for the mantis by Hesychius, who in a very intriguing entry states that the word refers to a ὑπὸ Πεισιστράτου καλαμαίᾳ ἐμφερὲς ζῷον, ἀπὸ τῆς ἀκροπόλεως προβεβλημένον, ὁποῖα τὰ πρὸς βασκανίαν. LSJ understand this as some kind of model rather than an actual insect specimen.[115]

2. Habits and popular beliefs

Scholion(f) on Theocritus X.18 correctly describes the mantis as being green in colour and as having long slender forelegs which are continually in motion. It is characterised in Suidas (s.v. *Mantis, Arouraia mantis*) as a form of *akris*, green in colour. Gow[116] supposes that the author chose the insect as a point of comparison in this context by reason of its thin appearance rather than because of its sinister reputation: this is in line with the explanation given by scholion (a) (ἴσως δ᾽ ἐκ τούτου τὴν ἰσχνὴν καὶ διεφθινηκυῖαν). E. K. Borthwick,[117] however, argues that Theocritus has in view here the female mantis' notorious propensity for devouring her mate. Scholia (g–i) all suggest that the comparison has something to do with the animal's colour, which they describe incorrectly as μέλαινα but this is clearly a rather unintelligent guess.

As has been noted above, the mantis was known particularly as an insect frequenting cornfields and thus acquired the names *kalamaia* and *arouraia*, the former of which is employed by Theocritus. Scholion (a) explains that it is a form of grasshopper found on corn stalks (ἐν τῇ καλάμῃ γινομένη), while scholion (i) incorrectly states that it actually feeds upon the ears. There is quite an accurate depiction of a mantis of indeterminate species perched on an ear of corn on a coin of the late fifth-century from Metapontum.[118]

Because of its habit of sitting motionless for long periods, the mantis is described by Suidas (s.v. *Mantis* and *Arouraia mantis*) as δυσκίνητος presumably particularly in comparison with the rapid activity of other types of *akris*: the word is said here to have been applied to sluggish and inactive people.

It was evidently the insect's curious posture adopted to await its prey that particularly excited the curiosity of the ancients and was responsible for its rather sinister reputation in popular belief and for its name *mantis*, as well perhaps as other names. As to the particular

---

[113] Cf. 11.1.2.xv above.   [114] Cf. 11.1.2.iii above.

[115] According to G. P. Shipp, *Modern Greek Evidence for the Ancient Greek Vocabulary*, p. 308, the word survives in dialectical Greek with the meaning 'malevolent ghost'. If it had this meaning in classical times it would be quite an appropriate appellation for the mantis with its sinister reputation.

[116] A. S. F. Gow, *Theocritus*, II pp. 197–8.

[117] In an unpublished article. Gow mentions this possibility only to reject it on the grounds that there is no evidence that this aspect of the creature's life was known in antiquity. Borthwick suggests that the name γραῦς σέριφος may also contain an allusion to this habit.

[118] Keller, *Münzen & Gemmen*, p. 48, pl. VII.41.

reason for the application of its commonest name, our sources display some uncertainty, three separate explanations being provided. The most frequently mentioned of these, given by Scholia (e) and (i) on Theocritus, Zenobius, and Hesychius, states that the mantis is an insect of ill omen, causing harm to any living creature that it looks upon (ἔι τινι ἐμβλέψειε ζώῳ, τούτῳ κακόν τι γίνεσθαι). Scholion (e) (=Zenobius and Hesychius) credits this explanation to Aristarchus, who rendered it in connection with the insect's appearance in Aeschylus' *Lycurgus*; scholion (i) is based on a different source and emphasises also the insect's malevolent appearance (ὅτι κακόχρους καὶ πρασίζουσα οἷς ἂν ἐντρανίσῃ ζῴοις κακὸν προμηνύει). The latter scholion also provides the alternative explanation that the mantis' occurrence is prophetic of famine, while Suidas (loc. cit.) says that omens were taken from its movements. It is likely also, although not explicitly mentioned in our sources, that, as Fernandez[119] suggests, the insect's holding of its forelegs in an attitude of prayer or contemplation was also a factor in the application of its name: it is this particular aspect of its habits that has been responsible for several of its modern European names.

3. Medicinal uses

There is only one recorded medicinal use for the mantis: this is found in Dioscorides (*Eup.* 1.158) who prescribes it for the treatment of tumours in the throat. The author follows this with the cryptic comment that the insect resembles the Indian *akris* (ἔστι δὲ ὅμοιος ἀκρίδι Ἰνδικῇ), but no creature of this name is mentioned in any other surviving classical work.[120]

Insecta: Ephemeroptera (Mayflies)

15. *Ephemeron/Hemerobion*

1. Identification

The insect known as ἐφήμερον (sc. ζῷον) is first described in the *Historia Animalium*, upon which all later sources are dependent for their information. The account appears in a passage of paradoxographic type, the whole of which is deleted as an interpolation by Aubert and Wimmer and Dittmeyer, alongside the wholly or partly fabulous snow worms of Media and the fire-dwelling *pyrallis* of Cyprus.[121] On account of its brief life span, the feature which especially captured the imagination of the ancients, the creature has been identified by Sundevall[122] as a species of mayfly (Ephemeroptera). This identification is accepted by Keller, D'Arcy Thompson and Leitner.[123] And indeed many species of Ephemeroptera do live for only a matter of hours in the adult state. Aubert and

---

[119] Op. cit., pp. 189–90. Cf. Davies and Kathirithamby, *Greek Insects*, pp. 177–80.

[120] Pliny does mention Indian Orthoptera (cf. 11.10.vii above) but there can scarcely be any connection here.

[121] Cf. 59–60 below.

[122] *Thierarten*, p. 199. He suggests that the species particularly concerned is *Palingenia longicauda* (Oliv.), the largest European form, which lives in the appropriate part of the world.

[123] Keller, *Ant. Tierwelt*, II pp. 454–5; D'Arcy Thompson, *Hist. An.*, V.19 note.

Wimmer,[124] however, argue that the creature cannot be a mayfly since it is described as having only four legs and since the account of its life history is inappropriate, and conclude that it is unidentifiable as the data fit no known species. Sundevall's suggestion that the impression of its having four legs is a result of the inconspicuousness of the middle pair is summarily rejected (Leitner puts forward a similar view; namely that the forelegs were mistaken for antennae). It may be said, nevertheless, that such errors of fact as are pointed to by Aubert and Wimmer are frequent in classical natural history, even in connection with insects native to Greece, and are only to be expected in a hearsay report from a remote part of the known world. It is therefore quite probable that, however distorted the data, some form of mayfly does lie at the foundation of the Aristotelian account.

This, however, does not mean that *ephemeron* is a simple equivalent of the English term 'mayfly'. It is a conspicuous fact that there is no evidence of any recognition of similar insects from more familiar parts of the classical world, or any information independent of the *Historia*. So that, although mayflies would certainly have been observed as a feature of the Southern European insect fauna, it is not necessarily to be assumed that these would have been termed *ephemera* or associated with the insects described in the *Historia*.[125]

2. Life history

In the account in the *Historia Animalium* (522b 18 ff.) the *ephemeron* is located in south Russia, as living around the river Hypanis in the Cimmerian Bosporus. At the summer solstice, we are informed, certain objects resembling sacks or pouches (θύλακοι), larger than grapes, are carried down river by the current. These then split open to reveal a winged insect with four legs, which flies until evening and then, as the sun sets, shrivels (ἀπομαραίνεται) and dies after a life of only a single day. The insect appears again at 490b 1 ff., where its short life and its unusual nature as a winged quadruped are emphasised: if the above passage be regarded as spurious, then presumably this reference should be deleted as well, especially as in the context it is inappropriate as an example of a many-limbed invertebrate. There is also a brief allusion to the insect's life span on the part of Theophrastus (*Metaph.* 29). Descriptions of the *ephemeron*'s life history derived from the *Historia* are given by Antigonus Carystius (85), Plutarch (*Mor.* 111c), Aelian (V.43), Pliny (XI.120) and Cicero (*Tusc.* I.94). All of these simply reproduce, with varying degrees of abbreviation, the Aristotelian account. Pliny translates the sack-like objects as *acinorum effigie tenues membranas*. Aelian omits mention of this stage altogether and simply has the adult insects born (τικτόμενον) at the first light of dawn. Similarly Pliny has them born at morning, in their prime at noon, and sinking into old age by evening. Of the above authors, Pliny employs the alternative name ἡμερόβιον found in Theophrastus while Aelian uses μονήμερον, having applied ἐφήμερον to the vinegar fly.[126]

---

[124] *Thierkunde*, I. p. 164.

[125] There is a curious scarcity of data on freshwater insects generally in classical literature, extending even to such conspicuous creatures as dragonflies (cf. 53–4 below). Sundevall, op. cit., p. 199, suggests that Aristotle's reference to *empides* shedding their skin (cf. 50.2 below) is an allusion to the development of small mayflies.

[126] Cf. 51.1.2 below.

IV

CICADAS, BUGS AND LICE

Insecta: Hemiptera

16. *Tettix/Cicada*

1. Identification: general

In general terms the above names present no problem of identification, covering as they do the Southern European species of cicada. These insects belong to a single family, the Cicadidae, and are of similar appearance and habits, making themselves a conspicuous part of the countryside by reason of their loud monotonous stridulation. Along with honey bees they may be said to have been the most popular insects in classical antiquity, one among the small number of living creatures that the ancients felt genuine affection for, and allusions to them are frequent in a wide variety of Greek and Roman authors.

In poetry cicadas are often closely associated with grasshoppers and crickets,[1] which although very different in appearance and taxonomically unrelated share with them the habit of stridulation. Gow[2] and Borthwick[3] discuss the possibility of actual confusion between the two groups of insects. Gow concludes that, although some writers were under the mistaken impression that the two groups' mode of sound production was identical, it is hardly credible that any could have actually confused them, because they are clearly different in appearance and song, and appear as distinct in visual representation.[4] However, as Borthwick points out, there is literary evidence[5] for such confusion, since it is evident that writers such as Theocritus and the poets of the Greek Anthology, when they wished to introduce an allusion to a singing insect, often selected one or other name quite arbitrarily, placing them in similar contexts and applying to them similar epithets (cf. for example Theocritus V.29 and VII.41, V.34 and VII.159; Anyte in *AP* VII.190 and IX.373).[6] As a result of such confusion, Borthwick suggests,[7]

---

[1] For which the ancients had the general name *akris/locusta*. Cf. 11.1 and 12.1 above.

[2] A. F. S. Gow, *Mnasalces*, CR, NS VI, 1956, p. 93.

[3] E. K. Borthwick, 'A Grasshopper's Diet—Notes on an Epigram of Meleager and a Fragment of Eubulus', *CQ*. XVI, 1966, pp. 103–6.

[4] Cf. Keller, *Münzen und Gemmen*, pl. VII and XXIII.

[5] Cf. also 11.4 above.

[6] Such confusion of nomenclature still continues today, as some modern translators of classical works inadvertently demonstrate, e.g. Rackham, *Nat. Hist.* III pp. 488–91.

[7] Art. cit., pp. 103–5, Cf. 11.8 above.

traditional motifs attached properly to the cicada, such as dew drinking, could be transferred in poetry to the *akris*.[8]

There is very little by way of physical description of cicadas in classical sources, attention being mostly directed to their song. Martial (I.115) describes them as black in colour and Theocritus (VII.158) as αἰθαλιώνες. Hesiod's epithet κυανόπτερος (Sc.393) must mean simply 'dark-winged'.[9]

1.2. Identification: the Aristotelian species

Aristotle divides cicadas into two species, distinguished mainly by size and time of appearance. These are a small form, which appears first and dies off last, and a large form which, though it emerges later than the other, does not survive as long. The strongly singing males of the large species he terms ἀχέται and the small, whose males he says sing only a little, τεττιγόνια (*HA* 556b 14 ff.). Once again Aristotle is here using popular terminology to provide himself with something approximating to a scientific nomenclature; that is, neither of the words he uses here was originally a specific name. Ἀχέτας which is the Doric form of ἠχέτης, meaning 'singing', is in its earliest usage an adjective applied to *tettix* (Hesiod *Op.* 582 and Sc.393; also in *AP* VII.213), and is later found as a synonym of the latter (Aristophanes *Av.* 1095 and *Pax* 1159; Ananios fr. 5 Bk).[10] Hesychius gives both Attic and Doric forms and defines the latter as the male cicada; that is as opposed to the mute female (as do the Scholiast on *Av.* 1095 and Suidas s.v. *echetes*). The atticist Pausanias (*ap.* Eustathius 396.1 ff.) defined ἀχέτας as 'τὸν μέγιστον τέττιγα', and the word appears in the catalogue of cicada names given by Aelian X.44 and Suidas s.v. *tettix* (but cf. Suidas s.v. *achetas*).

Sundevall's identification of Aristotle's larger variety as e.g. *Tibicen plebejus* (Scop.) and *Cicada orni* Linn., and his smaller as e.g. *Cicadetta montana* (Scop.) *Cicadatra atra* (Oliv.) and *Pagiphora annulata* (Brull.) has been generally accepted.[11]

1.3. Identification: Pliny's account

Pliny (XI.92) reproduces Aristotle's account of the two species, transliterating the Greek terms, but through misunderstanding of the not altogether clear way in which he expresses himself he makes the small species entirely mute. Later on (94) he records certain Latin terms not found elsewhere: *surcularia* for a large form of cicada, and *frumentaria* or *avenaria* for one which appears at harvest time, whose identity it is not possible to determine.[12]

---

[8] Nomenclatural confusion is not evident in Latin literature of the classical period, though in mediæval Latin we find *cicada* being used even for the house-cricket (cf. Bodenheimer, *Geschichte*, p. 175).

[9] M. Melon *La Cigale dans l'Antiquité Gréco-Latine*, Thèse de Licence, Ann. Acad. 1941–2. Univ. de Liège, p. 55, takes it to refer to the wings' iridescence.

[10] Fernandez, *Nombres*, pp. 121–2.

[11] Sundevall, *Thierarten*, p. 201; Aubert and Wimmer, *Thierkunde*, I. p. 162; Steier in PW V.1113.

[12] Leitner, *Zool. Terminologie*, p. 122, considers them Latin equivalents of Aristotle's two species.

1.4. Identification: names transmitted by the lexicographers

The cicada shares with grasshoppers and crickets the distinction of having an extraordinary number of additional names, most of which have been transmitted to us through lexicographical channels only. The largest number, mostly not found elsewhere, are contained in Hesychius, but there is also a catalogue of six names (including *achetas*) found in Aelian (X.44) and Suidas (s.v. *tettix*). Smaller lists are given by the atticist Pausanias (*ap.* Eustathius 396.1 ff.) and by the scholiast on Aristophanes *Av.* 1095.

Since for most of these names we have no examples of their actual usage, nor for the most part definitions which are genuinely informative, it is not possible to determine conclusively which of them—if any—are names of distinct species and which are simply dialect terms used instead of *tettix* in various different parts of the Greek world.[13] Aelian in reporting his catalogue of names claims that they all represent distinct species which οἱ δεινοί enumerate, but it must be doubted whether such apparent precision of nomenclature is not due simply to the misplaced ingenuity of lexicographers working upon words which in their origin were simply local synonyms of *tettix*. Certainly a multiplicity of dialect names for familiar animals is a well known phenomenon in modern languages. And the fact that most of the names in question can be derived from roots which have a very non-specific reference—namely to the sound produced by cicadas—provides strong support for this interpretation.

(i) *Τεφράς*. The 'ashen' cicada, named from its colour, is listed by Aelian and Suidas, and is perhaps one of the likeliest candidates for recognition as a genuine species name.

(ii) *Μέμβραξ*. A name found in the same sources as the above and of whose origin Aelian confesses ignorance. Fernandez[14] suggests that it derives from the same root, with a meaning 'to hum', as the hymenopterous name πεμφρηδών. If this is so, the name would refer to the sound production common to cicadas in general, and not, as Gossen[15] suggests, to some feature by which one member of the group might be distinguished from the rest. It would thus be most likely to be a synonym of *tettix*.

(iii) *Λακέτας*. Catalogued by Aelian and Suidas, the name derives from λακέω and would mean something like 'chirper'.[16]

(iv) *Κερκώπη*. This is the most well documented of cicada names, thanks to the preservation by Athenaeus (133 b) of a number of literary references to it, and the one most likely to refer to a distinct form. Aristophanes (fr. 51) wrote of both τέττιγες and κερκῶπαι being captured for food, and Alexis (fr. 92) characterised

---

[13] Gossen, *Tiernamen*, nos. 77–81, and *Zool. Glossen, passim*, has proposed a large number of arbitrary identifications for these names, based on the erroneous assumption that the ancients possessed a precise zoological nomenclature. He extends the range of the concept *tettix* to cover various allied groups of Hemiptera for which there is no reason to suppose that the ancients would have required names at all. In view of their general nature, there is no purpose in discussing these identifications in detail.

[14] Op. cit., pp. 233–4. [15] *Tiernamen*, no. 80. [16] Fernandez, op. cit., p. 122.

both as 'talkative' (λαλίστατος) creatures. Epilycus (fr. 4) also used the latter word (Libanius *Decl.* XXVI.34 = Alexis fr. 92).

Classical lexicographers display disagreement as to the meaning of the word. According to one view, represented by the scholiast on Aristophanes *Av.* 1095 and noted by Hesychius, it is the name given to female cicadas (the scholiast divides cicadas into three types: males (ἀχέται), females, and a silent type (σίγιον)). The alternative view was that the κερκώπη was 'a creature like the τέττιξ and the τιτιγόνιον' (Speusippus' *Homoia*, quoted by Athenaeus 133b; so Eusthathius *Comm. in Il.* 1282.40 ff. and Photius s.v.) or 'the smallest τεττίγιον, the καλαμαῖον' (Pausanias *ap.* Eusthathius 396 1 ff., who contrasts it with the larger ἀχέτας. Hesychius gives this as his primary definition, with the above as an alternative). The name is included without comment in the lists given by Aelian and Suidas.

Various explanations for the name have been put forward. Steier[17] associates it with the mythological dwarfs known as κερκώπαι, the point of comparison between these and the insects being their elusiveness. Fernandez[18] considers the view that the name has the meaning of 'tailed' and alludes to the female ovipositor, thus confirming the first of the definitions given above.[19] But he rejects this on the ground that Alexis credits the insect with song, whereas it was well known in antiquity that the females of cicadas are mute, and prefers to see a link with the grasshopper name κέρκα and a derivation from an onomatopoeic root. If we follow the line of Speusippus and Pausanias, it is possible that κερκώπη was the popular term for those smaller kinds of cicada that Aristotle chose to call *tettigonia*.

(v) 'Ακανθίας. Is doubtless related to the plant name ἄκανθα (covering various thorny species), as is stated in Suidas s.v. *achetas* ('. . . ἀκανθίας, because of its singing among ἄκανθαι').[20] It is also listed by Aelian (loc. cit. = Suidas s.v. *tettix*) and Hesychius ('a type of *tettix*'). Cf. the adjective ἀκανθοβατίς applied to a grasshopper in *AP* VII.198.

(vi) Τιτιγόνιον. Speusippus (loc. cit.) described the κερκώπη as 'a creature like the τέττιξ and the τιτιγόνιον, and in the *EM* the latter is itself defined as 'a creature like the τέττιξ'. Pausanias *ap.* Eustathius loc. cit. notes it as a type of cicada along with *achetas* and *kerkope*. Cf. also Epilycus fr. 4 ap. Photius s.v. Fernandez[21] sees the word as a diminutive of an original non-surviving name τιτιγών derived like τέττιξ itself from the verb τιτίζω and sees the Aristotelian τεττιγόνιον as a development from it under the influence of τεττίγιον, the true diminutive of τέττιξ found in Pausanias (cf. iv. above).

Eustathius (*Comm. in Il.* 1282.40 ff.) reproduces the statement of Speusippus quoted above, but with τριγονίῳ instead of τιτιγονίῳ. The former occurs also as

---

[17] In PW V.1113.   [18] Op. cit., pp. 45–6.
[19] Strömberg, *Gr. Wortstudien*, p. 16.   [20] Fernandez, op. cit., p. 165.   [21] Ibid., pp. 130–1.

a variant reading for τεττιγόνια in *HA* 556a 21. Fernandez[22] suggests that this is a genuine name rather than merely the product of copyists' errors, deriving it from τρίζω, though LSJ do not recognise it as such.

(vii) *Σίγιον*. According to the scholiast on Aristophanes *Av*. 1095, there are three types of *tettix*: the males or ἀχέται, the females or κερκώπαι, 'and another kind called σίγιον because they make no sound'. Fernandez[23] classes the name among those alluding to the cicada's song, in which case the supposed relationship with σιώπη would be merely an incorrect popular etymology.

(viii) *Βάβακος*. According to Hesychius, a local name for the cicada in Elis.

(ix) *Ἑρπυλλίς*. Defined simply as *tettix* by Hesychius, this word is probably formed from the plant name ἕρπυλλος.[24]

(x) *Ζειγαρά*. Another local synonym of *tettix*, according to Hesychius: ὁ τέττιξ παρὰ Σιδήταις.[25]

(xi) *Καλαμίς*. Hesychius records that 'the Ceryneans call the small *tettiges* καλαμίδες'. Probably therefore a synonym of *kerkope* and the Aristotelian *tettigonion* (cf. xii below).

(xii) *Καλαμαῖον*. Given as a synonym of *kerkope* by Pausanias (loc. cit.) and Hesychius, this is the neuter of an adjective καλαμαῖος meaning 'associated with corn stalks'.

(xiii) *Κίκους*. 'The young cicada' according to Hesychius, by which he presumably means the nymph, which one would not expect to be given a name alluding to the adult's song, as Fernandez[26] suggests.

(iv) *Κίλλος*. This Cypriot local name for the cicada in Hesychius is evidently descriptive, like τεφράς, of its dark colour.[27]

(xv) *Κίξιος*. This apparent synonym of *tettix* (Hesychius s.v.) is classified by Fernandez,[28] along with κίκους, σίγιον, σιγαλφός and ζειγαρά among cicada names derived from roots imitative of their song.

(xvi) *Κῶβαξ*. Defined by Hesychius as 'the large cicada', this is probably a local name for those species termed *achetas* by Aristotle. Like most cicada names, it may well have originated from the insect's song.[29]

(xvii) *Λιγάνταρ*. A name among the Spartans for 'a kind of *tettix*' (Hesychius), it is derived from the verb λιγαίνω which is related to the adjective λιγυρός often employed to describe the cicada's song.[30]

(xviii) *Σιγαλφός*. In Hesychius, οἱ ἄφωνοι, καὶ οἱ ἄγριοι τέττιγες. It is suggested by Fernandez[31] that the connection here made between this name and words suggesting silence is only the product of popular etymology, and that in fact once again the insect's song is in view.

---

[22] Ibid., p. 125.   [23] Ibid., p. 126.   [24] Strömberg, op. cit., p. 17; Fernandez, op. cit., p. 165.
[25] Cf. Fernandez, op. cit., p. 126.   [26] Ibid., p. 126.   [27] Ibid., p. 100.
[28] Ibid., p. 127.   [29] Ibid., p. 122.   [30] Fernandez, ibid., p. 122.   [31] Ibid., p. 126.

2.1. Life history: the Aristotelian account

Aristotle (HA 556 14 ff.) gives a detailed and generally accurate account of the cicada's life history, which Pliny (XI.93) reproduces in an abbreviated form. According to the text as we have it, after mating, which Pliny describes in a rather confused way (coitus supinis), the females give birth (τίκτουσι) in uncultivated land boring into the ground by means of an organ behind. Aubert and Wimmer[32] delete this sentence as an interpolation, since it contradicts what follows and seems to be based on a confusion with the habits of locusts. Assuming this to be so, the authentic text which follows says that they (Pliny omits this statement) bore into the canes used for supporting vines or into the stalks of the squill. Oviposition in the latter is also referred to by Plutarch (Mor. 767d). Cicadas in fact lay their eggs in slits made in twigs etc.[33] On hatching, the newly emerged nymphs—cicadas have an incomplete metamorphosis—descend to the ground and burrow into it. They spend this stage of their development beneath the soil feeding upon roots. This is what Aristotle is referring to when he speaks of the κυήματα laid by the females burrowing.[34] They are, he says, numerous in rainy weather. He goes on to describe how the nymphs—which he rather imprecisely calls σκώληκες (Pliny vermiculi), since they are undeveloped versions of the adult rather than true larvae—increase in size to become what are known as τεττιγομήτραι.

This fully developed nymph, the account continues, emerges from the soil by night at the time of the summer solstice, its outer covering splits open, and the adult cicada is revealed, which soon becomes dark in colour, firmer and larger, and begins to sing. These observations are all correct. The shedding of the nymphal skin is also mentioned in HA 601a 6 ff., where it is observed that on emergence the adults leave a little moisture behind them. Callimachus' reference in Aetia 1.29 ff., where he compares himself as poet to a cicada, to casting off old age (γῆρας ἐκδύοιμι) is evidently an allusion to this stage of the insect's life history,[35] though it is possible that he believed that, like snakes, they shed their skin more than once during their life: snakes were popularly believed to perpetually renew themselves in this way and thus to be more or less immortal (Aelian VI.51; cf. the description of the cicada as ageless in Anacreontea 34). Lucretius (IV.58; cf. V.803) associates the ekdysis of cicadas (cum teretis ponunt tunicas aestate cicadae) with shedding of skin on the part of snakes.[36]

2.2. Life history: popular beliefs

The earliest belief concerning the development of cicadas was simply that they were, as

---

[32] Op. cit., I. pp. 532–3. Plato Symp. 191c may embody a similar confusion with the habits of Orthoptera (Davies and Kathirithamby, Greek Insects, pp. 124–5).

[33] Cf. Imms, Gen. Textbook, p. 710.

[34] καταρρεῖ, so rendered in LSJ, rather than Peck's 'seep down', which implies that they are not active in the process.

[35] E. K. Borthwick, art. cit., p. 112.

[36] Melon, op. cit., p. 787, compares this concept of the immortality of the cicadas with the myth of Tithonus, which seems to contain a number of motifs drawn from actual beliefs concerning cicadas (cf. E. K. Borthwick, art. cit., pp. 105, 109; Davies and Kathirithamby, op. cit., pp. 126–7).

Anacreontea 34 describes them, earthborn (γηγενής); that is, that they were, in Aristotelian terms, spontaneously generated from the soil. This idea was founded no doubt on observation of the developed nymph's emergence from the ground, and it continued to exist alongside the 'scientific' view of the matter as found in Aristotle. Plutarch (*Mor.* 637b) definitely classes the cicada as a product of spontaneous generation, as do Sextus Empiricus (*Pyrrh.* I.41) and Lucretius (V.803), who uses it as an argument in support of his theory that at the beginning of time all living creatures were brought forth by the earth.

The supposed earthborn nature of the cicada caused the insect to be thought of by the Athenians as a symbol of their own claim to be an autochthonous people. The mysterious golden '*tettiges*'[37] which the men of Athens and Ionia are reported to have worn in archaic times as a hair ornament (Thucydides I.6; Aristophanes *Equ.* 1331, *Nub.* 984 & schol. ad. loc.; Eustathius 395.33; etc.) are said to have been a product of this association, though the actual nature of these items is a matter of dispute.[38] Interesting in this regard is the comment of Antisthenes (fr. 129) reported by Diogenes Laertius (VI.1.1) to the effect that the Athenians' boast of autochthony did not make them any higher born than creatures such as locusts.

Isidorus (*Or.* XII.8.10) preserves the interesting belief that cicadas are generated *ex cuculorum . . . sputo*. This latter, which we know today as 'cuckoo-spit', is the protective 'froth' produced by the nymph of the froghopper *Philaenus spumarius* (Linn.) and its relatives.[39] Probably this belief simply results, like the association with cuckoos, from the time of year at which the substance appears, although it is not impossible that the nymphs themselves were observed within the froth and interpreted as embryonic cicadas.

3.1. Habitat

Details of the preferred habitat of cicadas are given by Aristotle (*HA* 556a 21 ff., followed by Pliny XI.95), who notes first of all they do not occur in treeless localities. Reference to cicadas singing from trees are widespread in literature (cf. 6.2). As an example, he observes that at Cyrene they do not live on the flat country but only around the city, especially where olive trees grow, because these do not produce deep shade. They are absent from places where the temperature is low (cf. Aelian III.38) and from dense woodland.

3.2. Anomalies of distribution

The subject of curious distributional anomalies among cicadas evidently held a certain fascination for the ancients and therefore came within the ambit of the paradoxographers

---

[37] Concerning which cf. E. K. Borthwick, art. cit., pp. 107–9; A. B. Cook, *Zeus*, III pp. 246 ff. These authors also discuss, in relation to the respect held by the Athenians for cicadas, the significance of Athenian myths such as those of the earthborn Cecrops (whose name, it is suggested, might be related to the cicada name κερκώπη) and of Tithonus.

[38] A. B. Cook, op. cit., III pp. 253, 255, illustrates some brooches in the form of cicadas, and examples of Attic coins depicting them.

[39] Steier, art. cit., 1118.

who specialised in collecting unusual stories from around the classical world. It was reported that in Miletus the insect was present in some neighbouring districts but absent in others (HA 605b 26; Pliny XI.95), and that in Cephallenia there was a certain river on one side of which cicadas were numerous, while on the opposite side they were absent (or few, according to Pliny loc. cit.; HA 605b 28; Aelian V.9). Most popular of all was the story about the cicadas of the neighbouring districts of Rhegium and Locri in Italy. Those inhabiting Rhegium were said to be wholly silent, while the population on the opposite bank of the river Kaikinos (or Halex) in Locri were said to sing normally (Timaeus ap. Antigonus Hist.Mir. 1; Strabo VI.1.9; Pausanias VI.6.4; Pliny XI.95; Solinus II.40; Isidorus Or. XII.8.10). Aelian (V.9) puts things rather differently, saying that the insects never fly across the river, and that it will be found that the Locrian ones will keep silent in Rhegium and vice versa. Solinus cites an aetiological myth recorded by Granius, to the effect that the Rhegian cicadas kept silence because they were once commanded to do so by Hercules, whose sleep they were disturbing. Strabo puts forward as an explanation for the phenomenon the theory that since on the Rhegian side of the river the country is shady the cicadas there become wet with dew and are thus unable to expand the 'membranes' with which they sing, whereas over in Locri where it is sunny the insects have dry and horny membranes. This report is associated by Strabo and Antigonus with the legend about Eunomus of Locri (cf. 6.3).[40] The cicadas of Akanthos in Athamania were also said to be mute and gave rise to a proverb 'Ἀκάνθιος τέττιξ (Simonides fr. 220 B).

4. Feeding habits

The most general belief in antiquity was that cicadas' diet consisted entirely of dew. Aristotle records this once as an opinion of country-folk (HA 556b 15 ff.) and once as an unqualified statement (532b 10 ff.); and the idea is found very frequently in authors from Hesiod (Sc. 395) onwards (Anacreontea 34; Callimachus Aetia 1.29 ff.; AP VI.120, VII.196, IX.92,373; Aesopica 184 Perry; Athenaeus 46e; Aelian I.20; Philostratus Ep.71; Libanius Decl. 33, VII.85 F; Ps. Eustathius Hex. MPG 18,736; Virgil Ecl. V.77; Pliny XI.93).

According to a less frequently mentioned tradition, they were also credited with subsisting upon air (Philo de Vit.Cont. 35, Quod Omn.Prob. 8; Ambrose Ep. 28.5; upon air and dew, according to Plutarch Mor. 660 f.), or even on nothing at all (Plato Phaed. 259c; Philostratus Vit.Apoll. VII.11). Cicadas thus came to be employed as symbolic of the reputed neglect of the basic necessities of life on the part of poets and intellectuals[41] (cf. the quotations above from Plato, Philostratus and Libanius (also Decl. 32,VII,56F; 26,VI.537F), and Aristophanes Nub. 1360). Steier suggests[42] that the idea of a diet of dew

---

[40] M. Melon, op. cit., pp. 97–8, taking these stories together with the tale of Eunomus, suggests that they result from rivalry between the two cities.

[41] E. K. Borthwick, art. cit., p. 110.

[42] Art. cit., 1117; supported by Davies and Kathirithamby, op. cit., p. 123.

arose from observation of drops of clear excrement on the branches of trees frequented by
cicadas.

Aristotle (*HA* 532b 10 ff.) states that they have no mouth, but feed by means of a long
unforked 'tongue' like that of flies, and (*PA* 682a 18 ff.) that they draw up dew with this
organ in the same way as liquid is drawn up by a root. Cf. Pliny XI.93. Cicadas' actual
mode of feeding is by means of a piercing rostrum with which they suck fluid from
within the branches on which they rest.

5. Popular beliefs and attitudes

Aristotle tells us that cicadas were believed to have poor eyesight (cf. Porphyry *de Abst.*
III.23) on the grounds that they would climb onto a moving finger placed near them,
under the impression that it was a leaf (*HA* 556b 17 ff.; Pliny XI.94).

As has been said, cicadas were the most popular insects in antiquity, with the exception
of honey bees which were of direct service to man, because of their song (cf. 6.3) and
because, unlike most insects—or so the ancients thought—they did no harm either to
man or his crops. In *Anacreontea* 34 they are described as φίλτατος to farmers for this
latter reason. Far from harming the farmer's crops, they contribute positively to his
wellbeing with their song (*Aesopica* 299, 387). However, the fact that they apparently
spent all their time in song, enjoying themselves rather than doing anything constructive,
was regarded as an example not to be emulated. They were seen as typifying carefree,
indolent, improvident and impractical persons who disregarded the practicalities of life
and took no heed for the future. Hence the famous Aesopic fable of the idle cicada
contrasted with the industrious ant which stores up food in summer while the cicada is
wasting its time in song, so that it can survive during the winter (Babrius 140; Avianus
fab. 34).[43] But of course the hard-working ant would never have been proclaimed in
literature as beloved of Apollo and the Muses, for example, or 'almost like the gods'
(*Anacreontea* 34); the cicada's life was viewed, therefore, as being in some sense enviable,
even if not to be imitated by man.

5.2. Weather lore

Since the song of the cicada was thought of as typifying high summer—it is described as
prophetes of summer in *Anacr.* 34—it followed that their presence in unusual numbers was
seen as a sign of an especially hot and therefore unhealthy season (Theophrastus *de Sign.*
54).

6.1. The cicada's song: how produced

The fact that it is only the male cicadas which sing was apparently well known in
antiquity, though it was not universally accepted: there was a rival interpretation to the
effect that the mute cicadas were a separate species. Aristotle in his *Historia* shows
evidence of some confusion on the subject. In his main account (556b 14 ff.) he divides

---

[43] Cf. E. K. Borthwick, art. cit., pp. 107–11.

cicadas into two species according to size (cf. 1.2 above), and then states that these each occur in two forms: those with a 'division' (διῃρμένοι) at the *hypozoma* (the point at whch the thorax joins the abdomen), which sing, and those without, which are silent. Some way later (556b 12), he adds that the former are the males and the latter the females. However, at 532b 15 ff. earlier on he writes that it is the *achetai* alone which are provided with organs for singing, which the *tettigonia* do not possess. Pliny (XI.92), evidently confused by Aristotle's account, reproduces both of these views in succession. Aelian (I.20 and IX.26) states that it is the females which are mute (cf. also Aristotle *de Resp.* 475a 1 ff.); that this view was popularly held rather than being just a scientists' theory is indicated by the fragment of Xenophanes preserved by Athenaeus (559a).[44]

Cicadas produce their song by means of the rapid vibration of two membranes or tymbals, which are situated in resonating cavities on either side of the base of the abdomen.[45] Since this fact is far from being immediately obvious or deducible by observation, it is not surprising that there was considerable confusion in antiquity as to how the sound was produced. The commonest erroneous view seems to have been that cicadas stridulate in somewhat the same way as grasshoppers and crickets, whose mode of sound production is easily observable, either as Meleager explicitly says (*AP* VII.196), by striking saw-like legs against its body or its wings against its back (*AP* IX.264; cf. Ps. Eustathius *Hex.* MPG 18.736). Hesiod presumably had some such idea in mind when he described the insect as pouring forth song ὑπὸ πτερύγων, a phrase also employed by Alcaeus (fr. 39 B). It is somewhat unclear what Aelian (I.20) means by the phrase κατὰ τὴν ἰξύν and whether he is perpetrating the above error or expressing the more correct Aristotelian view in a rather strange way: the same may be said with regard to the statement of Archias (*AP.* VII.213) that the insect sings εὐτάρσοιο δι' ἰξύος, which Gow[46] takes as being a similar error to that of Meleager noted above. Bianor (*AP* IX.273) uses the odd expression διγλώσσῳ στόματι to describe how the song is produced, representing a view not paralleled elsewhere. Bodson[47] points out that Archilochus (fr. 143 B; cf. Lucian *Pseud.* 1), since he describes how a cicada sings more loudly when held by its wings, was evidently aware that it did not employ them in sound production, though this does not necessarily imply that the developed view found in Aristotle was current in his time.

Aristotle's opinion on the subject is given in *HA* 535b 7 ff. and *De Resp.* 475a 1 ff., where he compares the buzzing of bees, wasps and cockchafers with the song of cicadas. His theory is that both types of sound are produced by the friction of πνεῦμα inside the abdomen against a membrane (ὕμην) that is under the *hypozoma*. It is important to note that this account is not as accurate as, for example, D'Arcy Thompson[48] and Peck[49] imply: the 'membrane' described is evidently not to be understood as being the cicada's tymbals, but simply as an area of thin outer tissue between thorax and abdomen,

---

[44] Cf. also Hesychius s.v. κερκώπη, and 1.4.vii and xviii above.   [45] Cf. Imms, op. cit., p. 183.

[46] Art. cit., p. 93.   [47] L. Bodson, 'La Stridulation des Cigales', *AC*, XLV, 1976, pp. 75–94.

[48] *Hist. An.* IV.7 note.   [49] *Hist. An.*, II p. 74.

common to most other insects as well. Pliny's remarks in XI.266 are based upon the
above with variations (cf. Basil *Hex.* MPG 29.184), while in XI.93 he says somewhat
vaguely that *pectus ipsum fistulosum, hoc canunt achetae.* Strabo (VI.1.9) also speaks of
cicadas singing by means of a *hymen* (cf. 3.2) as does *Aesopica* 387.

6.2. The cicada's song: habits

There are frequent references to the cicada's singing as characteristic of the hottest part of
the summer (the time of the rising of Sirius) (Hesiod *Sc.* 397; Alcaeus fr. 39 B; Ps.
Aristotle *Mir.Ausc.* 835a 24; Martial X.48; Ovid *AA* 271; Juvenal IX.69; Nemesianus *Ecl.*
IV.42). Aristophanes (*Av.* 39–40) describes them as singing for one or two months. They
are said by Aelian (I.20) to sing only from midday onwards, whereas Hesiod (*Sc.* 396)
credits them with continuous song throughout the day. Allusions to their activity in the
midday heat are again frequent in literature (Aristophanes *Av.* 1096; Aristopho fr. 10; *AP*
VII.196, IX.264,273,373,584; Aristaenetus I.3; Plato *Phaed.* 230c,258e; Virgil *Ecl.* II.13, *G*
III.327; Basil loc. cit.; Ambrose *Hex.* V.76), and they are customarily depicted as singing
from the branches of trees, their preferred perch (Hesiod *Op.* 583 and *Sc.* 393–4; Alcaeus
fr. 39 B; *Anacreontea* 34; *Iliad* III.150; Aristophanes *Av.* 40; *Aesopica* 299; Lucian *Am.* 18;
Aristotle *Rhet.* 1395a 2=Demetrius *de Eloc.* 99; Theocritus VII.158, XVI.94–6;
Philostratus *Vit.Ap.* VII.11; *AP* VI.120, VII.190,196,200,213, IX.71,264).

6.3. The cicada's song: attitudes toward

It is interesting to note the difference in attitude toward the cicada's song that is displayed
by Greek and Roman authors respectively. Among Roman writers uncomplimentary
remarks are common, whereas these are not found in Greek sources. It is evident that the
Greeks in general found the insect pleasant to listen to.[50] Their song—the verb ἀείδειν is
commonly used—is described as a ἡδὺν νόμον (Aristophanes *Pax.* 1160) and compared to
that of the nightingale (παροδῖτιν ἀηδόνα, *AP* IX.373) where the adjective applied to the
sound, ξουθά, is one often used with reference to the nightingale. It is characterised as
clear or shrill (λιγυρός, Alcaeus fr. 39 B, Hesiod *Op.* 582, Aristaenetus I.3, Callimachus
Aet. I.29 ff.; ὀξὺ μέλος, Aristophanes *Av.* 1095), as continuous (πυκνός, Alcaeus &
Hesiod loc. cit.; Lucian *Bacch.* 7) and in the *Iliad* (III.150–2) by the term λειριόεις (also in
Timo fr. 30). It is described by Meleager (*AP* VII.196) and Archias (*AP* VII.213) as
resembling the note of a lyre (cf. Phaedrus III.16): with this may be compared the tale of
the *kithara* player Eunomus of Locri at the Pythian games, who won his contest with the
assistance of a cicada which, when one of his strings broke, settled on his instrument and
supplied the missing note (Strabo VI.1.9; Antigonus *Hist.Mir.* 1; *AP* VI.54, IX.584;

---

[50] The tendency of modern authors is to agree with the Romans that the cicada's song is monotonous and
hardly musical, and to criticise the Greeks' musical taste. N. Douglas, *Birds and Beasts of the Greek Anthology*, pp.
190–2 (cf. A. S. F. Gow, art. cit., p. 93), seeks to explain their attitude by saying that the Greeks disliked solitude
and silence, and that the insects provided company. M. Melon, op. cit., pp. 64, 75, reminds us that judgements
of what is musical are relative, and argues that cicada songs would have borne comparison with the kinds of
instrument with which the Greeks were familiar. Cf. also Davies and Kathirithamby, op. cit., pp. 116–17.

Clement Alex. *Protr.* 1; Julian *ep.* 41,421D). Their song is said to be a pleasure to farmers (*Aesopica* 299) and wayfarers (*Aesopica* 387), and in cultivated gardens (Achilles Tatius I.15).

Because of the continuous nature of the insect's stridulation, apparently carried on without respite throughout the day, they are often characterised as 'talkative' (λάλος, -ίστατος: Alexis fr. 92, Aristopho fr. 10, *AP* IX.122,273, Longus I.25; Julian *Ep.* 50). In the *Iliad* (III.150–2) garrulous old men are compared to cicadas (cf. Lucian *Bacch.* 7; Artemidorus *On.* III.49), as are the Athenians discussing their lawsuits by Aristophanes (*Av.* 39–41) (cf. *Nub.* 1360; Theopompus fr. 44; Dio Chrysostom 47.16). Plato (*Phaedrus* 258e) depicts them as ἀλλήλοις διαλεγόμενοι (cf. Theocritus VII.159, Ananios fr. 5 B). It is pointed out by Gow[51] that Theocritus' phrase κὔμμες ἐρεθίζετε τὼς καλαμευτάς is a reference to this constancy of song as being a challenge to the reapers and is not intended to suggest that the sound was an annoyance (V.111; cf. V.29).

Among Roman writers, the song is usually described in less approving terms, suggesting that it tended to be thought of as more of an irritation than a pleasure. Virgil, for example, writes of how *raucis . . . arbusta resonant cicadis* (*Ecl.* III.12–13; cf. Nemesianus *Ecl.* I.2; Baehrens *PLM* LXI.35), and similarly of how *cantu querulae rumpent arbusta cicadae* (*G.* III.328; cf. *Copa* 27). Martial characterises them as *inhumanae* (X.58) and elsewhere as *argutae* (XI.18; cf. Calpurnius V.56, *Culex* 152–3, Novius fr. 26 Ribb.). Cf. also *Laus Pisonis* 79–80 (*stridula convicia*) and Solinus II.40.

7. Predation

It was believed that swallows were especially fond of eating cicadas, as we are informed by Plutarch (*Mor.* 727e), Aelian (VIII.6) and Clement (*Strom.* V.5). This motif is found also in Longus (I.26) and *AP* IX.122. A verse in *AP* IX.372 depicts a cicada caught in a spider's web, and there is an epitaph by Archias (VII.213) concerning one that has been killed by ants.

8. Cicadas employed as food

Though the Greeks were fond of cicadas, it seems that this did not prevent them from eating them on occasions. Aristotle (*HA* 556b 7 ff.) gives details on this subject, saying that they are pleasantest to eat at the *tettigometra* stage, before the adult emerges from the nymphal skin. As for when they are adult, the males are best at first, but after mating the females, at which stage they are full of white eggs. Pliny in his parallel account (XI.92) states that they are an article of diet among *gentes ad orientem*, including the Parthians. Athenaeus has preserved for us a number of references from Greek comedy to the use of cicadas as food (131e = Anaxandrides fr. 41.59, 55a = Alexis fr. 162, 133b = Aristophanes fr. 51 (cf. also fr. 569)). From the latter of these it would appear that they were captured by means of a birdlimed twig as is described more clearly in *AP* IX.264 and 273.[52]

---

[51] *Theocritus*, II p. 112. M. Melon, op. cit., p. 72, agrees that Theocritus envisages the cicadas as spurring the reapers on.

[52] On the eating of cicadas, cf. F. S. Bodenheimer, *Insects as Human Food*, p. 62; Davies and Kathirithamby, op. cit., p. 128.

9. Cicadas in captivity

It is clear, most notably from items in the Greek Anthology, that it was well known in antiquity for singing insects to be kept as pets in homes, in some form of small cage known as an ἀκριδοθήκη.[53] It is grasshoppers and crickets (akrides) which are most often mentioned in this regard, but there is apparent evidence for the keeping of cicadas as well. There is an epitaph by Anyte (AP VII.190 = VII.364) on a young girl's pet akris and tettix. Pliny's statement about monumenta in XXXIV.57 seems to be based on a misunderstanding of AP VII.190.[54] There has been considerable disagreement as to whether it is in fact physically possible to keep cicadas—in contrast to crickets—in captivity. Gow[55] and Borthwick[56] tend towards the view of Fabre, who said that under such conditions the insects would refuse to sing and would quickly starve.[57] Douglas[58] and Steier,[59] however, accept the possibility of cicadas being kept for their song, and Sibson[60] gives an example of this being done, although not from Europe.[61]

10. Medicinal uses

Pliny (XXX.68) lists roasted cicadas among treatments for the bladder (cf. Dioscorides DMM. II.51), a prescription repeated by Paulus Aegineta (VII.3), who further suggests that sufferers from colic should take a number of them in drink along with pepper (cf. Galen XII.360 K).

11. Association with religion and mythology

Cicadas are commonly associated with the gods in Greek literature, though not in Roman, and seem in popular belief to have been thought of as possessing something of the divine themselves. In Anacreontea 34, a hymn addressed to them, they are spoken of as ageless and σχεδὸν θεοῖς ὅμοιος, and Plutarch (Mor. 727e) describes them as ἱερούς καὶ μουσικούς. Aristophanes (Av. 1095) applies the adjective θεσπέσιος. In Plato's Phaedrus (259c), a myth is related to the effect that cicadas were once men who, having encountered the Muses, sang continuously until they starved to death, whereupon they became insects to whom the Muses gave the gift of being able to sing all their life without sustenance and of going to the Muses upon their death. The story is alluded to by Philostratus (Vit.Ap. VII.11). The Anacreontic song describes them as beloved of Apollo and the Muses, who gave them their voice, and Alcaeus (fr. 2 B) depicts them as singing praises to Apollo. They are also associated with the Nymphs (AP IX.373) and with Athene (VI.120).[62]

---

[53] Cf. 11.8 above.   [54] M. Melon, op. cit., p. 118.   [55] Art. cit., p. 93.   [56] Art. cit., p. 105.

[57] Borthwick suggests that a motif applicable in fact only to crickets has been transferred to cicadas in poetry due to the popular association of the two, while Gow suggests the possibility that the insects in Anyte's epitaph, though described as παίγνια, may not necessarily have been kept in captivity.

[58] N. Douglas, op. cit., p. 190.   [59] Art. cit., 1119.

[60] R. B. Sibson, 'Some Thoughts on Cicadas', Prudentia, XI, 1979, p. 105.

[61] There is archaeological evidence of model cicadas as children's toys. Cf. Keller, Ant. Tierwelt, II. p. 404; Steier, art. cit., 1119.

[62] Cf. also 2.2 and notes 36–7 on pp. 96–7.

17. Koris/Cimex

1. Identification

The above names refer almost exclusively in classical literature to a single species of insect pest, the bed-bug, *Cimex lectularius* Linn., as has been generally recognised by authors from Sundevall[63] onwards.[64] There are, however, a few examples of extended usage to cover insects of similar appearance.

The bed-bug belongs to a small distinctive family of Hemiptera-Heteroptera, the Cimicidae. This includes three species worldwide that attack man, of which *C. lectularius* is native to Europe. The adults live in human habitations and are nocturnal, hiding by day in crevices in walls and floors, or inside beds and other furniture. They emerge at night to feed upon human blood, piercing the skin with their proboscis or rostrum. The eggs are laid in the crevices and other places where the adults conceal themselves, and the nymphs which hatch from them appear as smaller versions of the parents and feed upon blood throughout their development.[65]

1.2. Nomenclature

The Heteroptera, unlike many insect orders and sub-orders, does not constitute a popularly recognised group such as, for example, butterflies and moths, or beetles. There is therefore no reason to suppose, as Richter implies, that the terms κόρις and *cimex* would have been considered as applicable to the generality of species which modern entomology has classed as Heteroptera, but only to certain of them which are similar in appearance to the domestic *C. lectularius*.

Pliny (**XXIX.62, XXX.24**) mentions a non-domestic type of *cimex* found on the mallow plant (*eos qui agrestes sint et in malva nascuntur*) which was used in medicine for ear ailments. Richter[66] suggests that this might be a type of shield-bug (Pentatomidae), but one would expect something more similar to *C. lectularius* (for example members of the Anthocoridae, some species of which are known to bite man occasionally[67]).

Alciphron (*Ep.* II.5.2) uses the names κόρις and σής[68] to refer to the insects that devour old documents. This is evidently an example of the kind of loose terminology by which names of insect pests, though possessing a primary specific definition, may also be found applied to any small noxious creature.[69] There is no reason to suppose that the author has any particular insect species in mind, or that he has not selected the terms he uses somewhat arbitrarily.[70] In Petronius 98.1 the word *sciniphes*, latinised form of the Greek

---

[63] *Thierarten*, p. 228.

[64] Aubert and Wimmer, *Thierkunde*, I p. 166; Keller, op. cit. II p. 399; W. Richter in PW Supp.XIV.822.

[65] Smith, *Insects*, pp. 385–8.

[66] Art. cit., 822.   [67] Smith, op. cit., p. 389.   [68] Cf. 26.1 below.

[69] Κόρις belongs to the *ses–kis–ips–thrips–knips* complex of names which all mean essentially 'small insect pest', but which each tend to have a particular association with one form of pest. Cf. 36–7, 58 below. On the etymology of κόρις cf. Fernandez, *Nombres*, pp. 109–10.

[70] Since Alciphron is not intending to be entomologically precise, Gossen's identifications, in PW XII.1039 and Supp.VIII.238, are misguided.

σκνίψ,[71] appears in place of *cimices*. Isidorus (*Or*. XII.5.17) states that the insect's name derives from a certain plant having a similar smell.[72] A single synonym, πετηνίς, is listed by Hesychius.[73]

2. Life history and feeding habits

Data on the insect's development are provided by Aristotle, Pliny omitting it from his zoological section. He classes it (*HA* 556b 21 ff.) along with lice and fleas among those insects which feed on the juices of living flesh and which, upon mating, produce nits (κονίδες) out of which nothing further develops. The adult insects are produced by spontaneous generation out of the moisture from living animals (ἐκ τῆς ἰκμάδος τῆς ἀπὸ τῶν ζῴων συνισταμένης ἐκτός), a belief which is also referred to by Augustine (MPL 32.1372).

3. Relations with man

Often associated with lice and fleas, the bed-bug was well known as one of the major domestic pests of antiquity (Plautus *Curc*. 500; Livius Andronicus *com*. fr. 1; *AP* IX.113; Philostratus *Vit.Soph*. 588; Jerome *in Jl*. 2.22, *in Is*. 18.1, *in Ps*. 103), and, to judge from Pliny (XXIX.61, *animalis foedissimi et dictu quoque fastidiendi*), was thought of as particularly repugnant. In keeping with its known habits today, it is commonly mentioned as infesting beds (Aristophanes *Nub*. 634,699,725, *Pl*. 541, fr. 266; Lucian *Merc.Cond*. 17; *AP* V.184; Dioscorides *DMM*. II.34; Catullus 23.2; Martial XI.32; Petronius 98.1; *C.Gl.L*. II.573.19), biting the occupants during the night and averting sleep in the process (Aristophanes *Pl*. 541; Artemidorus *On*. III.8; Chrysippus *ap*. Plutarch *Mor*. 1044d–e). They are referred to as being a well known hazard in inns (Aristophanes *Ran*. 114–15).

Various suggestions for measures to be taken against infestation by bed-bugs are listed by Pliny, Varro (*RR*. I.2.25), and in most detail by the *Geoponica* (XIII.14.1–11). The majority of these involve certain herbal preparations to be smeared on one's bed (Varro; *Geoponica* 1,2,4–6). Pliny (XXVII.80) prescribes scattering fern leaves about, and the *Geoponica* (4) gives an 'insecticidal' preparation of a herb with ox-gall and oil for scattering over the insects themselves. Other substances used for treating beds were pitch (*Geoponica* 1), brimstone (*Geoponica* 3), gall of various domestic animals (Varro; *Geoponica* 2,6) and isinglass (*Geoponica* 3). Two of the treatments noted by the *Geoponica* are said to be effective against bugs inhabiting house walls (2,4). Also suggested are fumigation with burning leeches (Pliny XXXII.124,136; Cyranides p. 104; Ps. Theodorus Priscianus *Addit*. I.12) or centipedes (Pliny XXIX.58; *Geoponica* 8—with leeches 7), placing the foot of a hare or deer by the bedposts (*Geoponica* 9), a deluge of hot

---

[71] Cf. note 69 above and 58 below.

[72] It is unclear whether *C.Gl.L*. V.617.1 *Cimes sunt vermiculi dicti quod comedunt olus* belongs here. If so we have a parallel to the use of lice and flea names in a secondary sense to refer to plant pests (cf. 63–4 below).

[73] The name can hardly mean 'winged', unless ἀκρίς should be read here instead of κόρις. Cf. Gossen, *Zool. Glossen*, p. 88; Fernandez, op. cit., p. 77.

water (*Geoponica* 11), the placing of a vessel of cold water under the bed (*Geoponica* 10), or a trap under the bed consisting of an open vessel rubbed with goat grease (Ps. Theodorus loc. cit.).

4. Relations with domestic animals

Pliny (XXIX.62) notes that *cimices* are eaten by hens, and that they were said to render them protected against snake bite for one day.

5. Medicinal uses

Bed-bugs figure quite widely in authors who deal with the medicinal properties of animals, though Pliny (XXIX.61 ff.) is particularly sceptical about the reputed uses of these insects, implying that he would prefer not to have to include them at all. They were said to be effective in the treatment of eyes (Pliny 62), ears (loc. cit.) and nose (Q.Serenus 629); to relieve vomiting (Pliny 63; Marcellus XVII.34), strangury (Pliny 63; Dioscorides *DMM.* II.34, *Eup.* II.113) ληθαργία (Pliny 63; Q.Serenus 1003) and ὑστερικός πνίξ (Dioscorides loc. cit.); to be usable for conditions of the bladder (Marcellus XXVI.101) and uterus (Pliny XXX.131); to avert or cure various kinds of fevers (Pliny 63–4; Dioscorides loc. cit.; Q.Serenus 921,899); as a depilatory (Theodorus Prisc. *Eup.* I.11,41); as an antidote to snakebite (Dioscorides *Eup.* II.125); and, when burnt as a fumigant, to cause expulsion of swallowed leeches or make sucking leeches release their hold (Pliny 62; Marcellus XVI.95; Dioscorides *DMM.* II.34; Galen XII.363K, XIV.538K; Paulus Aegineta VII.3; Cyranides p. 104). Soranus (*Gyn.* II.29) condemns the use of fumigation with bed-bugs for ὑστερικός πνίξ as one of the unreasonable remedies of the ancients rejected by his school.

6. Veterinary uses

Bed-bugs are said to have been employed as a treatment for *dysouria* in horses and cattle (Pelagonius 153; Vegetius *Mul.* I.61, II.79; Chiron 480), and, burnt as a fumigant, to cause the ejection of leeches swallowed by livestock (Pliny XXIX.62; Columella *RR* VI.18.2, with description of the process; *Hippiatrica Ber.* 88.4; Vegetius IV.24; Gargilius *Cur.Bov.* 11; *Geoponica* XIII.17, XVI.19).

18. Tiphe/Tippula

Although one of the least frequently mentioned forms of insect in Latin literature, the *tippula* is nonetheless one of the simplest to identify. Plautus (*Pers.* 244) uses it to typify the idea of extreme lightness (*neque tippulae levius pondust quam fides lenonia*). The grammarian Festus (p. 503 L), citing this passage, states very clearly that 'the *tippula* is a small creature with six legs, so light that it runs over the water without sinking'. Equally clear is Nonius' (p. 180.9–10) description of the insect as a creature of extreme lightness which crosses water not by swimming but by walking. He quotes in illustration a line from Varro's *Bimarcus* (*Men*.50)—*ut levis tippula lymphon frigidos transit lacus*. Elsewhere in surviving literature it seems to appear only in Arnobius (*adv.Nat*. II.59), who includes it in passing in a list of insects. In the Latin glossaries *tippula* is defined as *vermis aquaticus* (V.397.44) and as *animal liberis vestigiis ambulans super aquas* (V.624.28). In II.198.28 it appears in the form *timulus* with the Greek equivalent given as ἐξυδρίς, explained as 'a six-footed creature which skims over the water surface' (ἑξάπουν ἐπινηχόμενον τῷ νώτῳ).[74]

There can be no doubt that the insects here described are the various members of the three allied families Gerridae, Hydrometridae, and Veliidae, comprising such well known aquatic insects as the Pond-skater *Gerris lacustris* (Linn.), the Water-measurer *Hydrometra stagnorum* (Linn.), and the Water-cricket *Velia caprai* Tam. These are distinctive insects which live on the water surface of lakes and ponds, moving about supported by the surface film.[75]

That the Latin *tippula* is related to the Greek τίφη is suggested by Keller[76] who derives the latter from τῖφος, 'pond or marsh', which would be very appropriate for a freshwater insect. Nonetheless, there is an apparent problem in that in normal Greek usage as it appears in our surviving sources τίφη is a synonym of σίλφη and refers to the cockroach.[77] Fernandez[78] argues that this is a secondary meaning which arose through confusion of the two words, and that originally *tiphe* had no connection with *silphe* but referred to an entirely different insect, although he does not attempt to define what this other insect might be. It is likely, however, that a unique example of *tiphe* in its original meaning is in fact provided by Aristophanes *Ach*. 920 ff., where we have a joke about the possibility of a lighted wick attached to a *tiphe* being introduced into the Athenian dockyard through a water conduit. This passage has puzzled modern authors as it puzzled the scholiasts, who could only conclude that the author was referring to the cockroach. Some commentators such as Rogers and Starkie[79] agree with the scholia, the latter suggesting a reference to the rose chafer or μηλολόνθη which children played with on the end of a string;[80] while others consider that Aristophanes means us to understand *tiphe* in its botanical sense as a hollow corn stalk with a light inside it.[81] Whereas the idea of a

---

[74] The name ἐξυδρίς is not recognised by LSJ or included by Fernandez.

[75] The only other surface living insects are the Gyrinid beetles (cf. 49.2 below), but these do not have legs that are clearly visible.

[76] O. Keller, *Über die Bedeutung einiger Thiernamen im Griechischen und Lateinischen*, Graz, 1878, p. 11. By 'water spider' he presumably means the water-skater and not the arachnid *Argyroneta* (*Oxford Latin Dictionary*'s 'water boatman', i.e., Notonectidae, is clearly a mistake). Keller is supported by Ernout and Meillet, *Dict. Ét.*, p. 692.

[77] Cf. 13.1.2.i above.

[78] *Nombres*, p. 239.

[79] B. B. Rogers, *The Comedies of Aristophanes*, Vol. 1, London, 1910, p. 142; W. J. M. Starkie, *The Acharnians of Aristophanes*, London, 1909, p. 254.

[80] Cf. 32.3 below. Starkie thinks that 'lighted wax' was attached to these insects, but there is no clear evidence for this.

[81] For example J. van Leeuwen, *Aristophanes Acharnenses*, Leiden, 1968 (original ed. 1901), p. 153; Chantraine, *Dict. Ét.*, p. 1123.

water–skater skimming along the water surface, thus keeping the light affixed to it dry, would fit the sense of the passage perfectly.

19. Kokkos/Coccus

1. Identification

The scale insect *Kermes vermilio* Planch, commonly known as 'kermes', was well known in antiquity because of the crimson or scarlet dye produced from it, though it was usually thought of as being a vegetable product rather than an animal. This is not surprising when the appearance of the adult female is considered, since it appears as an immobile scale-like object attached to the stalks and branches of the oak *Quercus coccifera* Linn., native to the Mediterranean coast and the Near East. Like all members of the group to which it belongs (superfamily Coccoidea), the kermes exhibits marked sexual dimorphism, the male, unrecognised in antiquity, being active and of conventional insect form. The eggs develop beneath the scale-like covering of the static female and hatch into active nymphs which allow dispersal of the species to take place. These nymphs in subsequent stages of their development become stationary, attached by their sucking mouthparts to the host plant. Those which develop into females remain in position , while those developing into males pass through a 'prepupal' and 'pupal' stage first.[82]

1.2. Other names

(i) *Cusculium*. The native Spanish name, according to Pliny (XVI.32). J. and C. Cotte[83] suggest that it has survived in Basque and Spanish terms for the foodplant and the insect.

(ii) Σκώληξ or σκωλήκιον. Pliny (XXIV.8) refers to a form of *coccus* found in Asia and Africa that swiftly turns into a small larva (*celerrime in vermiculum se mutans*), named *scolecium* and considered to be of low quality for dye production. Hellenistic papyri concerned with the subject of dyeing also include mentions of a low grade variety of kermes known as σκώληξ or σκωλήκιον (Ps. Democritus *Alch.* p. 42 Berth.). Forbes[84] interprets this data as indicating an awareness of the fact that when the young nymphs leave the body of the female they empty the scale of its dye. *Skolekion* would therefore be simply a term for scales that had reached an advanced stage of development, not having been harvested sufficiently early and Pliny would be in error in making the name referrable to a distinct geographically limited species. J. and C. Cotte,[85] however, understand Pliny's *vermiculi* to be internal parasites of the kermes.

(iii) *Vermiculus* is the term used in the Vulgate to translate the Hebrew *tola'at*, which

---

[82] Cf. R. J. Forbes, *Studies in Ancient Technology*, vol. IV, Leiden, 1956, pp. 102–3; Imms, *Gen. Textbook*, pp. 725–8.

[83] J. and C. Cotte, 'Le Kermes dans L'Antiquité', *Rev. Arch.* VII, 1918, pp. 105–7.

[84] Op. cit., p. 104.

[85] Art. cit., pp. 93–4.

literally means a small worm or insect larva[86] and is employed as a term for kermes-dyed fabrics. It appears in *Gloss.* II.351.68 as a synonym of κόκκος, and likewise in Isidorus (*Or.* XXX.28.1), who writes in a section *de coloribus vestium*: *κόκκον Graeci, nos rubrum seu vermiculum dicimus; est enim vermiculus ex silvestribus frondibus.*

2. Habitat and development

As has been noted, the popular view in antiquity was that the kermes was not an animal but a product of the tree upon which it was found. Theophrastus, the earliest writer to describe it, classes it among those miscellaneous products of trees—for example catkins—that are neither leaves, flowers nor fruit (*HP* III.7.3, 16.3), and calls it 'a kind of scarlet berry' (κόκκον τινὰ φοινικοῦν) growing on the oak known as *prinos*. Pliny (XVI.32) likewise thinks of it as something developing from the tree itself, referring to it as a 'grain' (*granum hic primoque ceu scabies fruticis*). The earliest classical reference of all, by Simonides (fr. 54 Bk), describing the coloured sail given to Theseus by his father also implies the belief that the kermes was an aberrant form of fruit (ὑγρῷ/πεφυρμένον ἄνθει πρινῶν ἐριθάλλων).

Some authors, however, display an awareness, even if they are somewhat confused in matters of fact, of the animal nature of these unusual objects. Pausanias (X.36.1) describes how there develops in the fruit of the tree to which he gives the name *kokkos* a small creature which, when the fruit has ripened, makes its way to the outside world and flies off as a developed insect resembling a gnat (κώνωψ). By the 'fruit' of the tree Pausanias evidently means the scale-like female insects. He informs us that it is necessary to gather them before the 'small creatures' emerge, as it is their blood which serves as a dye material. These are presumably just the emerging nymphs, erroneously described as being winged by analogy with other insects, though J. and C. Cotte[87] regard them as the imagines of internal parasites. Dioscorides (IV.48) appears to consider the scale itself as a living creature, since he describes it as resembling a small snail attached to its host tree; he also says that it is similar to a lentil.

3. Distribution

We find mention of the *kokkos* being 'harvested' in many parts of the classical world and beyond. Those from different regions were noted to vary in quality and were graded accordingly, though there is some difference of opinion evident as to the respective merits of the various regional forms. Thus Pliny (XVI.32) seems to rate the Spanish variety quite highly and states that the lowest quality is found in Sardinia, while Dioscorides (loc. cit.) classes the Galatian and Armenian as best, followed by the Asian and Cilician, with the Spanish at the bottom of the list. In IX.141 Pliny concurs about the Galatian *kokkos*, but ranks that from Lusitania in

---

[86] Used in this sense in *Ex.* 25.4, *Lev.* 14.4, where LXX renders κόκκινον. Cf. F. S. Bodenheimer, *Animal and Man*, p. 157.

[87] Art. cit., p. 94.

Spain as equal to it. In Ps. Democritus (loc. cit.), however, ὁ τῆς Γαλατίας σκώληξ figures as a cheap and inferior dye, unless the reference is to a particular grade of the Galatian product (cf. 1.2.ii above) and not to the product as a whole.

The Spanish variety is that which Pliny considers in most detail (XVI.32); he singles out that originating from around Emerita in Lusitania as being of special value (IX.141, XXII.3). Spain is also mentioned as a source by Dioscorides, as noted above, Strabo (III.2.6), who notes it as one of the exports of Tourdetania in the south of the country, and Solinus (XXIII.4). Apart from its mentions by Pliny (IX.141, XVI.32, XXII.3), and Dioscorides (loc. cit.), Galatian kermes is referred to by Pausanias (loc. cit.), *Papyrus Holmiensis* (1097) and Marcellus (XXXI.33). Other regions mentioned in this regard are Phocis (around Ambrossos; Pausanias, loc. cit.), Africa (Pliny XVI.32 and XXII.3), Sardinia (see above), Pisidia (Pliny XVI.32), Cilicia (Pliny XVI.32; Dioscorides loc. cit.), Syria (*Pap.Holm.* 922), and Armenia (Dioscorides loc. cit.).

Forbes[88] suggests that in the East a second species of scale insect, *Margarodes hameli* Br., living on grasses, was used as a source of dye, though without being distinguished from the kermes insect in antiquity. This was known to the Assyrians from very early times as native to the Ararat vallies in Armenia, and would have been one of the sources of the kermes referred to in the Old Testament. Presumably Dioscorides' Armenian variety of *kokkos* would have been this second species.[89]

4. Employment as a source of dye

As has been said, the primary use of the kermes insect in antiquity was as a source of a scarlet or crimson dye. This dye, as with that produced from the murex shell, was very highly valued. Pliny (XXXVII.204) places it at the head of a list of the most valuable plant products known to him, just as he classes murex purple as the most precious product of the sea after the pearl; and he notes that (XVI.32) it provided the Spaniards with the means of paying half their prescribed tribute.

The actual gathering of the scales is described only by Dioscorides (loc. cit.), who writes that in Cilicia the women collect them by picking them from the branches with their nails.[90] The product derived from the crushing of these was employed for the colouring[91] of woollen fabrics (most frequently mentioned, e.g. in Strabo XIII.4.14, Pausanias loc. cit., Solinus XXIII.4), silks and leather (Martial II.29). Mentions of scarlet garments (*coccina*) are frequent in Roman sources (e.g. Martial II.39,43, V.23,35, X.76, XIV.131; Horace *Sat.* II.6.102; Petronius 28,32; Pliny XXII.3, for military cloaks of generals), often as a sign of ostentation. The dye is said

---

[88] Op. cit., p. 102.

[89] F. S. Bodenheimer, op. cit., p. 139, suggests that the red dye referred to by Strabo (XI.14.9) as being produced in Armenia is that derived from this insect, but the passage clearly refers to a mineral substance.

[90] The reading is uncertain here. Στόμωματι (Wellmann) and στόνυχι have been proposed, implying that the women used some kind of instrument for the purpose. But, bearing in mind what Forbes, op. cit., p. 103, says about kermes gathering in more recent times, 'with their nails' is probably meant.

[91] On dyeing processes, cf. Forbes, op. cit., pp. 104 ff.

by Pliny (IX.141) to lack strength when freshly produced, or when it is over four years old. As well as being used alone, kermes dye was also compounded with other colouring materials such as murex purple. There are numerous references to a compound due known as ὕσγινον,[92] related to the terms ὕσγη (apparent synonym of πρῖνος in Suidas) and ὗς (Galatian name for the same, according to Pausanias), but whose exact nature is unclear (cf. for example Pliny IX.140—where it is the result of combining kermes and sea purple—XXXI.170, XXXV.45; Vitruvius VII.14.1; Ed.Diocl. XXIV.8–12; Isidorus XIX.17,15; AP VI.254; Nicander Ther. 511; LSJ s.v.).

5. Medicinal uses

The kermes insect was employed as a treatment for fluxes and dysentery (Marcellus XXVII.50), for eye ailments (Pliny XXIV.8), to promote the healing of wounds (Pliny loc. cit., Dioscorides loc. cit., Galen XII.32 K; Paulus Aeg. VII.3) and for damaged sinews (Dioscorides Eup. I.156).

20. Lakkos

As well as the kermes, a second dye-producing insect was recognised in antiquity; namely the lac insect, Laccifer lacca Kerr., a native of India, from which its product was imported from at least the first century A.D.[23] The females of this scale insect, which in their adult state are wingless and immobile like those of the kermes, live attached to the branches of various trees such as figs, enclosed in a dense resinous cell which they secrete around themselves.[94] The correct name for the insect and/or the dye produced from it, similar in colour to that from the kermes, is given only in the Periplus Erythraei Maris (6), where λάκκος χρωμάτινος appears in a list of items imported from India.[95] A full description of the insect, though inaccurate in relating its physical appearance, is preserved by Aelian (IV.46) from Ctesias' account of India. Ctesias depicted it as an insect of the colour of cinnabar, the size of a scarab (κάνθαρος), with very long legs and soft to the touch. They live on the trees that produce amber, feeding on their fruit, and are gathered by the natives, crushed and used to colour cloaks and other articles, such articles being imported into Persia.

21. Margarites chersaios

In his account of the different geographical varieties of pearl, Aelian (XV.8) mentions that as well as those produced in the shells of oysters there are also 'land-

---

[92] Cf. J. and C. Cotte, art. cit., pp. 95–8; LSJ s.v.

[93] So identified by Gossen, Tiernamen, no. 85; Scholfield, Aelian, I p. 167; R. J. Forbes, op. cit., p. 104.

[94] R. J. Forbes, op. cit., pp. 104–5; J. L. Cloudsley-Thompson, Insects and History, pp. 212–13.

[95] R. J. Forbes, op. cit., p. 105, who also identifies as referring to lac the term λακχά, mentioned as a dye substance by Ps. Democritus (Alch. p. 42 Berth.); but this latter seems rather to refer to a vegetable product, as stated in LSJ.

pearls', found in India, which are said (by Juba, fr. 275.70 J) to be generated in some way which is not specified from rock crystal. These items are also referred to by Origen (*in Matt.* X.7), who says that they are unique to India and suitable for necklaces and signet rings.

These so-called pearls are identified by Scholfield[96] as scale insects of the genus *Margarodes*,[97] which secrete pearl-like waxy scales as an outer covering and which have been used as jewellery in some parts of the world in recent times. By contrast, Rommel[98] views the 'land-pearl' as a vegetable production, most probably certain pebble-like items formed inside bamboo canes which are known to have been collected in India for decorative purposes.

Insecta: Anoplura and Mallophaga (Lice)

22. *Phtheir/Pediculus*

1. Identification: general

Lice are the most frequently mentioned of the four types of external parasites of man distinguished in antiquity, the other three being bugs, fleas, and ticks. In contrast to popular opinion today, they were not regarded as having any particular association with uncleanliness: infestation by them was considered as a perfectly normal and inevitable, if undesirable, fact of life in all sections of society.[99] Lice are a distinctive group of insects, so that, with a few possible exceptions which will be dealt with in due course, it may safely be said that the range of the classical terms φθείρ and *pediculus* is coextensive with the two orders of insects known by the name of lice today, these being the Anoplura or blood-sucking species and the Mallophaga which feed upon the surface of the skin.

The classical references to lice may be regarded as dividing them into two basic groups. In the first place, there are those parasitic upon man, whose specific identification is quite straightforward. There are two species of human lice, of which one—*Pediculus humanus* Linn.—is divided into two distinct subspecies; namely, the Head Louse—*P. humanus capitis* de Geer—which is mainly confined to the head but may occur on hairs in other parts of the body, and the Body Louse—*P. humanus humanus*—which is found only on the body or attached to clothing in contact with the body. *P. humanus* is the louse which transmits disease, but this fact was unknown in antiquity. The second species is the Pubic Louse—*Phthirus pubis* (Linn.)—which lives attached to hairs in the area suggested by its name, and also in other areas such as the beard, eyebrows and eyelashes.[100] These three forms were for the most part lumped together by the ancients, though in technical literature there are signs of an awareness of the existence of distinguishable forms.[101]

---

[96] *Aelian*, III p. 221.   [97] Cf. Imms, *Gen. Textbook*, pp. 727–8.

[98] In PW XIV.1700.

[99] Cf. H. Keil, 'The Louse in Greek Antiquity', *Bull. of the Hist. of Medicine*, XXV, 1951, pp. 305 ff.

[100] Smith, *Insects*, pp. 395–7.

[101] The three forms are cited under varying names by Aubert and Wimmer, op. cit., I p. 172, and Keller, op. cit., II p. 399.

In the second place, we have those lice which are parasitic on domestic birds and mammals. Here numerous distinct species are involved, few of which were recognised as such in antiquity: theoretically at least this second group covers the whole of the rest of the two modern orders. In view of this fact, there is little purpose in following the example of Gossen[102] in giving a lengthy catalogue of every possible species found on domestic animals (including those which the ancients believed did not exist!). But where it is possible to give likely identifications this will be done in the appropriate section below.[103]

1.2. Identification: terms for the louse's eggs

The louse's eggs, or nits, which the females attach to the hairs of their hosts, are as conspicuous as the insects themselves. Although it was known to the ancient natural historians that the former were the product of the latter, even if they did not realise that they were eggs, it is quite probable that the connection was originally unknown to the general public. This would account for the fact that, as in modern English, so in the ancient languages, the nits were given a distinct popular name. In Greek they were known by the term κόνις and in Latin lens, lendis (with later variants lendis—C.Gl.L. III.431.61, 454.29, 485.20—and lindines—Marcellus IV.22). Δόρκα, a synonym of konis, is recorded by Hesychius as a Cypriot word but is not found elsewhere.[104]

1.3. Identification: synonyms and distinct forms

(i) *Pedis.* The name of which *pediculus* was originally a diminutive evidently went out of use in favour of the latter at quite an early date. Examples of its use are found only in the early Latin comic dramatists and satirists (Lucilius fr. 883 M; Livius *com.* 1; Novius *com.* fr. 104 Fr.; Plautus *Curc.* 500, *Vid.* 114; cf. Festus p. 230 Li.) and in Varro (*RR* III.9.14).

(ii) *Peduculus.* An alternative spelling of *pediculus,* found in Petronius (57) and Pelagonius (*Vet.* 118), as well as in the *Corpus Glossarum. Peduclus* is found in Sextus Placitus V.a.11.

(iii) *Tinea or tinia.* A number of late medical writers use *tinea* (and the associated adjective *tineosus*), which in earlier Latin normally means a clothes-moth or book-louse, as an alternative to *pediculus* (Marcellus VI.25; Theodorus Prisc. *Faen.* 4; Cassius Felix; Sextus Placitus IX.a.24; *Antidotarium Brux.* 164–5), especially for head-lice (Cassius and Sextus use *pediculus* for body-lice only).

(iv) Ἄγριος φθείρ. The 'wild louse' is described by Aristotle (*HA* 557a 1 ff.) as a particular type of human louse which is harder than the more common form and

---

[102] In PW XII.1030–9.

[103] The names φθείρ and *pediculus* are also used for a number of small insect pests regarded as similar to lice in appearance: cf. 9 above and 25.1.2, 36.1, 64 below. Veterinary writers (for example Vegetius *Mul.* I.44.1, *Hippiatrica Ber.* 85.3) also use them to refer to internal parasites.

[104] Cf. Fernandez, *Nombres,* p. 109.

more difficult to remove from the body. It is probable that, as has been suggested,[105] Aristotle has here correctly recognised the distinctive nature of *Phthirus pubis*. Caelius (*Chron*. IV.14) describes them, under the name *ferales pediculi*, as less common than ordinary lice, broader and harder, and biting more fiercely, being capable of penetrating the body.

(v) Βραύλα. According to Hesychius, a synonym of *phtheir*.

(vi) Κικνίον. Defined by Hesychius as μικρὰ φθειρία.

(vii) Νίρνος or νίρμος. Another Hesychian synonym of *phtheir*: the lexicographer gives both spellings and describes the word as Achaean.

(viii) Σάθραξ. This name, defined as *phtheir* by Hesychius, appears etymologically to have a more or less identical meaning;[106] that is, to derive from a verb meaning 'to ruin or destroy'.

(ix) Κάρ and κάρνος. Synonyms of *phtheir* in Hesychius, related etymologically to the bed-bug name *koris*.[107]

(x) Ἄχωρ. A term normally referring to 'scurf or dandruff' (LSJ), but given as a Greek name for head-lice by Cassius (p. 10; cf. *C.Gl.L.* III.598.35).

2. Life history

In his discussions of the development of insects, Aristotle (*HA* 539b 10, 556b 21 ff.) includes lice among that group of creatures which are themselves spontaneously generated and which, although they mate, produce as a result only an 'imperfect' offspring from which nothing further is produced. In the case of lice what are produced as a result of copulation are the *konides* or nits. Aristotle was thus unaware of the fact that the nits are in fact eggs and that from them are hatched nymphs which are miniature versions of their parents.[108] The belief that lice were produced spontaneously from the human body was evidently a popular one (Diogenes L.II.81). Those on the head were believed to result from the action of moisture in the scalp (Aristotle *Probl.* 861a 10 ff.; Galen XII. 462 K),[109] while those on the body were said to be caused by putrefaction of the flesh (Aristotle *HA* loc. cit., *Probl.* 924a 8; Galen XIV.290 K; Pliny XI.114; Q.Serenus 56; Augustine MPL 32,1372; Isidorus *Or.* XII.5.14), or alternatively of the blood (Theophrastus *CP* II.9.6; Pliny XXVI.138) or of sweat (Philo *de Prov.* 2.69; Augustine MPL 42.236) or other bodily exudation (Caelius *Chron.* IV.14–18). Aristotle's description of the process occurs in the context of his discussion of reported fatalities from louse infestation and will be dealt with below.

[105] Aubert and Wimmer, op. cit., I. p. 172; Keller, op. cit., II p. 396; Gossen, art. cit., 1031.

[106] Fernandez, op. cit., p. 119.

[107] Ibid., p. 110.

[108] It has been questioned, for example by Gossen, art. cit., 1035, whether Aristotle can really have meant what he appears to be saying in these passages, that lice have no continuous life cycle, but there can be no doubt that this was in fact his belief.

[109] Aeschylus *Ag.* 560–2 seems to refer to the influence of moisture on the generation of lice. Cf. Davies and Kathirithamby, *Greek Insects*, p. 10.

3. Popular beliefs

It was popularly believed that the wool of sheep that had been killed by wolves was especially liable to give rise to lice: Plutarch devoted one of his *Quaestiones Conviviales* (*Mor.* 642b ff.) to a discussion of the reasons for this supposed phenomenon. From the same author we learn that clothing made from flax was said to be the least liable to infestation (*Mor.* 352 f.). It was held that persons with louse-infested heads were less liable to headaches (Aristotle *HA* 557a).

4.1. Human lice; normal infestations and their treatment

References, both specific and general, to infestation by lice are very frequent. They are the most commonly mentioned of the external parasites of man, and were regarded as the most difficult to contend with, medical writers devoting considerable space to measures for prevention and treatment. There is no specific statement to the effect that they suck blood, the nearest being Aristotle's note that they feed upon 'the juices of living flesh' (*HA* 556b 21), but they are simply referred to as biting and causing irritation (Appian *Bell.Civ.* I.101; Q. Serenus 56–8). Aside from general allusions to lice as pest species (e.g. Aristophanes *Pax* 740, *Plut.* 537; Archilochus fr. 137; Plutarch *Mor.* 208e; Plautus *Curc.* 500; Petronius 57; Symphosius *Aen.* 30b; unnamed in Aeschylus *Ag.* 560–2), references to all of the three forms of human lice can be distinguished, though only one of them, *P. pubis*, was recognised as constituting a distinct variety. Lice in the hair, i.e. *P. humanus capitis*, are considered on two occasions by Aristotle (*HA* 557a, *Probl.* 861a 10 ff.), who states that women and children are more troubled with them than men. They are also referred to by Pausanias (X.10.7), Eubulus (fr. 32), Lucilius (fr. 883 M), Claudian (XVIII.1.113–15,260), and frequently in Pliny (esp. IX.154, XI.114) and in the medical works of Galen and Marcellus. A number of medical writers (Galen XIV.415,771 K; Dioscorides *Eup.* I.48; Oribasius V.717; Celsus VI.6.15; Marcellus VIII.128) single out for particular mention infestations of lice occurring on the eyebrows. Galen explicitly describes them as broad in shape, which provides additional confirmation that *P. pubis* is the insect in view here. Presumably Marcellus' *vermes* on eyebrows (VIII.127) are the same creature loosely named.

Lice infesting the body had a rather sinister reputation as compared with the other forms, since they were reputed to be capable of causing death (see below). References to these are numerous. It was observed that they would depart from the persons of the dying (Plutarch *Mor.* 49c; Apollonius *Hist.Mir.* 27). The form primarily concerned here would have been P. *humanus humanus*, as with those references to lice in clothing (e.g. Plutarch 352 f., 642 b; Pliny XXVI.138). These are most frequently referred to by medical writers. Caelius (*Chron.* IV.14–18) devotes a distinct section to *phthiriasis* as a 'disease' in its own right which often occurs as a concomitant of other conditions. According to him, the creatures are generated from a 'reddish bile' discharged from the pores of the sufferer.[110]

---

[110] There is an example from a votive tablet from Epidaurus, referred to by Keller and discussed by Keil, art. cit., p. 310, of louse infestation being viewed as a 'disease' serious enough to require miraculous cure.

Suggested treatments for louse infestations are numerous, Pliny's catalogue being the longest. Some of his prescriptions are given specifically for the removal of nits or adult lice from the head. In the majority of cases *lendes* only are mentioned (Pliny in fact rarely uses the name *pediculus*, preferring various circumlocutions such as *foeda animalia*), while others are associated with *phthiriasis* (the Greek technical term for louse infestation) in general or specifically of the head and/or body. The majority of countermeasures consist of various liquids, ointments or powdery substances to be poured on, scattered on, or rubbed into the scalp, eyebrows or body, although Pliny does describe two products as being taken internally (XXX.144). Caelius (loc. cit.) sets out a comprehensive programme of treatment involving exercise, bathing and special diet, as well as the medicaments prescribed by other writers. Various herbal remedies predominate, various plant products being described by Pliny (XX.8,24,53,239,125, XXIII.18,94,154, XXIV.18,72–4,79, XXV.61,136, XVI.138), whose lists cover fifteen species including bay, ivy, hellebore, bryony and cedar, Galen (XII.16,30,462, XIV.323 K), Celsus (loc. cit.). Marcellus (IV.22, V.65–7) and Q.Serenus (64–7). Also suggested for use in combating lice are wine or vinegar (Celsus loc. cit., Marcellus IV.22; Caelius loc. cit.); oil (Pliny XXIX.121; Marcellus IV.65; Celsus loc. cit.; Caelius loc. cit.); honey (Pliny XXII.108; Dioscorides *DMM.* II.82); whey (Pliny XXX.144); goat's milk (Pliny XXVIII.166; Sextus Placitus V.a.11,b.22); calf gall (Pliny XXVIII.164; Marcellus IV.23); dog fat (Pliny XXIX.111); antler ash (Pliny XXVIII.163; Marcellus IV.69); vulture marrow (Sextus Placitus XXIII.9); snakes boiled up with various substances, or their shed skins (Pliny XXIX.111,121, XXX.144); pig's dung (Marcellus IV.22; Sextus IX.a.24); ash or chalk dust (Marcellus IV.65); seawater (Pliny XXXI.65; Q.Serenus 69; Caelius loc. cit.; Galen XII.462, XIV.415 K); sulphur (Pliny XXXV.177; Marcellus IV.67, VI.25); soda (Pliny XXXI.117, XXXV.177; Galen XII.462, XIV.323 K; Celsus loc. cit.; Marcellus IV.67; Caelius loc. cit.; Theodorus Eup.I.12); sandarach (Galen loc. cit.; Caelius loc. cit.; Dioscorides DMM. V.105; Celsus loc. cit.); alum (Pliny XXXV.189); and Samian earth (Pliny XXXI.117). According to Herodotus (II.37), the priests of Egypt were in the habit of shaving their whole bodies to prevent infestation.

However, in the absence of serious proliferation, the most straightforward way of dealing with lice was simply to pick them off individually by hand (Heraclitus fr. 56D; Athenaeus 586a; Plato *Soph.* 227b). Pausanias (X.10.7) introduces into one of his historical anecdotes, evidently as something his readers would recognise as a normal everyday practice, not in any way associated with uncleanliness, the picture of a wife picking lice out of her husband's hair.

4.2. Human lice: exceptional infestations and fatalities

One of the more mysterious aspects of classical entomology is the universal belief that, apart from their normal and fairly innocuous types of infestation of human beings, lice could also be the agents of a horrific disease in which they were generated in large numbers in abscesses beneath the skin, later bursting out and spreading, and capable of causing the death of their unfortunate victim. This process, which bears no relation whatever to the actual life history of lice, is described by Aristotle (556 b 21 ff.) as the normal mode of their generation. According to his account, in the first place small eruptions form on the surface of the body out of which, if they are pricked, the insects emerge. This condition was thought particularly likely to befall persons in prison. Thus in Artemidorus' manual on dreams (III.7) to see oneself swarming with lice is a singularly ill omen. According to Aelian (IX.19), the disease may be caused by the consumption of contaminated oil.

Furthermore, severe suffering and death from this form of *phtheiriasis* was not regarded as a mere theoretical possibility, since a considerable number of well known historical personages were reported as having been afflicted in this way. The ancients indeed were rather fond of gruesome stories about individuals being eaten alive by insects. There are also several accounts of persons being devoured by maggots or worms (cf. 48 below). Such an unenviable fate was regarded as being a particularly fitting one for some of the more notorious figures of ancient history. Wishful thinking directed against unpopular men of the past and a general taste for the horrific (evidenced also by the manifest interest in vivid accounts of snake bite symptoms) would thus have been major influences in the rise of such stories.[111] In the case of one frequently mentioned victim, Pherecydes, Diodorus bears witness to an alternative version of his death in which lice are not involved. Certainly it was a popular libel to imply that a certain individual had died in this manner: this is evident from what we hear of rumours concerning Socrates and Plato.

The full list of those reputed to have thus suffered and, in most cases, to have died from *phtheiriasis* comprise one mythological figure, Acastus (Plutarch *Sulla* 36), and from historical times Alkman (Aristotle loc. cit.; Pliny XI.14; Plutarch loc. cit.; Antigonus Carystius 88; Q.Serenus 59), Pherecydes (Aristotle loc. cit.; Plutarch loc. cit. and *de Comm.Not.* 11; Diogenes L. I.118; Aelian *VH* IV.28, V.2; cf. Diodorus X.3.4), Democritus (Marcus Aurelius III.3), Socrates (Marcus Aurelius loc. cit.), Plato (Diogenes L. III.40), Callisthenes (Plutarch *Sulla* 36 and *Alex.* 55; Diogenes L. V.5; *Com.Adesp.* fr. 280 Ed.; Suidas s.v.), Speusippus (Diogenes L. IV.5), Eunus (Plutarch *Sulla* 36; Diodorus XXXIV/V.2.23), Mucius Scaevola (Plutarch loc. cit.), and Sulla (Plutarch loc. cit.; Pausanias I.27; Pliny XI.114, XXVI.138; Q.Serenus 62). Among these various references, the most lurid account is that to be found in Plutarch's life of Sulla (loc. cit.), where we are informed that the whole of the victim's body was converted into lice, so that even with attendants employed day and night in removing them they were unable to keep up with the rate at which new ones were being produced.

What, if anything, of actual entomological fact might lie behind this widespread conviction, once the over-vivid imagination of narratives such as Plutarch's has been discounted, has been a matter of some debate. In the opinion of Sundevall, Keller and Keaveney and Madden,[112] the origin of Aristotle's statements and of the various popular

---

[111] Cf. T. Africa 'Worms and the Death of Kings', *Classical Antiquity*, 1, 1982, pp. 1–17; Davies and Kathirithamby, op. cit., pp. 174–6.

[112] Sundevall, *Thierarten*, p. 229; Keller, op. cit., II, p. 397; A. Keaveney and J. A. Madden, 'Phthiriasis and its Victims', *Symbolae Osloenses*, LVII, 1982, pp. 88–9. The activity of mites is also proposed as an explanation by T. Africa, art. cit., pp. 1–2.

stories was probably the activity of the scabies mite (*Sarcoptes scabiei* (Linn.)). This creature is a minute arachnid whose females burrow into human skin, where they lay eggs which hatch into larvae: in this way they produce severe irritation. Normally they are not present in very great numbers, but there is a particular form of scabies in which the affected areas become covered by thick crusts and in which mites are produced in great profusion.[113] Such a condition could well have given rise to the erroneous belief in the possibility of fatalities. The main objection to this theory is that the mites concerned are so small as to be almost microscopic. However, though it is perhaps surprising that such minute animals, no larger than Aristotle's 'smallest creature' the *akari*, itself a form of mite, should have been regarded as identical with normal lice (unless the size difference was felt to be sufficiently accounted for by the assumption that the creatures were only newly born), this remains the most likely suggestion to account for the classical data. Gossen[114] rejects the identification with the scabies mite in favour of the view that some louse-born disease lies behind the 'fatal *phtheiriasis*' of antiquity. However, although lice are of importance in the transmission of certain diseases—indeed it has been suggested[115] that the famous Athenian plague is to be identified as louse-borne typhus fever—it should be remembered that the connection between insects and disease was something of which the ancients were wholly ignorant.[116]

5. Lice and domestic animals

Lice were well known not only as pests of man, but also of his domestic animals, accounts of the various animals affected being provided by Aristotle (*HA* 557a 1 ff.), Pliny (XI.114–15), and the agricultural writers. The number of modern species involved here is very large and, as has been said, the provision of a comprehensive catalogue is neither necessary nor reasonable, in view of the fact that the ancients did not share our concept of species and for the most part lumped together all lice as identical. Only in one case is an animal louse noted as distinctive in appearance by comparison with its relatives.[117]

According to Aristotle (loc. cit.), lice increase in number when animals make a change of the water in which they bathe; and the same author devotes a section of his *Problems* (861a 10 ff.) to a discussion of this phenomenon (here, though, he seems to be speaking of human beings rather than animals). Pliny (XI.115), not fully comprehending Aristotle's meaning, manages to render his words in a manner which gives them a somewhat different connotation, as if lice are spontaneously generated in bath water (*Aquas quoque quasdam quibus lavemur fertiliores eius generis invenio apud auctores*).

---

[113] Smith, *Insects*, pp. 458–60.
[114] Art. cit., 1032.
[115] H. Keil, art. cit., pp. 313–4. Cf. Davies and Kathirithamby op. cit., pp. 169–70.
[116] In his article on mites, in PW Supp.VIII.355, Gossen further suggests that the skin condition known to the ancient medical tradition as ψώρα or *scabies* is to be identified as that produced by *Sarcoptes*, but this is denied by W. H. S. Jones, *Nat. Hist.*, VIII p. 582, and Keaveney and Madden, art. cit., p. 96.
[117] It may be noted that there are a number of mites, for example *Sarcoptes* and *Psoroptes* spp., which cause serious skin conditions in sheep, cattle, dogs and other animals; but, although these conditions would have been observed by the ancients, the creatures themselves evidently were not, being excessively small.

The animals from which lice were recorded are as follows (citations from Aristotle and Pliny are all from 557a 1 ff. and XI.114–15 respectively):

(i) *Cattle*. Lice on cattle are explicitly mentioned only by Aristotle. We are dealing here with such species as *Haematopinus eurysternus* (Nitzsch), *Linognathus vituli* (Linn.), and *Damalinia bovis* (Linn.)[118]

(ii) *Horses and donkeys*. According to Aristotle, followed by Pliny, donkeys are free from infestation by lice; but this is not in fact correct. Lice as pests of horses appear in a number of places in the *Hippiatrica* (*Ber.* 85.4–7, *Par.* 722, *Cant.* 70.4). They are spoken of as occurring either on the mane or over the whole body, and in the first of the above references they are termed ἄγριοι φθεῖρες. Prescribed treatments consist of the use of herbal preparations and the sprinkling of the animals' straw and fodder with brine. Among true lice, species such as *Haematopinus asini* (Linn.) and *Damalinia equi* (Linn.) occur on horses.[119]

(iii) *Sheep and goats*. Aristotle, followed by Pliny, states that lice do not occur on sheep or goats, although ticks do. The *Geoponica* (XVIII.16), however, describes both lice and ticks as pests of sheep and goats, and prescribes treatment with herbal preparations. It is the *Geoponica* which is correct here, the offending creatures being, e.g. *Linognathus pedalis* (Osborn) and *Damalinia ovis* (Linn.) on sheep, and *L. stenopsis* (Burm.) and *D. caprae* (Gurlt) on goats.[120] There is also a wingless parasitic fly, *Melophagus ovinus* (Linn.), which is found in the wool of sheep and sucks blood.[121]

(iv) *Pigs*. Aristotle describes the lice which occur on pigs as being distinctive by reason of their large size and hardness: these may be identified as *Haematopinus suis* (Linn.), which are indeed relatively large.[122] Isidorus (*Or.* XII.5.12) gives what are presumably these same insects the name *usia*, not found elsewhere (*Vsia et vermis porci . . . nam ubi momorderit, adeo locus ardet ut ibi vesicae fiant*).

(v) *Dogs*. The *Geoponica* (XIX.2.10) describes flies, ticks, and lice as being responsible for causing ulcers on the ears and between the toes of dogs. In earlier sources such as Columella, however, lice are not mentioned in this regard, their place being taken by fleas. Dog lice, such as *Trichodectes canis* (Deg.),[123] do of course exist, but conditions such as those described are caused rather by various types of mite: for example, the harvest mite *Trombicula autumnalis* (Shaw) affects the spaces between the dogs' toes.[124]

(vi) *Birds*. Bird lice comprise the majority of the Mallophaga and were well known in antiquity. Aristotle and Pliny mention them as infesting birds in general and, more

[118] Soulsby, *Helminths*, pp. 370, 372.
[119] Ibid. Gossen, in PW XII.1031, suggests that flies of the genus *Hippobosca* would have been classed as lice in antiquity, but cf. 8.1 above.
[120] Soulsby, op. cit., pp. 370, 373.
[121] Cf. 8.1 above.
[122] Soulsby, op. cit., p. 370.
[123] Ibid., p. 373.
[124] Ibid., pp. 495–6.

specifically, pheasants, which, we are told, may be killed by them unless they take regular dust baths. That lice can be severe pests of domestic hens is noted by Varro (*RR* III.9.14) and Columella (*RR* VIII.7.2). The former advises that young chicks are especially vulnerable and that care should be taken to manually remove the insects from them. The species *Menacanthus stramineus* (Nitzsch) is particularly harmful to chicks. Columella states that when hens are being fattened in small cages the feathers should be removed from their heads and from beneath their wings to prevent infestation. The head and neck of poultry are the particular target of *Cuclotogaster heterographus* Nitzsch, while *Lipeurus caponis* Linn. is found on the underside of the wing feathers.[125] Other parasites of fowl include *Goniodes gigas* (Tasch.), and *Menopon gallinae* (Linn.). Lice infesting wild birds are explicitly mentioned only by Theophrastus (*de Sign.* 16) with reference to ravens.

6. Uses in veterinary medicine

Pelagonius (*Vet.* 122) mentions the use of human lice (*peduculus humanus*) in the treatment of horses, as does Gargilius (*Cur.Bov.* 20) for strangury in cattle.

7. Exotic species

(i)   In *Egypt*. Herodotus II.37 (see 4.1 above).

(ii) In *Libya*. Herodotus describes (IV.168) the members of a certain Libyan tribe, the *Adyrmachidai*, as catching and biting each others' lice.[126]

(iii) In *Ethiopia*. According to Diodorus (III.29.5–7) the *Akridophagoi* of Ethiopia[127] never live beyond the age of forty owing to their becoming infested with a fearsome variety of winged louse. The crediting of fabulous varieties of normally wingless creatures with wings was clearly felt to make them appear especially sinister (cf. the reputed winged snakes and scorpions). These creatures were said to be generated inside the body and subsequently to break out onto the surface in large numbers accompanied by a stinging liquid, with eventual death as the inevitable result. The same account is given by Agatharchides (58); but Strabo (XVI.4.12), on the other hand, omits mention of the lice and says simply that the tribesmen die ἀποθηριουμένης αὐτῶν τῆς σαρκός.

---

[125] Ibid., pp. 371–3.
[126] F. S. Bodenheimer, *Insects as Human Food*, p. 40, suggests that the *Phtheirophagoi* tribe (Strabo XI.2.19, etc.) were so named because of this same practice, but the term seems rather to mean 'eaters of pine cones' (cf. LSJ Supp.; Davies and Kathirithamby, op. cit., p. 172).
[127] Cf. 11.10.vi above.

V

BUTTERFLIES, MOTHS AND WOOD-BORING LARVAE

Insecta: Lepidoptera (Butterflies and Moths)

23. Psyche/Papilio

1. Identification: general

These terms refer in their most general sense to any species of butterfly or moth, i.e., of diurnal or nocturnal Lepidoptera.[1] Unlike Latin, Greek also has a specific term for nocturnal forms (see next article) and ψυχή is never explicitly used for these outside lexicographical sources. However, since the ancients were chiefly interested in insects from the point of view of their importance as pests, rather than for aesthetic reasons, *psyche* and *papilio* refer primarily, not to the many attractive species of butterfly native to Europe, but to two pest species not distinguished in antiquity. These are the 'cabbage whites' *Pieris brassicae* (Linn.) and *P. rapae* (Linn.), well known as pests of cabbages, lettuces and related plants, and the insects designated as *psyche* by Aristotle.[2]

References to butterflies in imaginative literature are, surprisingly enough, extremely rare, a phenomenon which is commented upon by Keller[3] and Fernandez.[4] The latter suggests that, in view of the frequent appearance of other conspicuous diurnal insects in literature, their omission is deliberate and results from their sinister reputation as emblems of the human soul and thus creatures of ill omen. This explanation is not, however, entirely satisfactory in view of the fact that the ancients were quite happy to illustrate butterflies,[5] and also that the belief in question is said to have been more particularly associated with moths.[6] Is it perhaps possible that the ancients tended to think of butterflies as small brightly-coloured birds rather than insects, and that authors comprehended them under the former heading?[7]

---

[1] With the exception of the clothes-moth (26 below), which was too small to be reckoned as a lepidopteron.

[2] Sundevall, *Thierarten*, p. 202; Aubert and Wimmer, *Thierkunde*, I p. 510. As Gossen, in PW 2A.579, and Leitner, *Zool. Terminologie*, point out, there are also moths, for example *Mamestra brassicae* (Linn.), whose larvae attack cabbages.

[3] *Ant. Tierwelt*, II. p. 436.

[4] *Nombres*, p. 200. Cf. also Davies and Kathirithamby, *Greek Insects*, pp. 100–1.

[5] Keller, *Münzen und Gemmen*, p. 141, pl. XXIII. Gossen, art. cit., 573–5, claims that two of these are intended to represent recognisable species, but they are too stylised for this to be so.

[6] Cf. notes 36–8 below.

[7] Butterflies are described as ὄρνεα by Schol. Nicander *Ther*. 760, and as *aviculae* by Isidorus (*Or*. XII.8.8). *Papilio* is quaintly defined in *C.Gl.L*. V.231.3 as *avis quae numquam crescit*.

1.2. Identification: early stages

The early stages of Lepidoptera, which appear most frequently in agricultural sources, are not usually explicitly linked with their appropriate adult stage, although the zoological tradition from Aristotle was well aware of the details of their life history. The standard Greek term for 'caterpillar' is κάμπη, sometimes replaced by the more general term σκώληξ meaning 'worm' or more scientifically 'larva' (as in Aristotle, where it refers to newly hatched caterpillars). The Latin equivalent is eruca or uruca,[8] replaceable by the generalised vermis or vermiculus, or rarely (Ovid Met. XV.373) tinea.[9] In the Latin glossaries (for example II.338.10, V.423.20) eruca is often explicitly defined as the 'cabbage caterpillar' (Pieris spp.). In Aristotle the pupal stage is termed χρυσαλλίς, latinised by Pliny to chrysallis.[10]

The appearance of κάμπη (Vulgate eruca) in the LXX (Am. 4.9, Jl. 1.4, 2.25) is the result of the translators' shortage of Greek equivalents for the numerous Hebrew words for locusts and stages in their development.[11] It is, however, understood as a caterpillar by patristic commentators (e.g. Jerome MPL25. 1031 = Isidorus Or. XII.5.9; Gregory Magnus Mor. XXXIII.39).[12] Perhaps as a result of this, kampai are in a number of other places (e.g. Ptolemy Tetr. II.8.83; Hephaestio Apotel. I.20.37; Eusebius Praep.Ev. MPG 21.101) lumped together with locusts as plagues of crops on the same level.[13]

1.3. Identification: other names and named varieties

(i) Σκυταλίδες and σκυταλωτούς. Two related names[14] defined in the Etymologicum Magnum (721.45) and Etym.Gudianum as referring to 'a type of caterpillar'. In view of the meaning of these names, they presumably refer to larvae resembling sticks or twigs, i.e., certain of the Geometridae with their very effective camouflage.[15]

(ii) Σητοδοκίδες. According to Hesychius' definition, 'butterflies or winged creatures'. Fernandez[16] relates this word to the preceding, following Strömberg[17] in relating the first element to σής,[18] but identifying the second with δοκίς, a beam or plank.

---

[8] A Gallic synonym dolva is noted by Eucherius (Instr. II.12). Cf. Ernout and Meillet, Dict. Ét., s.v. Delpa, which appears in Polemius Silvius' list of insect names, is equated with this by A. Thomas, 'Le Laterculus de Polemius Silvius', Romania, XXXV, 1906, p. 167.

[9] For other uses of tinea, cf. 24.2 and 26 below.

[10] The term may result, as Keller, Ant. Tierwelt, II p. 436, suggests, from the colouration of some butterfly pupae.

[11] Cf. 11.1.2.i,iii,xxi,xxiii–iv above. It translates Hebrew gazam, a stage in the locust's life history, according to I. Aharoni, art. cit., p. 475.

[12] The Hebrew for foliage feeding larvae is tola'at, rendered in LXX by σκώληξ, for example Deut. 28.39, Jon. 4.7.

[13] Schol. Juvenal VI.276 erroneously equates uruca with curculio (36 below), and furthermore confuses it with ericius, as Gossen, art. cit., 574 has failed to notice.

[14] Fernandez, op. cit., p. 38.

[15] As Gossen, art. cit., 580 suggests, although no particular species can be said to be primarily in view.

[16] Op. cit., p. 38.

[17] Gr. Wortstudien, p. 21.

[18] The connotation of σής (=Latin tinea) is 'worm' rather than 'moth'. Unlike tinea, it is not in surviving literature used for anything as large as a typical caterpillar. Cf. 23.1.2 above, 24.2 and 26 below.

Presumably, then, we have here another term for stick-like larvae, and Hesychius is in error in referring to the adults.

(iii) Σκῆν ὃ τινες μὲν ψυχὴν, τινὲς δὲ φάλαιναν according to Hesychius. Fernandez[19] follows Immisch[20] in applying this name to the pupae of Lepidoptera rather than to the adults, relating it to σκῆνος 'tent'. It is perhaps more likely, as Gossen[21] suggests, to refer to the beehive pest known as κλῆρος.[22]

(iv) Μεμφίδες. Defined by Hesychius as αἱ τῶν πτηνῶν ψυχαί, this term is included by Fernandez[23] as a genuine insect name, though he adds that the text is 'evidently corrupt' and offers no etymology. The word is recognised by LSJ, but Latte brackets the entry in his text of Hesychius and suggests that it is an error for πεμφίδες.[24]

(v) Ὕπερον and πηνίον. Following his account of the life history of *Pieris* spp., Aristotle (*HA* 551b 6 ff.) goes on to say that 'the *hyperon* and *penion* too come from similar caterpillars, which walk with an undulating motion: they go forward with one part of their body, then bend themselves, and so move forward; each of the resulting animals derives its own proper colour from its caterpillar'. Hesychius defines *hypera* in vague terms as σκώληκες τινες, while Photius and Suidas, explaining the other name, provide us with a quotation from Aristophanes' *Clouds*, the lost first version, κείσεσθον ὥσπερ πηνίω βινουμένω, the insects here typifying ξηρότητα καὶ ἀσθένειαν.[25]

As has been generally recognised, Aristotle's account is a good description of the characteristic 'looping' movement of the larvae of Geometridae, a family which includes such pest species as *Abraxas grossulariata* (Linn.), *Erannis defoliaria* (Cl.), and *Operophtera brumata* (Linn.).[26] His statement about colour could refer, as Aubert and Wimmer suggest,[27] to the unique correspondence of colour and pattern between larva and adult of *A. grossulariata*, or else to some more generalised and erroneous belief that, for example, brown larvae give rise to brown moths. Both names, which in popular parlance were probably synonyms, clearly refer to the insects' physical appearance, the former meaning 'pestle or object so shaped' and the latter 'bobbin or spool', and would normally have applied, as Sundevall states,[28] to the larvae and not the adults as in Aristotle.

---

[19] Op. cit., pp. 203–4.

[20] O. Immisch, 'Sprachliches zum Seelenschmetterling', *Glotta*, VI, pp. 200–1.

[21] *Zool. Glossen*, p. 105.

[22] Cf. 39 below.

[23] Op. cit., p. 242.

[24] Which means 'ghost' in Lycophron, thus providing a possible parallel with ψυχή if this is recognised as a genuine insect name.

[25] According to the lexicographers, Speusippus in his *Homoia* equated the name with κώνωψ and ἐμπίς (50 below), but this must have been an error on his part.

[26] These are the three species cited by Sundevall, op. cit., pp. 204–5. The former is a garden pest, while the others defoliate trees.

[27] Op. cit., I p. 510.

[28] Op. cit., pp. 204–5, supported by Fernandez, op. cit., pp. 36–7. Aristophanes' allusion to a thin dried-up appearance supports this view. The terms cannot have referred to the pupae, as D'Arcy Thompson, *Hist. An.*, V.19 note, and LSJ suggest.

1.4. Identification: unnamed varieties

In addition to those of the 'cabbage whites', caterpillars are associated with the following types of plant. Not all of these are necessarily Lepidoptera.[29]

(i) *Orchard trees.* There is considerable confusion in classical literature between foliage feeding and wood-boring larvae attacking fruit trees.[30] Some authors (Aetius XIII.54; *Geoponica* X.18) distinguish between the two by naming the former κάμπαι and the latter σκώληκες, but often this distinction is ignored, and in any case the same remedies are usually prescribed for both. Palladius (I.35.6) and the *Geoponica* (XII.8) offer prescriptions covering both larvae from trees and those from garden vegetables (cf. 5 below).

Theophrastus (*HP* IV.14.10, *CP* V.10.1), followed by Pliny (XVII.230), speaks of the fruit of pears, apples, medlar, and pomegranate being eaten by *skolekes*, and states elsewhere (*CP* V.10.5) that those from different species of tree differ not only in appearance but also in the fact that if transferred from one to the other they do not survive. Aelian (IX.39) refers to insects generated in apple trees which destroy the fruit, but which are also useful in helping women to conceive. It is stated by a number of authors (Pliny XVII.266, cf. XV.117; Palladius III.25.15; *Geoponica* X.18.7) that the fruit of apple trees may be protected against *urucae* or *kampai* by application of green lizard gall; Pliny adds that to have a menstruating woman circle the trees is also effective. Augustine (MPL 32.1365, 34.288, 38.412) also refers to caterpillars of apples and pears. In the Greek Anthology, Antiphanes (*AP* IX.256) describes an apple tree devastated by the attacks of a πτιλόνωτος κάμπη. The apple pests of Aelian and Pliny would include among others the Codling moth *Cydia pomonella* (Linn.).[31]

(ii) *Pulses.* Theophrastus refers in a number of places (*HP* VIII.6.5, 10.1.5, *CP* III.22.3) to *kampai* which are generated among chickpea whenever hot weather follows rain, the latter being said to wash a protective saltiness from the leaves. This account is followed by Pliny (XVIII.154). These same *kampai* are referred to also by Aelian (IX.39). Also referred to are *skolekes* generated under the same conditions among peas, *Lathyrus sativus*, and *L. ochrus* (*HP* VIII.10.5, *CP* III.22.3). Such larvae are apparently recognised as lepidopterous by Aristotle (*HA* 552a 19), who speaks of certain 'winged creatures' as the adult stage of *skolekes* generated among pulses (Aristophanes Byzantinus (36) seems to make their nature as Lepidoptera explicit, but only because he has erroneously lumped together several distinct Aristotelian life histories at this point). Aelius Promotus (pp. 776.28–9) prescribes 'the *skolex* from chickpea' as an amulet for semitertian fever.[32]

---

[29] Other foliage feeding caterpillars are dealt with under 25 below.

[30] Cf. 30 below.

[31] So identified by Gossen, in PW 2A.583, who also identifies Antiphanes' species as *Malacosoma neustria*, ibid., 575.

[32] *Leptidea*, mentioned by Gossen, ibid., 573, is not a pest species.

(iii) *Spurge.* Hippocrates (*Superf.* 28) describes certain *kampai* with tails or 'stings' (κέντρα) found on a sea spurge, which may be dried in the sun and drunk in wine as a remedy for disorders of the womb. As has been noted,[33] the conspicuous larva of the Spurge Hawk-moth *Hyles euphorbiae* (Linn.) with its horn-like process at the end of its abdomen would fit the description very well.

(iv) *Thorns.* According to Pliny (XXX.101,139), certain *vermiculi asperi lanuginosi* found on thorn bushes were employed as amulets against quartan fever, or to prevent food from sticking in the throats of babies. There are a number of conspicuous furry caterpillars with the appropriate foodplant, for example *Gastropacha quercifolia* (Linn.), *Malacosoma neustria* (Linn.) and *Euproctis chrysorrhoea* (Linn.).

(v) *Corn.* Pliny (XXVIII.78) lists *urucae et vermiculi* among pests in cornfields.

(vi) *Rue.* A form of *kampe* or *uruca* devouring this herb is referred to by Nicander (*Alex.* 413) and Pliny (XIX.156).

(vii) *Thistle.* A *vermiculus in cardone herba* is recommended as an amulet against toothache by Marcellus (XII.21).

(viii) *Grass.* Pliny (XXX.125) mentions a *vermiculus in gramine* as an amulet to prevent miscarriage.

(ix) *Teasel.* A *vermiculus* from the small teasel *Dipsacus pilosus* (*veneris labrum*) is prescribed by Pliny (XXV.171, XXX.24) as a treatment for hollow teeth. Similarly he mentions the use of *vermiculi* generated in the flower head of the larger teasels *D. fullonum* and *D. silvestris* (*gallidraga*), as it ages, in amulets to relieve toothache (XXVII.89).[34]

(x) *Acanthus.* Marcellus (XIV.6) mentions *vermiculi* from species of *Acanthus* as providing a remedy for coughs.

2. Life history

According to Aristotle's account (*HA* 551a 13 ff.) the *psyche* arises from *kampai* generated upon the leaves of cabbage and related vegetables. To begin with these latter are smaller than a millet grain, but they soon become small *skolekes*, which in three days develop into small *kampai*. They then form themselves into an immobile *chrysallis*, fixed down by threads like those of a spider, out of which the adults shortly emerge. As is explained elsewhere (*GA* 758b 9–759a 3, 733b 14), Aristotle regards the pupal stage as corresponding to the eggs of higher animals, the preceding stages being, strictly speaking, the egg in process of development. He seems to be somewhat unclear as to whether his initial 'millet grain' (the true egg, in our terminology) is actually produced by the adult insect. Although in the passage above he clearly suggests that they appear on the foliage

---

[33] Gossen, ibid., 574.

[34] Suggested identifications are offered by Gossen, ibid., 579–80, and Steier, ap. Leitner, op. cit., p. 247.

by spontaneous generation, he states a little earlier (*HA* 550b 26 ff.) that all insects which mate produce *skolekes*, with the exception of 'a certain kind of butterfly' which gives rise to a hard object like a safflower seed.[35]

Apart from such minor areas of confusion, these details provide a quite accurate description of the development of *Pieris* spp., particularly *P. brassicae*, since its relative does not feed in an exposed position.

Pliny (XI.112) makes an addition to Aristotle's account by stating that the 'millet grains' are initially formed from dew settling on the foliage in spring and being condensed by the rays of the sun. This follows the account given by Aristophanes of Byzantium (*Epit.* 36), who applies to the larvae the non-Aristotelian name κραμβίδες. Aristotle himself does in fact state (*HA* 551a 1) that some insects arise spontaneously 'from the dew which falls upon leaves, during spring in the normal course of events, but quite often in winter', but he gives no specific examples.

A general belief that cabbage caterpillars are spontaneously generated from their food plant is found in Aelian (IX.39), who calls them *krambides*, Sextus Empiricus (*Pyrrh.* I.41) and Isidorus (*Or.* XII.5.9,18; cf. Augustine MPL 32.1365). Fulgentius (*Mit.* II.16) speaks of refuse thrown out in gardens giving rise to *vermiculi*. Theophrastus (*HP* II.4.4, *CP* V.7.3) and Plutarch (*Mor.* 636c) follow Aristotle's three stage metamorphosis from caterpillar to pupa to adult, but say nothing about the origin of the former (Plutarch does not use the word *chrysallis* but simply says that the caterpillar becomes brittle through dryness and bursts open to release the *psyche*). Elsewhere, however, Theophrastus (*HP* IV.14.10, VII.5.4) does explicitly include cabbage caterpillars among spontaneously generated pests.

General references to the life history of butterflies or moths other than *Pieris* spp. are found in Ovid (*Met.* XV.372–4), who speaks of *agrestes tineae* weaving leaves together with white threads when they are about to turn into *papiliones*, and in the poem *Phoenix* (107–8), which is indebted to Ovid for most of its phraseology but has the larvae attaching themselves to stones prior to their transformation.

3. Habits

Pliny (XVIII.209) notes that one of the clearest signs of spring is the first appearance of butterflies, which due to their *infirmitas* are particularly sensitive to inclement weather. However, he continues, he himself had observed them making initial appearances only to be destroyed by unexpected late frosts. The reference here is to those butterflies which hibernate and which are thus the earliest in the year to be seen, most notably the Brimstone *Gonepteryx rhamni* (Linn.) and also various Nymphalidae such as *Aglais urticae* (Linn.) and *Inachis io* (Linn.). These would not in fact be killed by late frosts, but would simply return to hiding.

Tertullian (*de An.* 32) characterises what he calls *papiliunculi* as creatures which delight in dryness; that is to say, which are active in hot sunny weather.

[35] There is no reason to suppose, as does Gossen, art. cit., 573, that Aristotle is intending to refer to a species other than *Pieris*.

4. Popular beliefs and attitudes

The term ψυχή or 'soul' was clearly applied to butterflies and moths as a result of a popular folk belief which visualised the human soul as an insubstantial winged entity in the form of these insects.[36] The presumed origin of this association is discussed by Immisch,[37] Keller[38] and Fernandez.[39] It is suggested that it was originally moths rather than butterflies to which the belief in question chiefly applied, their nocturnal habits, flitting like ghosts through the night, bringing about a link in the popular imagination with the spirits of the dead.[40] Keller cites the Homeric description of souls fluttering around in swarms when summoned out of Hades by Odysseus. Attention is also drawn to their fluttering and weak flight, and to their metamorphosis, the escape of the adult from the immobile pupa being seen as a picture of the flight of the soul from a dead body.

The evidence for this belief in Greek and Roman antiquity is almost entirely archaeological,[41] for example Latin epitaphs in which the soul is described as a *papilio* (*C.I.L.* II.2146, VI.26011). Literary references to an association between Lepidoptera and the spirits of the dead are very rare, although Ovid's phrase *feralis papilio* is generally understood as having this implication (*Met.* XV.374). Far more explicit is the Aesopic fable (Phaedrus *App.* 31) of the wasp and the butterfly, in which the latter describes with regret its former existence as a distinguished human being and compares its origin as a human soul rather incongruously with the wasp's spontaneous generation from a dead mule.[42]

Martial refers to butterflies twice, as typifying lightness and delicacy (VIII.33), and weakness and insignificance (XII.61).

5. Importance as pests

The caterpillars of *Pieris* spp.[43] were well known to the ancients as the most significant pests of garden vegetables in general and cabbages in particular. Theophrastus (*HP* VII.5.4) in his enumeration of garden pests associates *kampai* and *skolekes* (Pliny, XIX.177, follows this nomenclature with his *urucae et vermiculi*) with the cabbage in particular. Presumably he does not intend us to understand two distinct forms of pest here.[44]

---

[36] The use of the equivalent Latin term *anima/animula* for lepidoptera is attested only in glossaries (*C.Gl.L.* V.384.44, *papilio animal quomodo quasi apes tenues quas dicunt animula*). This usage is discussed by J. André, *Rev. Phil.*, **36**, 1962, p. 23, who also considers the possible synonym *vappo*, found only in Probus, *Gramm. Lat.* IV.10.30-1, where it is defined as *animal volans, quod vulgo animas vocant*.

[37] Art. cit., pp. 193-206.

[38] Op. cit., II pp. 436-40.

[39] Op. cit., pp. 201-2. Cf. also Davies and Kathirithamby, op. cit., pp. 99-101, 103-7.

[40] Cf. 24.1,1.2.i,3 below. Keller and Gossen, art. cit., 574, suggest that the particular species *Acherontia atropos* (Linn.), the Death's-head Hawkmoth, with its skull marking had an especial part to play in the development of folklore surrounding moths.

[41] Cf. Keller, op. cit., II pp. 437-40; Davies and Kathirithamby, op. cit., pp. 103-7.

[42] Cf. 40.4 below.

[43] More particularly *P. brassicae*, which feeds in numbers conspicuously exposed, while *P. rapae* feeds concealed.

[44] Perhaps he is following the Aristotelian terminology where *skolekes* are immature *kampai*.

Among non-technical writers, Pherecrates (fr. 145) refers to cabbage plants full of *kampai*, while Aristophon (fr. 10) speaks of them as a pest of vegetables in general. Martial (XI.18) mentions the *uruca* as a garden pest. The ἐν τοῖς κήποις ψυχαί of Schol. Nicander *Ther.* 763 are presumably the adult 'cabbage whites'.

A considerable variety of counter-measures, likely and unlikely, are prescribed by Pliny (XIX.177), Columella (*RR* XI.3.63–4), Palladius (I.35) and the *Geoponica* (XII.8, unless otherwise stated). It is suggested that before sowing the seeds be immersed in houseleek juice (Pliny, Columella, Palladius), fig ash solution (*Geoponica* XII.17.1) or the blood of other caterpillars (Palladius). Alternatively, chickpea or squill (Pliny, Palladius) may be sown among the plants, or a mare's skull (Pliny) or river crabs (Pliny, Palladius) may be hung up in a convenient place. Once the plants have grown, they, together with any caterpillars that may have appeared, may be sprinkled with a solution of houseleek (Pliny), wormwood (Pliny), fig ash (Palladius, *Geoponica*), vine ash (*Geoponica*) or dead larvae from another garden (Palladius, *Geoponica*) or with a mixture of cattle urine and *amurca* (Palladius, *Geoponica*). Or they may be fumigated by the burning of garlic stalks (Palladius, *Geoponica*), fungi from beneath nut trees (*Geoponica*), sulphur and bitumen (Palladius, *Geoponica*), or bat dung (*Geoponica*). Or branches of cornel (Pliny) may be applied to them. Columella also offers the more mundane suggestion that the caterpillars be collected by hand or shaken from the plants in the early morning when they are still torpid from the cold of the night. In addition he mentions the belief, noted also by Aelian (VI.36), Palladius and the *Geoponica*, that if a menstruating woman circles the garden three times barefoot all the insects would fall to the ground and die.

For pest species other than *Pieris* cf. 1.4.i–ii, v–vi above.

6. Caterpillars as harmful to domestic animals

The *Hippiatrica* (*Ber.* 93) refers to skin conditions in horses caused by their rolling on *kampai* or rubbing against trees infested with them; these are to be treated by the application of vinegar, soda, and copper sulphate. According to the *Geoponica* (XVIII.17.5), sheep which have swallowed *skolekes* along with their green fodder may be treated by bleeding or by medication with human urine.[45]

7. Medicinal and other uses

The majority of references to caterpillars in medical contexts explicitly prescribe those of *Pieris* spp. Nicander (*Ther.* 87–90) states that the 'garden *kampe* with the green back'[46] provides when crushed in vinegar an ointment to deter the attacks of venomous animals; similar details are given by Paulus of Aegina (VII.3) and Dioscorides (*DMM.* II.60), who also suggests such an ointment as a preventive against wasp stings (*DMM.* II.134). *Vrucae brassicae* are reported by Pliny (XXX.24) to cause decayed teeth to drop out (this is also stated by Galen XIV.430 K), and unspecified caterpillars are said to be used in a Magian amulet against quartan fever.

[45] Larvae of *P. brassicae* are toxic to animals. Cf. Soulsby, *Helminths*, p. 522.
[46] I.e. *P. brassicae*.

In the illuminated manuscripts of Dioscorides discussed by Kadar,[47] the section on caterpillars is illustrated by a pair of unusually accurate drawings. One clearly depicts a red-backed yellow-spotted hawk-moth larva with its prominent posterior horn, which is most likely to be identified, as Kadar suggests, with that of *Hyles euphorbiae*, a species described by Hippocrates (1.4.iii above). The other is presumably the larva of *Pieris brassicae*.

For the use of caterpillars other than *Pieris* in medicine, cf. 1.4.i–iv, vii–x above.

Cabbage caterpillars are mentioned as a component of a compound fish bait in the *Geoponica* (XX.31).

24. *Phallaina*

1. *Identification: general*

Φάλλαινα is the general Greek term for the nocturnal members of the order Lepidoptera, which we know as moths. Strictly speaking it constitutes a subdivision of ψυχή, which covers the whole order, but in normal usage the latter name is restricted to the diurnal species,[48] being referred to moths only in lexicographical sources (for example, Schol. Nicander *Ther.* 760). Latin has no equivalent term, and therefore *papilio* is used indiscriminately for all Lepidoptera.

The etymology of φάλλαινα and its variant form φάλλη found only in Hesychius is uncertain and the various suggestions put forward are discussed by Fernandez.[49] He rejects the theory of Immisch[50] that the name is, like *psyche*, the product of folk belief in favour of an interpretation which would link it with φαλός, 'white'. This view is supported by Chantraine,[51] who notes the fact that moths, whatever their actual colour, appear pale in lamp light.

As with *Pieris* spp. in relation to butterflies, a particular pair of moths stand out as being of pre-eminent importance due to their status as pests. These are the species described by Aristotle, the two wax moths *Galleria mellonella* (Linn.) and *Achroia grisella* (Fab.), which infest beehives. The adult females enter the hives by night and lay eggs on the combs. From these emerge larvae which devour the wax, producing tunnels lined with matted silk, and whose activities may cause the bees to desert the hive.[52] In popular belief these wax moths were not clearly distinguished from their relatives, with the result that all nocturnal Lepidoptera tended to be regarded as a threat to the beekeeper.[53]

---

[47] *Zoological Illuminations*, p. 60, pl. 68, 88, 93.

[48] There are of course diurnal moths such as *Zygaena* spp., but the ancients would have viewed these as butterflies rather than φάλλαιναι.

[49] *Nombres*, pp. 204–7.

[50] Art. cit., pp. 196–7.

[51] *Dict. Ét.*, p. 1175. Also supported by Davies and Kathirithamby, *Greek Insects*, p. 108.

[52] So identified by Sundevall, *Thierarten*, pp. 205–6; Aubert and Wimmer, *Thierkunde*, I p. 164; Keller, *Ant. Tierwelt*, II p. 442.

[53] It may be noted that the large hawk-moth *Acherontia atropos* (cf. 23 note 40) is popularly known for its occasional habit of entering beehives to feed on honey.

1.2. Identification: other names

(i) Ἠπίολος. This term is used by Ps. Aristotle (*HA* 605b 13 ff.) instead of *phallaina* for moths in general, but is not found elsewhere. Fernandez and Chantraine[54] support the theory of Immisch[55] that it is related to ἠπίαλος 'fever', and ἠπιάλης 'nightmare', and results from a folk belief that moths are the carriers of these latter. There are parallel beliefs in modern European folklore, although no direct evidence exists from antiquity.[56]

(ii) Πυραύστης. This is a literary name for moths, referring to their habit of being attracted to light. Aelian (XII.8) describes the *pyraustes* as a creature which flies into lamps as a result of its delight in fire and is consequently burned to death. He quotes a proverbial expression δέδοικα μωρὸν κάρτα πυραύστου μόρον, which was used by Aeschylus (fr. 288). With certain variations in wording, the same expression is quoted by the lexicographers (Suidas, Eustathius *in Il.* 1304.8, *in Od.* 1547.63, 1848.35) and paroemiographers (Zenobius V.79; Apostolius XVIII.18; *Mantissa Prov.* II.72). It is variously stated to be applicable to the short-lived, to those destroyed for the sake of some small pleasure, or to those who die ignobly through their own folly.

(iii) Κανδηλοσβέστης. A name with similar significance to the last,[57] this is preserved only by the scholiast on Nicander *Ther.* 763, who gives it as a synonym of *phallaina*. A feminine form κανδηλοσβέστρια is found in Schol. Oppian *Hal.* I.404.

(iv) Ψώρα. A synonym of *phallaina* mentioned only by Schol. Nicander *Ther.* 760, 763. It is probably derived from the adjective ψωρός 'rough', with reference to the texture of moths' scale-covered wings.[58]

2. Life history
The wax moths are the only species whose life history is detailed in surviving sources. Although both adults and larvae were observed as pests of beehives, there was generally no clear realisation of the true connection between them. Our earliest source, the non-Aristotelian book IX of the *Historia Animalium*, has two somewhat contradictory accounts of the matter. The first (*HA* 605b 13 ff.) lists among beehive pests certain creatures resembling the moths which fly around lights (this is the only place where wax moths are explicitly distinguished from their relatives), these damaging the combs and producing a good deal of down or fluff, and also certain κάμπαι known as τερηδόνες,[59] generated in the hive, but draws no connection between the two. Later (625a 5–14),

---

[54] Fernandez, op. cit., pp. 196–7; Chantraine, op. cit., p. 415.
[55] Art. cit., p. 193. Davies and Kathirithamby, op. cit., p. 110, suggest that in Hesychius s.v. ἠπίαλος is a synonym of ψυχή.
[56] Immisch, ibid., suggests that in *Aeneid* VI.284 the false dreams are being portrayed as moths at rest by day beneath the leaves of a tree.
[57] Fernandez, op. cit., pp. 154–5.
[58] Ibid., pp. 39–40.
[59] For other uses of this term, cf. 30 and 37 below.

however, it is stated that neglected combs become covered with cobwebs, i.e., spontaneously, and that from these are generated σκωλήκια which subsequently develop wings and fly off.

According to Pliny (XXI.65–6) and Columella (*RR* IX.14.8), who are followed by Palladius (IV.15.4) and Isidorus (*Or.* XII.8.8), the larvae, named by Pliny *teredines* and by Columella *vermes . . . quos alvorum tineas appellamus*, are generated from droppings produced by the moths. Pliny confuses the picture further by introducing a second form of *teredo*, generated *in ipso ligno*. The term *tinea* is used for the larvae also by Virgil (*G.* IV.241 ff.), and its Greek equivalent σής[60] is given this meaning by Hesychius.

3. Habits

The well known attraction of moths to light clearly intrigued the ancients, two of these creatures' names being derived from this phenomenon (1.2.ii–iii above). Nicander's κρανοκολάπτης,[61] which appears to be based upon some sort of large moth, is compared to 'the *phallaina* which darts about around lamps in the evening' (*Ther.* 760–1). The scholia (760–3) expand this latter description, referring to the scale-covered wings of the *phallaina* and its ashen colour.[62] It has been suggested that it was the nocturnal habits of moths which were the main influence in the association of Lepidoptera with the souls of the dead.[63]

4. Importance as pests

Both adults and larvae of the wax moths are described as serious pests of beehives by ancient writers on apiculture, though they are generally unclear about these being two stages in the life history of a single insect. Although in fact it is only the larvae which damage the combs, it was popularly believed that the adult *papiliones* or *phallainai* were equally responsible. Thus both Ps. Aristotle (loc. cit.) and Pliny (loc. cit.) credit the adults with producing silken webs, chiefly composed, according to the latter, from the scales of their wings. And Pliny and Columella (loc. cit.) state that they gnaw the combs and scatter droppings.

In order to combat the depredations of these pests, it is generally recommended that at the end of the year (Virgil *G.* IV.241 ff.) and at ten day intervals in late summer (Columella IX.14.1–2,7) the hives should be thoroughly cleansed of debris, and that they should then be fumigated by burning thyme or cow dung (Ps. Aristotle loc. cit.; Virgil loc. cit.; Pliny XXI.81; Columella loc. cit.; Palladius loc. cit.).

As a further precaution, it is suggested that light traps be set up to capture the adult moths as they fly around the hives. These are described by Columella (IX.14.9) and Palladius (V.7.7) as consisting of a tall narrow bronze vessel *simile miliario* with a light at

[60] Cf. 23 note 18 above.

[61] 7b.1.3.x above.

[62] There is little purpose in attempting, as Gossen, in PW 2A.579, does, to link such references with particular moth species, though it is such members of the family Noctuidae as he names which are generally thought of as 'typical moths', nondescript in colour.

[63] 23.4 and note 40.

the bottom: moths attracted by the light enter at the top, are restricted by the vessel's shape, and are thus forced into the flame. Columella advises that these be set up in late summer, when the *papiliones* are at their most numerous; and Pliny (XXI.81), whose account is briefer, adds that the best nights are those which are clear and moonless. Aelian (I.58) describes a somewhat different method of capture in which lamps are placed in front of the hives with vessels of oil beneath them into which the moths fall. Clearly this very effective use of light would result in the capture of many other species of Lepidoptera besides the wax moths, but it seems that no clear distinction was drawn between the latter and their harmless relatives, all moths being under suspicion as potential pests.[64]

5. Medicinal uses

Pliny (XXVIII.162), in a list of poisons and their antidotes, includes among the former 'the moth that flies to lights' (*papilio quoque lucernarum luminibus advolans inter mala medicamenta numeratur*), but this is probably a confused reference to the moth-like κρανοκολάπτης described by Nicander.[65]

25. Ips/Convolvolus

1. Identification: general

The above terms along with their several equivalents are used, mainly by agricultural and botanical writers, to indicate insect pests attacking the vine. Physical descriptions of these creatures are never given, and they are usually described purely in terms of their effects. Their nomenclature is somewhat ill defined. In some cases the terms used seem to apply to a particular form of insect rather than simply to vine pests in general, although only the *Geoponica* states explicitly that more than one species is involved. The pests most commonly referred to seem to be the larvae of two Tortricid moths, *Eupoecilia ambiguella* (Hubn.), a European species, and *Lobesia botrana* (D. & S.), whose range extends beyond Europe to Asia Minor, the Middle East and North Africa. These larvae feed on the developing fruit, spinning webs over the clusters, and pupate among the foliage.[66] Also mentioned, so it seems, are the much larger foliage-feeding larvae of the hawk-moth *Hyles lineata* (Fab.) which has a similar range to *Lobesia*.[67] Gossen[68] notes the existence of certain coleopterous pests such as the weevil *Rhynchites betuleti* (Fab.), but there are no explicit references to these in our surviving sources.

---

[64] Pliny (XXI.65) seems to lump all moths together as enemies of bees when he characterises the species as *hic ignavus et inhonoratus luminibus accensis advolitans*.

[65] Cf. 7b.1.3.x above.

[66] So identified by Gossen, in PW 2A.582; Leitner, *Zool. Terminologie*, pp. 99, 251; Bodenheimer, *Animal and Man*, p. 78. Another Tortricid pest, mainly a foliage feeder, is *Sparganothis pilleriana* (D. & S.), referred to by Leitner, op. cit., p. 99. Cf. Stapley and Gaynor, *Crop Protection*, pp. 190–2.

[67] Cited by Bodenheimer, op. cit., p. 78, and *Geschichte*, pp. 39–40.

[68] In PW 10.1487; *Zool. Glossen*, p. 12.

1.2. Identification: Greek terminology

(i) "Ἴψ or ἴξ. This pair of names belongs to the *ses-kis-thrips-knips* complex and would originally have meant nothing more specific than 'small insect pest'.[69] However, as is the usual pattern with these words, they came to be attached in normal usage to vine pests in particular. It is possible to interpret the two either as variants of one word, or as having distinct etymologies.[70] The form ἴξ is found only in a fragment of Alcman (fr. 43 Bk. *ap.* Ammonius *Diff.* 244), who characterises the insect as ποικίλον[71] ἴκα τὸν ὀφθαλμῶν ἀμπέλων ὀλετῆρα[72] and in the lexicographical tradition (Hesychius, *E.M.*, *Et. Gud.* s.v.).

Ips appears first of all in Theophrastus, who uses it for the especial pest of vines in *CP* III.22.5–6, and again in passing in *HP* VIII.10.5.[73] When he describes the use of bituminous earth (*Lap.* 49) to protect vines against pests, this name is applied to the offending insects. This usage is followed by Strabo (XIII.1.64) and Lydus (*de Ost.* 14). *Ips* occurs in one of the *Geoponica*'s chapters on vines (V.53), but elsewhere more general terms are used.

Since *ips* and *ix* imply a creature quite small and nondescript in appearance, the two Tortricid species would seem to be primarily in view in these passages.

(ii) Σκνίψ. In a parallel passage to Theophrastus *Lap.* 49, Galen (XII.186 K) describes the insects in question as '*skolekes* which our vine growers call *sknipes*'.[74]

(iii) Φθείρ. In another parallel passage to Theophrastus *Lap.* 49, Posidonius (fr. 235 *ap.* Strabo VII.5.8) uses the verb φθειρίαω to describe infested vines. In *Geoponica* V.30 φθεῖρες and κάμπαι are distinguished as two types of vine pest, the former being presumably the Tortricids and the latter *Hyles*.[75]

(iv) Σκώληξ. The *ips* or *sknips* is described by Lydus (loc. cit.) and Galen (loc. cit.) as a form of *skolex*, i.e., a 'worm' or larva. In parallel passages to Theophrastus *Lap.* 49, Dioscorides (*DMM.* V.160), Oribasius (XIII.G.2) and Paulus (VII.3) name the insects referred to simply as *skolekes*. The term appears alongside *kampai* in *Geoponica* V.48. In LXX *Deut.* 29.39 the Hebrew *tola'at*, referring here probably to *Lobesia botrana*,[76] is translated as σκώληξ.

---

[69] Cf. 26, 36–7, 58 below. *Ips* is used for the clothes-moth in Homer. Cf. 26.1 below.

[70] Fernandez, *Nombres*, pp. 115–17.

[71] The adjective presumably means 'wily' and not 'many-coloured' as Gossen, *Zool. Glossen*, p. 38 suggests.

[72] Possible emendations of this fragment are discussed by S. Daris, *Actes du XVᵉ Congrès International de Papyrologie* II, 1979, pp. 9–13. Daris argues, supported by W. G. Arnott (in a footnote to Daris's article), that ἴξhere refers to a species of bird, citing a papyrus of the fragment with an incomplete marginal gloss which appears to define the creature as an ὄρνεον; but he leaves out of account the similarity between ἴξ and the undoubted insect name ἴψ. Ammonius evidently regarded Alcman's ἴξ as an insect, since he groups it with ἴψ, κίς, and θρίψ, and the other lexicographical sources, using terms like θηρίδιον, are in keeping with this.

[73] There is no reason why the word should be rendered 'wood-worm' here, as it is in LSJ and in Hort's translation.

[74] For other uses of *sknips*, cf. 58 below.

[75] For other cases of *phtheir* used for plant pests, cf. 64 below.

[76] Bodenheimer, *Animal and Man*, p. 78.

(v) *Κάμπη*. The term 'caterpillar' is used in *Geoponica* V.30 and V.48.6 in distinction from φθεῖρες or σκώληκες. It implies a creature of reasonable size[77] and so is likely to distinguish the large larvae of *Hyles lineata* from the much smaller and less conspicuous Tortricids. The latinised form *campe* is used by Palladius (I.35.6).[78]

(vi) *Κέρκος*. Defined by Hesychius as θηρίδιον τὰς ἀμπέλους βλάπτον. Gossen[79] and Fernandez[80] take it as having the same meaning in Galen (*Inst.Log.* XII.9), where it appears as the name of an undefined pest.

(vii) *Άψοος*. According to Hesychius, θηρίον τι κατεσθίον ἀμπέλους.[81]

1.3. Identification: Latin terminology

(i) *Convolvolus* or *involvolus*. These terms correspond to the Greek *ips*, similar pest control methods involving bitumen being prescribed for both, and have the same primary reference to Tortricid larvae. They refer to the insects' habit of spinning together foliage with silk and pupating in rolled leaves. The form *convolvolus* is that used by Cato (*Ag.* 95) and Pliny (XVII.264), while *involvolus* appears in Plautus (*Cist.* 728-30), where it is characterised as a leaf-rolling pest (*nequam bestiam et damnificam . . . quae in pampini folio intorta implicat se*). The grammarian Festus (p. 100 L) has the spelling *involvus*, defined as *vermiculi genus qui se involvit pampino*. Isidorus (*Or.* XII.5.9), citing Plautus, classifies the creature as a form of *eruca* or caterpillar.

(ii) *Volucre*. This term is applied by Columella (*Arb.* 15), followed by Pliny (XVII.265), to a creature which gnaws at vine leaves and grapes (*praerodit teneros adhuc pampinos et uvas*). It would appear that it is simply a synonym of the preceding.[82] The word does not necessarily imply that the creature to which it refers possesses wings, since there are other places where it apparently means nothing more than 'insect'.[83]

2. Life history

There is no evidence that the adult stages of any of the lepidopterous larvae mentioned above were recognised in antiquity. The generally held belief was that, like other plant pests, the insects were spontaneously generated from the vine itself. Such a belief is either stated or implied by Theophrastus (*CP* III.22.5–6), who states that the process takes place under the influence of moisture, Dioscorides (loc. cit.), Galen (XII.186 K), and Cato (loc. cit.). Elsewhere Theophrastus (*HP* VIII.10.5) includes them among pests which do not reproduce or develop into anything else, but which die when they have exhausted their

---

[77] On *kampe* and *skolex*, cf. 23.1.2 above.

[78] Palladius is not necessarily being specific here, though he could be distinguishing *campae* from *animalia* in I.35.4.

[79] In PW 10.1487.

[80] Op. cit., p. 45. In Hesychius it also means 'field mouse'.

[81] Latte reads 'Αψός here. Fernandez, op. cit., p. 243, offers no etymology.

[82] It is so understood by Leitner, op. cit., p. 251, who suggests that it also comprehends coleopterous pests.

[83] Cf. 8.6.v, 10 above.

food supply. Other theories were that they were formed out of mist (Lydus *de Ost.* 14), or that they were generated from the rotting of the stakes used to support the vines (*Geoponica* V.53.6).

3. Importance as pests

Vine pests under their various names are generally characterised in Greek sources as attacking young buds or shoots at the beginning of the year (Alcman loc. cit.; Posidonius loc. cit.; Galen loc. cit.). Similarly Columella (loc. cit.) speaks of them as damaging the developing foliage and fruit. How seriously their depredations were regarded is indicated by Strabo's statement (XIII.1.64) that the Erythraeans of Mimas honour Herakles under the special title *Ipiktonos* as the deity protecting their territory against this pest.

The most commonly mentioned countermeasure in Greek sources is the application to the vine at the beginning of the year, before the insects make their appearance, of a special form of bituminous earth known as *ampelitis* (Theophrastus *Lap.* 49; Posidonius loc. cit.; Dioscorides loc. cit.; Galen loc. cit.; Oribasius loc. cit.; Paulus loc. cit.). Similar to this is the recommendation by Latin writers (Cato loc. cit.; Pliny XVII.264) that a mixture of bitumen, sulphur, and *amurca* be boiled up and applied around the vine stock and under the branches. This mixture could also be used to fumigate the vines (Pliny loc. cit.; Palladius I.35.6). Other substances recommended for application to vines are bear fat (*Geoponica* V.30.1), and fig juice with oil (*Geoponica* V.48.5), while the following are listed (*Geoponica* V.48.1–4) as alternative fumigants: dung, stag's horn, ivory dust, lily root, all-heal, peony, burdock, and women's hair. It is suggested that the stakes supporting the vines be treated with smoke (*Geoponica* V.53.6), that cow dung be applied to the roots when planting (*Geoponica* V.9.1), that burnt vine twigs plus wine and vine sap be placed in the middle of the vineyard (*Geoponica* V.30.4), that peony or burdock be planted alongside the vines (*Geoponica* V.48.4), or that water in which specimens of the pests had been boiled be sprinkled over the plants (Palladius I.35.6). It was also believed that protection against pests could be gained by rubbing the pruning knife used on the vines with a beaver pelt (Columella *Arb.* 15; Pliny XVII.265), or ground garlic (Palladius I.35.6; *Geoponica* V.30.1, 48.6), or bear blood (Columella *Arb.* 15; Pliny XVII.265) or fat (*Geoponica* V.30.1), or goat fat (*Geoponica* V.30.3), or frog blood (*Geoponica* V.30.3), or *kantharides*[84] found in roses (Palladius I.35.4; *Geoponica* V.30.2).

4. Pests of the olive

Associated with Theophrastus' account of vine pests is his description of similar insects affecting the olive. Firstly, he distinguishes two varieties of caterpillar observed to affect olives at Miletus around flowering time (*HP* IV.14.9, *CP* V.10.3), one of which destroys the flowers, and the other the leaves. Their development may be impeded or assisted by the weather conditions prevailing at the time. These *kampai* are also referred to in *CP* III.22.6, where they are said to be generated under the influence of moisture.

---

[84] Cf. 33 below.

Secondly, we have certain *skolekes* which attack the fruit (*HP* IV.14.10, *CP* V.10.1) as it develops. These may be generated beneath the skin and devour the flesh, but alternatively they may confine their attentions to the stone, in which case the fruit remains usable. These, again, are affected by weather conditions, and will not appear if rain follows the rising of Arcturus. *Skolekes* may also be produced in the ripe olives later in the year, and these have a more serious impact upon the final yield of oil. Columella refers to such larvae as *vermiculi* in *RR* XII.52.21, saying that olives which drop as a result of their activities may still be used after immersion in hot water to draw out any ill taste.

Thirdly, Theophrastus describes (*HP* IV.14.10, *CP* V.10.2) a 'disease' known as ἀράχνιον or 'spider's web', which grows over the fruit under the influence of moisture and may destroy it. All these details are reproduced by Pliny (XVII.229–30).

Various suggestions have been made as to the identity of these pests. Clearest of all are the *skolekes* feeding upon the flesh of the developing, and later on of the ripe, fruit, these being the first and subsequent generation larvae of the olive fly, *Dacus oleae* (Gmel.).[85] Early larvae cause the fruit to drop, but those appearing later in the year simply result in a yield of poor quality oil.[86] The caterpillars feeding on the flowers are probably those of the moth *Prays oleae* (Bem.), whose first generation larvae spin webs among the flowers and young fruit.[87] Its second generation larvae bore into the fruit to attack the stone, and so may well be identifiable with those of Theophrastus' *skolekes* described as feeding in this way. The webs of *Prays* could perhaps be the *arachnion* of Theophrastus, the latter being identified by LSJ as spun by larvae of *Malacosoma neustria*, and by Bodenheimer[88] as the product of actual spiders.

26. Ses/Tinea

1. Identification: general

The Greek σής is one of a small group of chiefly monosyllabic insect names[89] whose basic meaning seems to be that of 'small insect pest', i.e., a creature too small to be readily identified with any group of larger invertebrates.[90] To a certain extent, therefore, they are interchangeable, but there is a tendency for each of them to become particularly attached to the pests of some particular plant, food product, or man-made material.

Thus *ses* is most commonly applied to those pests devouring clothing and other fabrics, especially woollens, which we know as the larvae of clothes-moths,[91] and secondarily to

---

[85] So identified by Gossen, in PW 2A,580; Bodenheimer, *Animal and Man*, p. 78.

[86] Stapley and Gayner, op. cit., p. 202; Fröhlich and Rodewald, *Pests*, p. 192.

[87] So identified by Bodenheimer, op. cit., p. 78. The problem here is that the larvae would scarcely seem large enough to qualify as *kampai*; cf. 25.1.3.v above. Bodenheimer identifies the foliage feeding κάμπαι as larvae of *Palpita unionalis* (Hubn.). Cf. Stapley and Gayner, op. cit., p. 201.

[88] *Geschichte*, p. 72.

[89] Cf. 25.1.2.i above; 36–7, 58 below.

[90] In the absence of any instruments of magnification, the physical structure of such creatures could not be examined, and it tended to be imagined that their structure was more or less undifferentiated. Tertullian (*de An.* 10) is able to assert that clothes-moths feed without possessing any mouthparts.

[91] The ancients did not recognise them as related to the larger Lepidoptera.

those attacking books and papers, which we know as booklice (e.g. in Lucian *Adv.Ind.* I; Strabo XIII.1.54; Alciphron *Ep.* II.52; *AP* XI.78,322,347).[92]

The clothes-moths are a small group of closely related Lepidoptera whose larvae feed upon a variety of organic materials such as wool, hair, fur, feathers and horn. The most important pest species, whose major populations are domestic, are *Tineola biselliella* (Humm.), *Tinea pellionella* (Linn.), *T. pallescentella* Staint., and *Trichophaga tapetzella* (Linn.).[93] In addition to its normal Greek designation, there are examples of clothes-moths being termed ἴψ (*Od.* XXI.395; hence Ammonius *Diff.* 244)[94] or θρίψ (Schol. Aristophanes *Ach.* 1111).[95]

Booklice are minute omnivorous insects belonging to the small order Psocoptera. Domestic species such as *Trogium pulsatorium* (Linn.) live among books and papers, feeding upon fragments of animal and vegetable matter, for example moulds growing on the paper, or the paste of book bindings.[96] There is one example in Greek of their being given the alternative name of κόρις (Alciphron loc. cit.).[97]

In a similar way, the Latin equivalent *tinea* (or *tinia*), which may be etymologically related to σής,[98] has a primary reference to clothes-moths and to booklice (for example Horace *Ep.* I.20.11 ff.; Martial XI.1.14, XIV.37, VI.61; Ovid *ex Pont.* I.1.72; Vitruvius II.9.13, VI.4.1; Pliny XIII.86; Ausonius *Epist.* IV, *Epig.* I.1, *Com.Prof.* XXII.3–4), but may also be used for other small insect pests.[99] There are also examples of clothes-moths being termed *teredines* (Pliny XV.33, VIII.197)[100] and booklice *vermes* (Ausonius *Epig.* I.13–14).[101]

1.2. Identification: other names

(i) *Βρωτήρ* and *βρωστήρ*. Synonym of *tinea* and *ses* in *C.Gl.L.*III.361.2, 503.29, with reference to the creatures' voracity.[102] Both spellings are found as replacements for LXX σής in Aquila *Is.* 50.9 and *Hos.* 5.12.

---

[92] *Ses* is also used to refer to wood-borers (37.1.2 below) and the larva of the wax-moth (24.2 above).

[93] One or other of these species are cited by Sundevall, *Thierarten*, p. 205; Aubert and Wimmer, op. cit., I p. 170; Keller, op. cit. II, p. 442; Gossen, in PW 2A.584–5.

[94] Cf. 25.1.2.i above. The v.l. κεραΐς is included by LSJ.

[95] Cf. 37.1.2 below.

[96] So identified by Gossen, in PW XII.1083, and Leitner, *Zool. Terminologie*, p. 238. Cf. Imms, *Gen. Textbook*, pp. 646 ff. Sundevall, op. cit., pp. 230–1, prefers to identify them as the silverfish *Lepisma saccharina* Linn., which also damages papers, but this is a relatively large and distinctive insect. It is not impossible, however, that this well known domestic insect was comprehended under the names *ses* and *tinea*, since no other term for it has come down to us.

[97] Cf. 17.1.2 above.

[98] Fernandez, *Nombres*, pp. 119–20.

[99] *Tinea* is also used for wood borers (37.1.3 below), corn pests (36.1 below), pests of stored meat (48.5 below), other lepidopterous larvae (23.1.2, 24.2 above) and, in later Latin, where nomenclature becomes increasingly vague, for lice (22.1.3 above) and intestinal worms (for example Chiron 224–5).

[100] Cf. 37.1.3 below.

[101] Cf. for example *HA* 557b, Isidorus XII.5.11, where the clothes-moth is viewed as a type of larva or 'worm', cited under 26.2 above.

[102] Fernandez, op. cit., p. 105.

(ii) *Τριχόβρως*. Found in literature only in Aristophanes *Ach.* 1111, as the name of a creature devouring the feather plumes of a helmet. The lexicographical references are all associated with this passage. The scholiast ad loc. and Suidas define it as a synonym of *ses*, also called *thrips*, a *skolex* eating hair or fur. Hesychius and Pollux (II.24) give similar details. Such a diet would be appropriate to clothes-moths.[103]

(iii) *Τριχοτρώκτης*. A name with the same meaning as the above,[104] found only in Hesychius, where it is defined as 'small creatures devouring hair'.

(iv) *Δερμηστής* or *δερμιστής*. The name clearly means, as Fernandez[105] points out, an insect which devours skins or hides. This would fit some of the Tineinae, as well as, and perhaps more especially, certain beetles of the family Dermestidae, for example, *Dermestes maculatus* DeGeer and *D. lardarius* Linn. Though it is not found in any surviving work of literature, a number of examples of its usage have been collected by the lexicographer Harpocration, who discusses the word in some detail. He notes that the term was used by Sophocles in his *Niobe* (fr. 449), by Lysias (fr. 104 S), and by the historian Aristides of Miletus (*FHG* IV.326), and that there was some uncertainty as to the kind of creature being referred to, Didymus declaring it to be a *skolex* and Aristarchus a snake. Hesychius, Suidas and *EM* all give abbreviated versions of Harpocration's discussion, the former synonymising the word with *ses*.

(v) *Σακοδερμηστής*. This name, recorded only by Hesychius and Suidas, is clearly[106] a compound of the preceding and *σάκος* 'shield'. Presumably therefore it was intended to refer to an insect attacking the leather of shields,[107] the same range of species being involved as with the preceding. According to the lexicographers, it was used by Sophocles in his *Troilus* (fr. 635), and there was a similar disagreement among the experts—presumably Didymus and Aristarchus again, although they are not named—as to its significance in this passage, some interpreting it as 'the *skolex* that eats skins' and others as 'a snake with a shield'.[108]

(vi) *Σάραξ*. Given as a synonym of *tinea* in *C.Gl.L.* II.429.50.[109]

2. Life history

It appears that no accurate knowledge of the clothes-moth's life history was available in antiquity, and there is no indication that the adult stages were recognised. The general belief, as with all insect pests, was that they were spontaneously generated from the

---

[103] Gossen, *Zool. Glossen*, p. 117, is clearly in error in regarding them as lice.

[104] Fernandez, op. cit., p. 105.

[105] Ibid., pp. 106–7.

[106] Ibid.

[107] As LSJ understand it.

[108] Gossen, *Zool. Glossen*, p. 99, has misunderstood the data.

[109] According to G. P. Shipp, *Modern Greek Evidence for the Ancient Greek Vocabulary*, p. 492, the word has survived as a name for wood-boring pests. I.e., one of the other meanings of *tinea* (cf. note 99 above) could be intended here.

material upon which they feed. Aristotle (*HA* 557b 1 ff.) describes the *ses* as a minute creature generated in woollens, a type of *skolex*, which is produced more abundantly if the fabric is dusty, or if a spider happens to have been enclosed with it, since the latter will take up any moisture in the fabric and dry it out. Aubert and Wimmer[110] tentatively suggest the emendation ἄρα χνοῦς for the very curious ἀράχνης, but Pliny certainly had the latter in his text.[111] The creature is briefly referred to again in *GA* 758b 22 ff.

Aristotle's account is paraphrased by Pliny (XI.117) as follows: *Idem pulvis in lanis et veste tineas creat, praecipue si araneus una includatur; sitiens enim et omnen umorem absorbens ariditatem ampliat.* A similar view of the insects' development is given by Gregory Magnus (*Mor.* V.38) and Isidorus (*Or.* XII.5.11), who defines the *tinea* as a *vestimentorum vermis.*

The booklouse is referred to in the same section of the *Historia Animalium* (557b 9–10) as the *ses*, though without being named, and is credited with the same mode of development, when Aristotle states that 'other small creatures are generated in books, some of them similar to those generated in clothes'.[112] Similarly Pliny (XI.117), after describing the origin of the clothes-moth, adds *hoc et in chartis noscitur.*

3. Popular beliefs and attitudes

The *ses* or *tinea* appears on a number of occasions in literature (Pindar fr. 222; Menander fr. 540) as a symbol of decay, associated with other destructive agents such as rust (cf. NT *Matt.* 6.19–20, *Lk.* 12.33). Σής (Vulgate *tinea*) is found frequently in the LXX in similar contexts, sometimes (*Job* 13.28, *Prov.* 25.20, *Sir* 42.13, *Is.* 33.1, 51.8) explicitly linked with the damaging of clothes, sometimes (*Job* 32.22, *Prov.* 14.30) as a generalised agent of decay.[113] Ovid (*ex Pont.* I.1.67–74) uses the image of the *tinea* gradually gnawing away at books to describe his state of mind while in exile, and similar imagery is used by Philo (*de Caini Post.* 56, *de Somn.* I.77) and Gregory (*Mor.* XI.48).

According to Pliny (XXVIII.33), it was believed that fabrics used at funerals were not subject to attack from clothes-moths.

4. Importance as pests

The clothes-moth figures in classical literature mainly as the characteristic pest of clothing and other fabrics (Menander fr. 540; Lucian *Sat.* 21; Horace *Sat.* II.3.117 ff.; Martial II.46), with woollens being singled out as particularly vulnerable (e.g. Aristotle loc. cit.; Aristophanes *Lys.* 729–30; Petronius 78). Gregory (*Mor.* XI.48) characterises them as *sine sonitu perforans vestimentum.* Under alternative names they appear also as devourers of feathers, furs, etc. (1.2.ii–v above), and under the name ἴψ[114] as the creatures which

[110] Op. cit., I p. 538.
[111] Aristotle may perhaps have been under the impression that silken galleries and cocoons produced by clothes-moth larvae were the work of spiders.
[112] Cf. 6 above.
[113] It normally translates Hebrew '*ash* (*Job* 4.19, 13.38, 27.20) or *sas* (*Is.* 51.8), whose meaning is identical, but its appearance on other occasions (*Job* 32.22, *Prov.* 14.30, 25.20, *Is.* 33.1, *Mic.* 7.4) is due to mistranslation or expansion of the text.
[114] Cf. note 94 above.

Odysseus fears may have been gnawing away at the horn of his bow (*Od.* XXI.393–5).

In order to deter clothes-moths, it is recommended by Cato (*Ag.* 98) that the chest in which fabrics are stored should be painted with boiled down *amurca*. It is also suggested that certain plants with deterrent properties be placed among the stored clothes, these being wormwood (Pliny XXVII.52), anise (Pliny XX.195), *Helichrysum* spp. (Pliny XXI.169), wild lettuce (Theophrastus *HP* IX.11.11), and the fruit of the citron (Theophrastus *HP* IV.4.2; Pliny XII.15). According to Pliny (XXIX.101), a snake's shed skin was held to be similarly effective.

Similarly, the booklouse appears as the characteristic destructive agent[115] of books and papers (cf. the passages cited under 1 above; Juvenal VII.26; Statius *Silv.* IV.9.10; Symphosius in *Anth.Lat.* 440.16). The idea of unwanted or neglected writings suffering their depredations is common in poetry. Against these, the recommended precaution was to treat books with cedar oil (Pliny XIII.86; Ausonius *Epig.* I.13–14).

27. Bombyx

1. Identification: general

The identity of the silkmoths of antiquity has been a subject of considerable debate, but it is generally recognised that we are dealing here with two distinct species groups. In the first place, we have either one or two species native to Europe, first referred to by Aristotle as being reared on the island of Cos; and, in the second place, the Chinese silkmoth, unknown in the classical world until the latter part of the second century BC, and even then only by its product, plus an indefinite number of other exotic species, whose silk was imported from the East along with that of the Chinese moth. To the ancients, however, all silkmoths, whatever their native country, were of one kind, with the result that in later literature details appertaining originally to the native species come to be applied indiscriminately to their eastern relatives.

Two identifications have been offered for the native Coan variety, once it had been realised that the Chinese silkmoth was unknown to authors as early as Aristotle.[116] The most generally accepted is *Pachypasa otus* (Drury), the largest European member of the Lasiocampidae.[117] The larva of this moth is impressive in size and appearance, with lateral warts on all its segments and bright reddish protuberances on the second and third, and its foodplants include cypress, oak, and juniper.[118] It would therefore fit both Aristotle's physical description and Pliny's list of foodplants. The silk which it spins for its cocoon is of reasonable quality, and is known to have been employed in more recent times.[119]

---

[115] Cf. note 96 above. As Sundevall points out, there are other insects besides Psocoptera which may cause damage to papers and books, for example Anobiid beetles boring through them. Popular parlance would lump all such pests together, naming them with reference to the damage caused, rather than to their physical appearance.

[116] Sundevall, *Thierarten*, pp. 202–4, and Aubert and Wimmer, op. cit. I, p. 162, are unaware of this fact.

[117] So identified by Keller, *Ant. Tierwelt*, II pp. 443–5; Gossen, in PW 2A.575–6.

[118] Gossen, ibid.; J. P. Wild, *Textile Manufacture in the Northern Roman Provinces*, Cambridge, 1970, p. 11.

[119] Peck, *Hist. An.* II p. 177.

W. T. M. Forbes has suggested that a second species was also used, the emperor moth *Saturnia pyri* D. & S.[120] Its larva is large and spiny, and feeds upon various plants including fruit trees, ash, and blackthorn. It pupates in a cocoon of glossy silk, brown in colour, which is inferior in quality to that of *P. otus*.[121] Forbes argues that Pliny's list of foodplants is composite and covers both species, but all in fact are possible foodplants of *P. otus*. His supposed nomenclatural distinction between the two species is rightly rejected by Fernandez.[122] It is quite probable, therefore, that the Coan silk industry involved *P. otus* only.[123]

The Chinese silkmoth presents no problem of identification, but is the well known *Bombyx mori* Linn., the species used in commercial silk production today, which even in classical times had long been domesticated.[124] In addition to *B. mori*, there is archaeological evidence for the importation of silk from other species native to India or China.[125] These include *Antheraea mylitta* (Drury), *A. pernyi* Guer., and *Philosamia cynthia* (Drury). The silk produced from these is not of so high a quality as that of *B. mori*, but is not distinguished from it in surviving literary sources.

1.2 Identification: life history stages and alternative names

(i) Σκώληξ. Aristotle's term for the first stage of the silkmoth's metamorphosis, the young larva, rendered by Pliny as *vermiculus*. This usage is followed by Clement, Basil and Ambrose, who are dependent upon Aristotle. *Skolex* or *vermis* are commonly used in a more general sense by those sources (Tertullian, Procopius, Isidorus, *C.Gl.L.*) which ignore the distinct stages in the insect's development.

(ii) Κάμπη. In Aristotle's terminology, the 'caterpillar' (*uruca* in Pliny) is the second stage in the silkmoth's development, the mature larva. This usage is followed by Clement, Basil and Ambrose.

(iii) Βομβυλίς. Aristotle's term for the third stage, the silken cocoon containing the pupa. It is latinised to *bombylis* by Pliny. It appears in the form βομβύλιος in Clement and Basil, latinised to *bombylius* by Ambrose.[126] The etymology of this term and the related βόμβυξ is confused by the existence of a parallel group of names referring to hymenoptera. *Bombyx*, normally a silk-moth name, is defined as a form of wasp by Hesychius, while Pliny manages to confuse his Assyrian silkmoth with the Aristotelian hymenopteron βομβύκιον. *Bombylios* also appears as a hymenopteron name in Ps. Aristotle and elsewhere.[127] Fernandez[128] suggests that

---

[120] W. T. M. Forbes, 'The Silkworm of Aristotle', *CPh*, XXV, 1930, pp. 22–6.
[121] Ibid.   [122] *Nombres*, p. 203.
[123] It has also been argued, by R. Pfister, *Textiles de Palmyre* I, Paris, 1934, p. 55, that *P. otus* is too rare, but in fact it is quite widespread.
[124] Cf. for example Bodenheimer, *Geschichte*, pp. 11 ff.
[125] R. Pfister, op. cit., pp. 55–6; J. P. Wild, op. cit., pp. 11–13.
[126] Βομβύλιος appears as a v. l. in Aristotle, and in Athenaeus 352 f. However, βομβυλίς is attested by Aristophanes of Byzantium.
[127] Cf. 42 below.
[128] Op. cit., pp. 131–6. The same view is taken by Chantraine, *Dict. Ét.*, pp. 184–5.

the two groups of names are of entirely independent etymology, but that they have become morphologically indistinguishable through confusion due to their phonetic similarity. On this view, the hymenoptera names would be of onomatopoeic origin and refer to the insect's buzzing, while the silkmoth names could either derive from a root meaning 'to swell' or from a Coan dialect name of non-Greek origin.[129]

(iv) Νεκύδαλος. The identity of Aristotle's fourth stage is disputed,[130] but, unless we credit him with a major blunder in what is essentially a well informed account, it is reasonable to assume that he intends us to understand the adult moth here.[131] It has been suggested that the term derives from νέκυς and results from a similar association of ideas to that which gave rise to the butterfly name ψυχή, of which it could perhaps be a synonym rather than a specific term restricted to the silkmoth alone.[132] In Pliny's rendering of Aristotle, the term is latinised to necydallus. Clement appears to be under the mistaken impression that it is a synonym of the preceding, while Basil and Ambrose deviate from the Aristotelian account at this point and do not use the word at all. Hesychius' entry s.v. is exceedingly confused.[133]

(v) Βόμβυξ. Latinised to bombyx. This is the most frequently used term for the silkmoth in all its stages, commonly appearing in those sources which view the insect in vague terms as a 'worm' and overlook the details of its life history. It is not found in Aristotle, though it must have been in use in his day since he employs the derivative βομβύκια to refer to the insect's silk.[134] Pliny misunderstands Aristotle's account and makes the bombyx an additional fifth stage of the insect's life history (XI.76): in XI.75 and 77–8 it is his general term for silkmoths. Elsewhere bombyx is used both with reference to the Coan insect and to its Chinese namesake (Hesychius; Clement; Pollux VII.76; Tertullian adv.Marc. I.14, de Pall. III.6; Servius in G. II.121; Isidorus Or. XII.5.8, XIX.22.13, 27.5). It is the standard name for silkmoths in the Latin glossaries, defined as e.g. vermes qui texunt (V.348.22) or vermis qui sericum facit (IV.602.37).[135]

Bombyx is also used to refer, not to the insects themselves, but to the silk derived from them or to silken fabrics (Alciphron I.39; Propertius II.3.15; Martial VIII.33), though these are more commonly termed βομβύκινα (Latin bombycina) (for

---

[129] Fernandez suggests Carian. This does not imply, as R. J. Forbes, Studies in Ancient Technology, vol. IV, Leiden, 1956, p. 53, suggests, actual acquaintance with far eastern silk at this date.

[130] It is identified as the cocoon by O. Immisch, Sprachliches zum Seelenschmetterling, pp. 203 ff., Fernandez, op. cit., pp. 202–3, and Chantraine, op. cit., p. 742.

[131] This is the view of Keller, op. cit., pp. 443 ff., 608, Gossen, art. cit., 575, and W. T. M. Forbes, art. cit.

[132] Fernandez, op. cit., p. 202; W. T. M. Forbes, art. cit.

[133] The confusion probably results simply from a presentation of the Aristotelian data in an abbreviated form. Aristophanes' epitome is similarly confused.

[134] Or its cocoon, as in LSJ.

[135] It is variously spelt bumbix, bambis, bombites, and bombycini. II.570.21 results from confusion with similarly named hymenoptera, while II.31.11 is an inexplicable error. A number of glosses (for example IV.313.12, V.170.42) give the definition aranea, since the silkmoth was sometimes compared to a spider.

example Hesychius s.v. *Bombyx*; Martial VIII.68; Pliny XI.76; Juvenal VI.260; Isidorus *Or.* XIX.22.13).

(vi) *Papilio.* A term which ought strictly to apply only to the adult stage of the silkmoth, but which is used by Pliny (XI.77–8) throughout his non-Aristotelian account of the Coan moth's development.

(vii) Σήρ. Latin *ser.* In the plural this is normally the term applied to the Chinese as a race, or occasionally to silk fabrics (Clement; Julian *ep.* 80), but Pausanias (VI.26.6) applies it to the silkmoth itself, the same usage being found in *C.Gl.L.* V.390.23. The name is borrowed from the Chinese, and applied originally to silk as a material rather than to the people involved in producing it.[136] Silk thread or silk fabrics are commonly termed σηρικά, σηρικὸν νῆμα, *serica, serica vestis*, etc. (e.g. Strabo XV.1.20; Lucian *Salt.* 63; Galen X.942; Propertius I.14.22, IV.8.23; Pliny XXI.11; Isidorus *Or.* XIX.22.13, 27.5).[137]

2. *Life history*

The silkmoth is notable for the unusual number of independent and contradictory versions of its life history which are recorded in antiquity. The major versions originally applied specifically to either the Coan or the Chinese variety, but, since the two were regarded as identical in form, there is a tendency for details to be interchanged.

(i) *The Aristotelian version.* According to Aristotle (*HA* 551b 10 ff.), the silkmoth of Cos begins its existence as a large *skolex* possessing what resemble horns.[138] This changes first of all into a *kampe*, and this in turn into the *bombylis*, out of which comes the *nekydalos*. The insect passes through all these stages in six months.[139] Aristophanes of Byzantium, summarising this account, adds cryptically that the original *skolekes* are 'hidden underground', but neither state explicitly how these themselves come into existence, though later writers describe them as spontaneously generated. Pliny (XI.76) reproduces these details but through some misunderstanding adds an extra stage to the metamorphosis (. . . *ex ea necydallus, ex hoc in sex mensibus bombyx*). Clement (*Paed.* II.10) appears to distinguish Ἰνδικοί σῆρες from τοὺς περιέργους βόμβυκας and gives a version of the Aristotelian life history in connection with the latter. He describes the *kampe* as furry and stops at the *bombylis* stage, of which he seems to regard the *nekydalos* as a synonym. Basil (*Hex.* MPG 29.184) and Ambrose (*Hex.* V.77, MPL 14.252) apply the Aristotelian life history to the Chinese silkmoth, described by a common geographical imprecision as 'Indian', but deviate from it after the third stage. According to them, the *bombylios* changes its form by

---

[136] Fernandez, op. cit., p. 237.

[137] *Lanarius*, found only in Polemius Silvius' list of insect names, may also be a term for the silkmoth, corresponding to the mediæval Latin *lanificus*, used for example by Albertus Magnus.

[138] Presumably a reference to the protuberances of the *P. otus* larva.

[139] The larval stage of *P. otus* lasts from July to May. If the ancients did not rear it in captivity, they may have reckoned its development from the time of its emergence from hibernation in spring.

developing 'feathers' described as 'broad loose leaves' (χαύνοις καὶ πλατέσι πετάλοις ὑποπτεροῦται: laxis et latioribus foliis videtur pennas assumere). This very curious statement appears to be the result of an attempt to harmonise the data given by Aristotle with the popular belief accepted by Strabo, Virgil and others that Chinese silk was obtained from a plant. Ambrose makes the connection explicit by using the Virgilian terminology to describe the obtaining of the silk from the insects (foliis . . . pennas assumere. Ex his foliis mollia illa Seres depectunt vellera).

(ii) *Pliny's independent version.* Pliny (XI.77) follows his account drawn from Aristotle with a second, extremely confused, version of events which is also referred by him to the Coan silkmoth. According to this, small *papiliones* are spontaneously generated under the influence of vapour exhaling from the ground, out of blossoms of oak, ash, cypress and terebinth (foodplants of *P. otus*) which have been stripped off by rain. These develop fur (*villis inhorrescere*) in order to protect themselves against the cold, and furthermore make for themselves *tunicas . . . densas* by scraping off the down from leaves with their feet. This down is compressed into *vellera*, carded by means of the insects' claws, *mox trahi in tramas* (mss *inter ramos*), *tenuari ceu pectini*,[140] and finally wrapped around the insect to form a nest (*nido volubili*). It is at this stage, the account continues, that they are collected and placed in earthenware vessels, where they are kept warm and fed upon bran, and where they develop a feathery down (*sui generis plumas*), *quibus vestitos ad alia pensa dimitti*.[141]

(iii) *Pausanias' version.* The earliest account of the development of the Chinese silkmoth is provided by Pausanias (VI.26.6–9) and is almost as confused as the preceding. Here the insect is described as resembling a spider, eight-footed and twice as large as the largest κάνθαρος,[142] which produces its silk in the form of a fine thread which it winds around itself with its feet (this presumably refers to the spinning of the cocoon and corresponds with Pliny's *nidus volubilis*). The Chinese, we are told, rear these creatures in specially made houses and feed them for four years upon millet.

In the fifth year, however, their diet is changed to fresh reeds, upon which they gorge themselves to such an extent that they burst open and die, whereupon the greater part of their thread is discovered inside them. It has commonly been supposed[143] that Pausanias obtained his information from Marcus Aurelius' embassy which arrived in China in AD 166, although if this is so the Chinese must have been deliberately misinforming their guests.[144]

(iv) *Procopius' version.* More accurate information on the Chinese silkmoth did not become available until the insect itself was first imported in the time of Justinian.

---

[140] The reading *inter ramos* could well be right here.

[141] This phrase is most unclear. Rackham translates '. . . they are sent out to other tasks'. Perhaps Pliny has combined two accounts here, one describing the insect's development in general and ending with the construction of the cocoon (*nido volubili*), and another referring to its collection and rearing in captivity.

[142] Cf. 31 below.

[143] J. G. Frazer, *Pausanias' Description of Greece*, vol. 4, London, 1913, pp. 110–12; Z. Kadar, 'Serica', *Acta Classica Universitatis Debreceniensis*, III, 1967, p. 96.

[144] Not, of course, impossible in view of the desire of the Chinese to retain a monopoly in silk production.

Procopius (VIII.17) describes how the monks who offered to obtain living stock for the emperor informed him that silk is produced by certain *skolekes* which each lay numerous eggs (the adult stage is ignored). These eggs are erroneously said to be buried in dung and warmed until they hatch. Once the monks had returned from their successful expedition, we are told how the larvae hatching from the eggs they brought were successfully reared on their characteristic foodplant mulberry.

In his parallel version Theophanes (*ap.* Photius *Bibl.* 64) describes how at the beginning of spring the newly hatched larvae were given mulberry to eat, and how reared on this they developed wings (here the pupal stage is ignored) and produced eggs of their own.

3. *Production of silk*

There was considerable confusion in antiquity as to exactly how, and for what purpose, silkmoth larvae produce their thread. Aristotle makes no explicit statement on the matter, and Pliny, reproducing his account, erroneously writes that they spin webs in the manner of spiders. This latter theory is found explicitly in Servius (loc. cit.), Tertullian (*de Pall.* III.6 *per aerem aliquando araneorum horoscopis idonius distendit*), and Clement (loc. cit.), and is implied by Heliodorus (*Aeth.* X.25 τῶν παρ' αὐτοῖς ἀραχνίων νήματα). Tertullian adds that having spun their web the insects proceed to devour it, and that it may be unwound from within them if they are killed: this statement has affinities with Pausanias' account (2.iii above). Isidorus (*Or.* XII.5.8, XIX.22.13) simply says that they *ex se fila generat*. The only writers actually to describe the construction of the cocoon in which the larvae pupates, the true purpose of silk production, albeit in a somewhat garbled fashion, are Pausanias (loc. cit.) and Pliny (XI.77).

A completely different origin was sometimes ascribed to Chinese silk, since it was believed initially that it was produced from a species of tree in a similar way to cotton. In the earliest statement on the subject, Nearchus (*ap.* Strabo XV.1.20) after referring to cotton added that silk was derived in a similar way ἐκ τινων φλοίων, although the more usual version was that it was a form of down taken from the leaves. Virgil (*G.* II.121) alludes to this belief when he characterises the *Seres* as those who *vellera . . . foliis depectant tenuia*, and the idea continues to be reproduced long after the animal origin of the fibre has been pointed out by for example Pollux (VII.76) and Pausanias (Seneca *Hipp.* 389, *Herc.Oet.* 666–7; Solinus 50.2; Ammianus Marcellinus XXIII.6.67; Pliny VI.54, XXXVII.204; Dionysius Periegetes *Orb.Descr.* 752 ff. and Eustathius *Comm.* ad. loc.; Claudian I.179–80; Ausonius *Tech.* X.24; Martianus Capella VI.693; Isidorus *Or.* XIV.3.29).

Basil and Ambrose (2.i above) seem to be attempting to harmonise these disparate versions when they describe the insects as sprouting *folia* from which the *Seres* somehow obtain their silk. Similarly Isidorus, who also includes the two separate versions in their original form, has a third conflated version (*Or.* XIX.27.5) in which he states that in the *Seres'* country *vermiculi . . . nasci perhibentur, a quibus haec circum arbores fila ducuntur.*[145]

---

[145] A similar conflation could perhaps be embodied in Pliny XI.77, with its talk of *folia* and *vellera.* This passage of Pliny also has interesting similarities with that in Pausanias, for example the references to a cereal diet.

4. Silk manufacture

According to Aristotle, silk manufacture in the West was invented by an inhabitant of Cos named Pamphile at an unspecified date. He says very little about how the industry was conducted on the island, giving no indication as to how the silkmoths were actually reared. It has been suggested[146] that *P. otus* is not amenable to domestication, and that therefore it must have been reared in the open on suitable trees near the owners' houses. However, although such larvae could not be handled so easily as the long domesticated and passive larvae of *B. mori*, it is quite possible to rear any lepidoptera in captivity, if attention is paid to supplying the correct conditions.[147]

Having related the insect's life history, Aristotle describes how the women of Cos unwind the cocoons and reel off the thread, out of which they weave silk fabrics. This implies the traditional method whereby the cocoons are dropped into boiling water to kill the pupae inside and to release the fibres from the gum which fastens them together; the silk may then be reeled off as a continuous thread.[148] This immersion of the cocoons in water probably underlies Pliny's words at the end of his very confused independent account of the Coan moth (2.ii above), where he speaks of the 'down' produced by the insects being moistened before being drawn into threads by a rush spindle (*quae carpta sint lanicia umore lentescere, mox in fila tenuari iunceo fuso*).

There is no evidence as to how long before the time of Aristotle the Coan silk industry was in operation. Attempts have been made to identify literary references to silk from the preceding period, but none of these is indisputable. Richter[149] suggests that the appearance of diaphanous garments in Greek sculpture in the latter part of the fifth century BC is evidence for the use of silk, and that the fabric known as ἀμόργινον and referred to by for example Aristophanes (*Lys.* 753 ff.) is also to be so identified; but the latter at least seems rather to be a species of flax produced from the plant of the same name.[150] After the time of Aristotle, Coan or any other silk is conspicuous by its absence in surviving literature[151] until the early Imperial period, when there is a sudden outburst of references in Latin poetry to *Coa* or *Coae vestes* (Horace *Od.* IV.13.13, *Sat.* I.2.101; Ovid *AA* II.298; Propertius I.2.2, II.1.5, IV.2.23, 5.23,56; Tibullus II.3.53, 4.29; Persius V.135; Juvenal VIII.101). The fabrics so named are noted for their extreme lightness and fineness, and for their diaphanous quality. Juvenal's is the latest reference to Coan fabrics as a contemporary item of fashion, although details about silk manufacture on Cos go on being reproduced as late as the time of Isidorus (*Or.* XIV.6.18, XIX.22.13).

---

[146] J. P. Wild, op. cit., p. 11; S. M. Sherwin-White, *Ancient Cos*, Göttingen, 1978, p. 382.

[147] There would be a problem in rearing *P. otus*, in that the larvae hibernate, but this would not be insuperable. Pliny's account in XI.77 could be taken to imply that they were taken into captivity at some point between emergence from hibernation and pupation.

[148] J. P. Wild, op. cit., pp. 26–7.

[149] G. M. A. Richter, 'Silk in Greece', *American Journal of Archaeology*, XXXIII, 1929, pp. 27–33.

[150] LSJ; J. P. Wild, op. cit., p. 13; S. M. Sherwin-White, op. cit., pp. 379–80. Richter suggests they were using imported silk from the East, though not that of *B. mori*. Similarly the μηδικά of Herodotus and Xenophon have been identified as silk, but the only classical authors to make the connection are too late to be reliable on the point; cf. S. M. Sherwin-White, ibid.

[151] And in Coan inscriptions. Cf. S. M. Sherwin-White, ibid., p. 381.

If the Coan weavers were still using *P. otus* silk in the period under discussion, the fabrics so produced would have been clearly inferior in quality to Chinese silk. Indeed it has generally been supposed that it was competition with the latter which eventually caused the demise of the Coan industry after the time of Juvenal.[152] However, since the trade routes with China had been opened up by the end of the second century BC,[153] it is not impossible that the Coans were in fact using imported silk in this period,[154] which would correspond far better with the high praises ascribed to *Coae vestes*, or that *Coa* had become a general term applicable to silk in general. Propertius, for example, seems to use the terms *Coae vestes*, *Serica* and *Arabius bombyx* more or less indiscriminately.

The earliest reference to Chinese silk in surviving literature is that supplied by Strabo (loc. cit.), and if the sentence in question is derived from Nearchus[155] it would be the only one to predate the opening up of trade routes between China and the West. The first references to the use of Chinese silk fabrics are found in Propertius (I.14.22, II.3.15, IV.8.23). Silk was imported either by an overland route, or by sea from India up the Persian Gulf and the Red Sea,[156] as described in the *Periplus Erythraei Maris* (39,49,64). Such trade relations with China culminated in the sending of a Roman embassy by Marcus Aurelius in 166.[157]

The few details given by Pausanias and Procopius concerning the rearing of *B. mori* in its native country and in the West have been noted above. Those writers who are under the impression that Chinese silk is a plant product normally refer to a moistening of the unworked fibres prior to drawing out the threads. The method used for unravelling the cocoons would have been the same as that described above for the Coan silkmoth.[158]

Chinese silk was imported not only in the form of finished cloth but also in that of raw silk and silk thread, all three being specified in the *Periplus*. From the latter cloth was produced in a number of centres in the eastern empire, e.g., Palmyra, Alexandria, and Pergamum.[159] Thus Procopius (*Anec.* 25.14–15) speaks of silks as having been manufactured 'from ancient times' at Berytus and Sidon. Pliny's reference (XI.75,78) to an Assyrian *bombyx*[160] presumably refers to fabrics produced in this region, as does Propertius' *Arabius bombyx* (II.3.15) and Lucan's *Sidonium filum* (X.141). As has been mentioned,[161] some of these fabrics would have been composed of silk from other species besides *B. mori*, including some from India rather than China, but this fact appears not to have been recognised. Chinese silk is sometimes termed 'Indian', but this is merely geographical imprecision.[162] By the time of Justinian, therefore, imported silk had come to play an important part in the economy of the eastern empire;[163] hence the significance

---

[152] J. P. Wild, op. cit., p. 13; S. M. Sherwin-White, op. cit., p. 382.

[153] Ibid., p. 379.   [154] As Richter, art. cit., supposed they were doing in earlier times.

[155] The preceding sentence about cotton is ascribed to Nearchus.

[156] R. J. Forbes, op. cit., pp. 54–5; Z. Kadar, art. cit.   [157] Z. Kadar, ibid., pp. 95–7.

[158] J. P. Wild, op. cit., pp. 26–7.   [159] R. J. Forbes, op. cit., p. 54.   [160] Cf. 42 below.

[161] Cf. 27.1 above.

[162] However, Servius (*in* G.II.121) writes that *apud Aethiopiam, Indos et Seres sunt quidam in arboribus vermes et bombyces appellantur.*

[163] R. J. Forbes, op. cit., pp. 55–8. Cf. R. S. Lopez, 'Silk Industry in the Byzantine Empire', *Speculum*, XX, 1945, pp. 1 ff.

of the eventual introduction of *B. mori* to the West. At this period silk was entering the empire only by land, through Persia, the supply being uncertain and liable to interruption, particularly in times of war. Justinian had made efforts to find an alternative route bypassing Persia, but without success.[164] In 553/4, however, according to the account given by Procopius (*Hist.* VIII.17.1–8), some Indian monks who had lived in China offered to obtain for the emperor a supply of the silkmoth's eggs. This they were successful in achieving, and it thus became possible for the first time to have an entirely native silk industry.[165]

28. *Pityokampe*

The 'pine caterpillar', otherwise known as πιτυίνη κάμπη or *pinorum eruca*, is mentioned on a number of occasions by medical writers as an insect of medicinal value. It is always closely associated with the *kantharis* and *bouprestis*,[166] and is regarded as having the same properties, being useful as a caustic agent in the treatment of skin disorders (Dioscorides *DMM.* II.61; Galen XI.756, XII.564; Pliny XXIX.95): they are said to be prepared for use in the same way.

As with *kantharides*, medicinally prepared pine caterpillars are said to be dangerous if taken internally, and they are therefore classed not only as drugs but also as a potential poison (Galen XI.767; Marcian *Dig.* 48.8.3.3). The symptoms of πιτυοκάμπη poisoning are described by a number of sources (Ps. Dioscorides *Peri Del.* 2; Paulus Aegineta V.34; Aetius XIII.53). Antidotes are said to be the same as those for *kantharides* (Aetius loc. cit.), including fresh milk (Pliny XXVIII.128) and vine products (Pliny XXIII.62,82).

According to Paulus (V.1.2), food should not be cooked underneath trees, especially pines, since these abound in salamanders and θανάσιμοι κάμπαι which, when warmed by the fire, are liable to drop off onto the food below. Most probably the creature involved here is the larva of the pine processionary moth *Thaumetopoea pityocampa* (D. & S.). This larva is a serious pest of pines and possesses urticating hairs which cause inflammation in man.[167]

29. *Xylophoron*

Ξυλοφόρον or 'wood carrier'[168] is the name applied by Aristotle (*HA* 557b 13 ff.) to a form of σκωλήκιον whose body is enclosed in a cobweb-like tunic to which are attached dry twigs. These twigs give the impression of having adhered to the creature as it moves along, but are in fact an integral part of its protective coat. The head, which is mottled,

---

[164] R. J. Forbes, op. cit., pp. 56–7.
[165] According to Theophanes (loc. cit.), the eggs were smuggled out by a certain Persian inside a hollow cane.
[166] Cf. 33–4 below.
[167] Gossen, in PW 2A.575; Fernandez, *Nombres*, p. 143; Leitner, *Zool. Terminologie*, p. 203.
[168] The v.l. ξυλοφθόρον, accepted by LSJ, is clearly out of keeping with the insects' habits. Cf. Fernandez, *Nombres*, p. 87.

and feet project outside the tunic. The covering is just as much a part of the animal as is a snail's shell, so that if it is removed the animal dies. As is the case with *kampai*,[169] the insect eventually becomes immobile and turns into a *chrysallis*, but it has not been observed what winged stage emerges from it. An abbreviated version of this account is provided by Pliny (XI.117), who characterises the insect as a *tinearum genus tunicas suas trahentium*.[170]

Since Aristotle gives no indication that he is describing an aquatic insect, it is unlikely that, as some authors have suggested,[171] he is speaking of the larva of the caddis-fly (Trichoptera). If this possibility is ruled out, there can be no doubt that this insect should be identified with the moths of the family Psychidae.[172] The larvae of these moths, known as 'bag-worms', construct protective cases for themselves which they carry about as they move around their foodplants. These cases are composed of silk covered with fragments of leaves, grass, twigs, or other objects, and vary according to species. The larvae pupate inside the case, having attached it to a suitable surface.[173]

Insecta: Lepidoptera and Coleoptera

30. *Kerastes/Cossus* and *Karabos/Lucavus*

1. Identification: general

This section covers the larger insect larvae known as pests boring into the wood of living trees and into cut timber, together with their adult stages where these were recognised. Normally these are clearly distinguished from the very small wood boring insects commonly known as θρῖπες,[174] but within the group itself the nomenclature is vague, so that in many cases it is impossible to determine to which order of insects a particular larva belongs.

Most of these larvae would be either Lepidoptera or Coleoptera. As regards the former of these orders, the most significant species are the Goat moth *Cossus cossus* (Linn.) and the Leopard moth *Zeuzera pyrina* (Linn.). The larva of the Goat moth is distinctive among wood boring larvae in being reddish brown in colour, others being whitish. It bores large tunnels within the trunk of various trees, including fruit trees, causing severe damage and often death. The smaller whitish larva of *Z. pyrina* is also a pest of fruit trees, but tunnels in the branches rather than attacking the heartwood. Less important, and smaller still in size, are the larvae of certain species of clearwing moth (Sesiidae) which have orchard trees as their foodplants. There are a number of likely references to these

---

[169] Cf. 23.2 above.

[170] There are other examples of the use of *tinea* for Lepidoptera other than the clothes-moth. Cf. 23.1.2 above.

[171] A. Ernout and R. Pepin, *Pline l'Ancien Histoire Naturelle* XI, p. 154; P. Louis, 'La Génération Spontanée chez Aristote', *Revue de Synthèse*, 89, 1968, p. 296.

[172] The identification accepted by the majority of authors; Sundevall, *Thierarten*, pp. 206–7; Aubert and Wimmer, op. cit., I, p. 169; D'Arcy Thompson, *Hist. An.*, V.32 note; Gossen, in PW 2A.580–1; W. Capelle, 'Zur Entomologie des Aristoteles und Theophrast', *RhM*, CV, 1962; Leitner *Zool. Terminologie*, p. 238.

[173] Imms, *Gen. Textbook*, pp. 1108–9.

[174] 37 below.

insects among ancient authors, but there is no clear evidence that their adult stages were recognised.

Among the Coleoptera, two families are chiefly in view. In the first place, we have the stag beetles (Lucanidae) including the well known *Lucanus cervus* (Linn.), whose larvae, white with a dark head, feed upon the rotting wood of trees and their roots. The adults are distinguished by their impressive antler-like mandibles. Secondly, there is the much larger family of long-horn beetles (Cerambycidae), the adults of which are distinguished in most species by their long antennae, sometimes equalling or exceeding the length of the body. There are a number of large and impressive species, for example *Ergates faber* (Linn.) and *Cerambyx cerdo* Linn. The larvae of most species bore into the wood of dead or decaying trees, but others attack living trees, and some, notably *Hylotrupes bajulus* (Linn.), cut timber in buildings.[175] Since the larvae may take several years to complete their development, adults of non-domestic species may emerge from timber after it has been used in the construction of buildings or for furniture.[176] The life history of such beetles was clearly recognised by the more well informed of the ancients, descriptions being provided by Aristotle and Theophrastus. The adults of some species were also of interest in their own right, due to their distinctive appearance.

Mention may also be made of the wood boring larvae of the horntails or woodwasps (Siricidae) belonging to the Hymenoptera, although there are no recognisable references to these in ancient authors.

1.2. Identification: names applying chiefly to larval stages

(i) Σκώληξ (Latin *vermiculus*). The most general term for wood boring larvae. In *HP* V.4.5 Theophrastus, followed by Pliny in XVI.220, divides the pests of timber into three groups; namely, the purely marine τερηδών,[177] the θρίψ[178] and the σκώληξ. He and Pliny use the term throughout to refer to the kinds of larvae considered above, although there is one place[179] where he also uses it to describe what, according to his later definition, should more properly be termed *thripes*. The same usage is found in Aristotle (*HA* 551b 17 ff.); and in *HA* 614b 1 and *de Mir.* 831b 5 ff., where larvae dug from trees by woodpeckers are so named. A number of non-scientific authors use *skolex* in the same sense (e.g. Apollodorus *Bibl.* I.9.12; Aelian V.3). In LXX *Prov.* 12.4, 25.20 the *skolex* is the characteristic destructive agent of timber, as *ses* is of clothing.[180]

Among agricultural writers, there is some confusion between foliage feeding and wood boring larvae attacking trees, but we are informed that *skolex* (*vermiculus*) is the correct name for the latter.[181] Certain of these larvae attached to

---

[175] There is archaeological evidence of the cerambycid *Hesperophanes fasciculatus* as a domestic pest in Roman times: cf. *Britannia*, II, 1971, pp. 159, 162.

[176] Imms, *Gen. Textbook*, p. 894.

[177] The shipworm *Teredo* spp., a form of mollusc.

[178] 37 below.

[179] *HP* V.1.2.   [180] Cf. 26.3 above.   [181] Cf. 23.1.4.i above.

particular species of tree are described as distinctive in appearance. Those infesting the medlar are described (Theophrastus *HP* III.12.6; Pliny XVII.221) as large and distinct from those of other trees, while those associated with the service tree are characterised as red and hairy. Judging from the colour, the latter is presumably *C. cossus*, though there are no 'hairy' wood borers.

Another red *skolex* or *vermis* from trees unspecified is mentioned by medical writers as a remedy for ear ailments (Galen XII.649 K; Pliny XXIX.135; Marcellus IX.84). This again is likely to be *C. cossus*.

Jerome (*adv. Jovin.* MPL 23.308) describes certain *vermes albos et obesos qui nigello capite sunt*, known also as *vermes Pontici* or ξυλοφάγα, as being eaten in Phrygia and Pontus. This description would fit very well the larvae of *Lucanus* spp.[182]

(ii) *Εὐλή*. This term, normally applied to fly larvae,[183] is used by Aelian (XVI.14) with reference to larvae extracted from θριπηδέστων φυτῶν.[184] He is presumably referring to the digging out of larvae with a metal implement, as described by agricultural sources (cf. 3 below).

(iii) *Eruca*. Normally meaning 'caterpillar',[185] this term is used once by Pliny (X.206) to refer to larvae hollowing out a tree.

(iv) *Τερηδών* (Latin *teredo*). According to Pliny and Theophrastus (i above), this is properly the name of the marine shipworm, but even Pliny himself does not adhere to this rule, and elsewhere the word is frequently used with reference to terrestrial woodborers. The word itself means 'borer' or 'perforator', and is etymologically related to *tarmes*.[186] When referred to terrestrial insects, it is usually associated with pests of cut timber rather than with those of living trees. Pliny uses *teredo* on three occasions (XVI.65,189, XXIII.135) to designate timber pests in general, while in XI.4 he characterises the insect as boring through wood *cum sono teste*, a detail which Theophrastus attributes to his *kerastes*. According to Suidas and Photius (= Pausanias Gr. fr. 21), the *teredon* is a *skolex* found in the timbers of houses,[187] and a similar definition is given by Eustathius (*Comm.* 437.18 ff., 1403.37, 1532.15).[188]

A number of Latin glossary entries (IV.182.49 etc.) define *teredo* as *vermis in ligno*, and it is also found equated with *cossus* (V.186.1, 654.3) and *termes* (V.485.56, 516.36, 580.61). According to Isidorus (*Or.* XII.5.10), *teredo* is the Greek equivalent of the latter.[189]

---

[182] Keller, op. cit., II. p. 407.

[183] Cf. 48.2 below.

[184] The adjective is used in its general sense of 'worm-eaten'.

[185] Cf. 23.1.2 above.

[186] Fernandez, *Nombres*, pp. 114–15.

[187] They quote Aristophanes *Equ.* 1305, where the shipworm is meant. It is unclear whether the τερηδών of Polybius VI.10.3 is terrestrial or not (cf. *AP* XII.190, where θρύψ and τερηδών seem more or less synonyms).

[188] The reference to feeding on bones is an error resulting from the medical sense of the word.

[189] Columella (*RR* IV.24.5) names as *teredines* pests which hollow out vine stocks. Pests attacking roots are so named by Pliny (XXI.42): cf. Dioscorides *DMM*. I.1,127. For other entomological uses cf. 24.2, 26.1 above, 48.5 below. The word is also used for internally parasitic worms (for example *Hipp. Ber.* 41).

(v) *Tarmes, termes.*[190] Plautus (*Most.* 825) uses this term, which is etymologically related to *teredo* and θρύψ,[191] to refer to domestic wood borers. The word is explained by Servius (*in G.* I.256) and Isidorus (loc. cit.), who says that it is the native Latin equivalent of the preceding.[192] The same equation is made by *C.Gl.L.* V.485.56, 516.36, 580.61.[193]

(vi) *Cossus* (*cossis* in Pliny XIII.134, XVII.220). According to Pliny's account (XVII.220), the *cossus* is a large edible variety of *vermiculus* found in oaks, belonging to the same group as the *skolekes* of Theophrastus. It is referred to elsewhere as a devourer of wood or timber (XIII.134, XI.113), and as usable for medicinal purposes (XXX.115). The grammarian Festus (p. 36.11 Li) associates it with the Roman surname *Cossus*, explaining that the insect name was once applied to men of wrinkled appearance (*cossi ab antiquis dicebantur natura rugosi corporis homines, a similitudine vermium ligno editorum qui cossi apellantur*). In the Latin glossaries it is generally defined as *vermes in ligno*, two entries (V.186.1, 654.3) adding *quos vulgo teredones vocant*, while another (II.119.32) gives the Greek ξύλου σκώληξ.[194] Traditionally this insect has been identified as the larva of the Goat moth, as the latter's scientific name testifies.[195] Fabre,[196] however, argued with good reason that the ill smelling *C. cossus* larva could never have been regarded as suitable for human consumption, and that a far more probable candidate was one or other of the large cerambycids such as *Ergates faber* or *Cerambyx* spp. The larva of the stag beetle *Lucanus* has also been suggested,[197] but, as Leitner[198] points out, it is unnecessary to choose between the two since the ancients would not have distinguished creatures so similar. The wrinkled appearance of such larvae would explain the entry in Festus, while their association with decaying or felled trees would fit the customary linking of the *cossus* with timber as opposed to growing trees.

(vii) 'Ρόμοξ or ρόμος. Defined by Hesychius (cf. Arcadius Gramm.59.24) as σκώληξ ἐν ξύλοις, this is etymologically a simple 'worm' name[199] and thus presumably a synonym for the *skolex* of Theophrastus.

(viii) Δήξ. According to Tzetzes (*in* Hesiod *Op.* 418), 'a type of *skolex* generated within timber'.[200]

---

[190] *Tarmes* is the original form. Cf. Ernout and Meillet, *Dict. Ét.*, p. 677.

[191] Fernandez, op. cit., pp. 114–15.

[192] His comment *quos tempore inoportuno caesae arbores gignunt* is derived from Plautus.

[193] For other senses of *tarmes* cf. 48.5 below.

[194] *Cossus* or *costus* is also used for parasitic worms of cattle (for example Vegetius *Mul.* I.52; Isidorus *Or.* XII.5.12).

[195] F. S. Bodenheimer, *Insects as Human Food*, p. 42.

[196] J. H. Fabre, trans. A. T. de Mattos, *More Beetles*, London, 1922, pp. 174–87.

[197] Keller, op. cit., II p. 407.

[198] *Zool. Terminologie*, p. 102.

[199] Fernandez, op. cit., p. 24.

[200] Cf. Ibid., p. 108.

1.3. Identification: names applying chiefly to adult stages

(i) Κεράστης. Theophrastus, in his rather confused account of the life history of *skolekes* (*HP* IV.14.5, V.4.5) says that the *kerastes* is the adult stage of some, if not all, of those infesting both living trees, particularly figs, and cut timber. In the former passage he adds the interesting comment φθέγγονται δὲ οἷον τριγμόν. These details are reproduced by Pliny (XVI.220, XVII.221), who latinises the name to *cerastes*. It is clear that Theophrastus is speaking here of the metamorphosis of larvae into 'horned beetles', the 'horns' being presumably the long antennae of Cerambycidae[201] rather than the mandibles of the stag beetle. The former identification is supported by the reference to stridulation,[202] since many cerambycids are able to produce sound by friction of two parts of the thorax, or of hind legs against wing cases.[203] Those causing damage particularly to figs would be *Hesperophanes griseus* (Fab.) and *H. fasciculatus* Fald.[204]

(ii) Κάραβος (v. l. καράμβιος in *HA* 551b 17 and Aristophanes Byzantinus).[205] Described by Aristotle as a kind of beetle (*HA* 531b 25) with distinctive antennae (*HA* 532a 27) which develops from larvae found in dry wood (*HA* 551b 17 ff.), this is almost certainly to be identified as a cerambycid like the preceding.[206] Hesychius gives the definition 'the *skolekia* in dry timber'. The word is more commonly used to refer to the marine or freshwater crayfish, and the insect has perhaps acquired its name through being regarded as similar in appearance, the crayfish having long backward-pointing 'antennae'.[207]

(iii) Κεράμβυξ. This insect is the subject of a myth related by Nicander (fr. 38 *ap.* Antoninus Liberalis XXII.5) concerning a certain Kerambos who was transformed by the Nymphs into a ὑλοφάγος κεράμβυξ. Antoninus goes on to say that it resembles a large black κάνθαρος[208] whose head, together with its curved jaws or horns, is shaped like a tortoiseshell lyre, and that it also goes under the name of ξυλοφάγος βοῦς: it is used as a plaything by children, who remove the head and carry it around. The word also appears in Hesychius, defined as 'an animal like the *kantharos*'. It is related to κέρας, with reference to the insect's horns.[209]

On the basis of this very accurate description, there can be little doubt that Goossens[210] has correctly identified this insect as belonging to the Lucanidae, whose curved mandibles have exactly the form depicted here.

---

[201] So identified by Steier, *Aristoteles & Plinius*, p. 47; Leitner, op. cit., p. 77.

[202] Τρίζω is used of the stridulation of Orthoptera in *Mir. Ausc.* 844b 26.

[203] Imms, op. cit., p. 893.

[204] Bodenheimer, *Animal and Man*, p. 78, *Geschichte*, p. 72. Cf. Fröhlich and Rodewald, *Pests*, p. 74.

[205] The forms -αμβιος and -αβιος are simple variants of κάραβος rather than separate terms for the larvae as Aubert and Wimmer, *Thierkunde*, I pp. 165–6, suggest. Cf. Chantraine, *Dict. Ét.*, p. 496.

[206] So identified by Sundevall, *Thierarten*, pp. 196–7; Keller, op. cit., II p. 408; D'Arcy Thompson, *Hist. An.*, V.19 note; Gossen, in PW 10.1486.

[207] On the etymology of κάραβος, which is uncertain, cf. Fernandez op. cit., pp. 228–9.

[208] Cf. 31 below.   [209] Fernandez, op. cit., pp. 78–81.

[210] R. Goossens, 'Identification de l'Insecte appelé κεράβυξ', *AC*, XVII, 1948, pp. 263–7. The same view is taken by Keller, op. cit., II p. 408, and Gossen, art. cit., 1487.

(iv) *Κεράμβηλον*. In a very interesting entry Hesychius explains that this name is applied to 'a small creature which, when tied onto fig trees, drives away the *κνῖπες*[211] by the noise it makes; or, according to others, *κάνθαροι* with horns'. It is likely that both definitions in fact refer to the same insect. Although the word is closely related etymologically to the preceding,[212] it would seem that we are dealing here with one or other of the larger cerambycids, whose stridulation would be the 'noise' referred to.[213]

(v) *Lucavus*.[214] Pliny describes a type of beetle (XI.97, XXX.138), so named by Nigidius, which possesses long forked and toothed horns (*cornua praelonga bisulca dentatis forficibus in cacumine, cum libuit, ad morsum coeuntibus*): these horns are hung around the necks of infants as an amulet (cf. their use as a plaything in iii above). This is doubtless to be identified as one or other of the Lucanidae.[215]

2. Life history

The general popular belief in antiquity was that all wood boring larvae were spontaneously generated from the decay of the wood upon which they would feed. The *cossus* of Pliny (XI.113) and Marcellus' (IV.38) corresponding *vermis*, the *termites* of Isidorus (*Or.* XII.5.10), and the *vermes albi* of Jerome (1.2.i above) are all described as being produced in this way. Similarly Augustine refers on a number of occasions (MPL 32.1365, 34.288, 38.412, 42.236) to *vermes* generated from wood. The agency of decay is explicitly mentioned by Marcellus and Jerome.

This belief continued to be held even in cases where the corresponding adult stages were recognised. Thus Aristotle (*HA* 551b 17 ff.) describes how the long-horn beetle *karabos* emerges from larvae found in dry wood, after these have become immobile (i.e. pupated), but leaves us to understand that the larvae themselves are of spontaneous origin. That such connections between adults and larvae were recognised in popular as well as scientific circles is illustrated by Nicander's characterisation of his *kerambyx* (1.3.iii above) as 'wood feeding', since such a description is strictly only appropriate to its larvae.

Theophrastus displays an awareness of the true facts but is unwilling to entirely discard the popular theory, with the result that his account is confused and somewhat contradictory. He describes the *skolekes* both of living trees (*HP* IV.14.5, *CP* V.10.5; Pliny XVII.221) and of cut timber (*HP* V.4.5; Piny XVI.220) as falling into two groups. In the first place we have those which are spontaneously generated from the decay of the wood. These are produced under the influence of moisture, their development being promoted by subsequent dry weather (*CP* III.22.5). They do not reproduce themselves,

---

[211] Cf. 46 below.

[212] Fernandez, op. cit., pp. 78–81.

[213] So identified by Gossen, art. cit., 1487 and Davies and Kathirithamby, *Greek Insects*, p. 94; cf. Keller,op. cit., II p. 408.

[214] V. l. *lucanus*.

[215] Keller, op. cit., II p. 407; Gossen, art. cit., 1487. It is suggested by A. Thomas, 'Le Laterculus de Polemius Silvius', *Romania*, XXXV, 1906, p. 171, that *cervus* in Polemius Silvius' list of insect names is also a term for the stag beetle.

but perish once they have exhausted their food source (*HP* VIII.10.5). In the second place we have those which are the offspring of the *kerastes* or long-horn beetle, which is described as digging out 'a kind of mouse hole' by turning round and round.[216] These larvae, unlike the rest, do not simply perish but develop into adult *kerastai* (*HP* VIII.10.5). In one place Theophrastus states that, in figs at least, all *skolekes*, including those spontaneously generated, develop into long-horn beetles (*HP* IV.14.5 = *CP* V.10.5), but this is in contradiction of what is said in *HP* VIII.10.5.[217]

3. Importance as pests

Wood boring *skolekes* have a significant place in botanical and agricultural sources as pests of orchard and other trees. According to Theophrastus (*HP* IV.14.2; Pliny XVII.220), all trees are attacked by them to some extent, but they are less liable to develop in those containing bitter juice. Most liable to infestation are trees that have been weakened by mechanical injury or by drought (*CP* V.10.5). The species particularly mentioned as suffering from larvae are fig (*HP* IV.14.3–5, II.5.5; Pliny XVII.220; Palladius IV.10.29), apple (*HP* IV.14.2; *CP* III.22.5, V.10.5; Palladius III.25.15; *Geoponica* X.18), pear (*HP* IV.14.2; Pliny XVII.220; Palladius III.25.5), medlar (1.2.i above; Palladius IV.10.20), service tree (1.2.i above; Palladius II.15.3), pomegranate (Palladius IV.10.4), plum (Palladius XII.7.15), and olive (*HP* IV.14.3).

In order to prevent their development, it is recommended to treat the roots with red earth, *origanum*, ox gall or pigeon dung (Palladius III.25.5; *Geoponica* X.18.9, 90.1,4–5); or to plant cuttings in a squill bulb or plant squill nearby (*HP* II.5.5; Pliny XVII.87; *Geoponica* X.18.2, 90.1); or to fix pine stakes around the tree (*Geoponica* X.90.1); or to plant with seedlings a cutting of terebinth or mastic (Pliny XVII.256; Columella *RR* V.10.9, *de Arb*.XX.3); or to make use of a written charm (Aetius XIII.54). Once larvae have appeared, they may be extracted by means of a bronze stylus or other implement, treating those parts where it has been necessary to remove bark with cow dung (Palladius IV.10.20,29,4; *Geoponica* X.10.10); use of such an implement is also effective in preventing further infestation (*Geoponica* X.18.3). Alternatively, the tree as a whole, or its affected parts, may be treated with lizard gall (*Geoponica* X.18.7), ox gall (Palladius III.25,15), pig dung (Palladius III.25,15, IV.10.4; *Geoponica* X.90.3), human or animal urine (Palladius IV.10.4,20,29, III.25.15, XII.7.5; *Geoponica* X.90.3), *amurca* (Palladius IV.10.20,29), pitch or bitumen (Palladius IV.10.29, XII.7.15), red earth (Palladius XII.7.15), lime (Palladius IV.10.20,29), or a decoction of lupine or fig root (Palladius IV.10.20; Aetius XIII.54). Or the trees may be fumigated with bitumen (Aetius XIII.54), or by burning specimens of the offending larvae (Palladius II.15.3).

Skolekes were also well known, along with the smaller *thripes*, as pests of cut timber, including the structural timbers of buildings (*HP* V.4.5, VIII.10.5; Pliny XVI.220; LXX *Prov*. 12.4, 25.20; Suidas s.v. τερηδών). In the legend of the seer Melampus (Apollodorus

---

[216] Hort's translation gives the incorrect impression that the *kerastes* is properly only a pest of trees.

[217] *CP* V.10.5 introduces further confusion by stating that the *kerastai* which produce offspring in fig trees are themselves generated from the olive.

Bibl. I.9.12; Schol. *in Odyssey* XI.287; Eustathius *in Od.* 1685) warning is given by two *skolekes* of the imminent collapse of a building whose roof beams they have been gnawing through. Theophrastus records that some timbers are more susceptible to attack than others, those which have a strong smell or a bitter taste or are especially hard being avoided (*HP* V.4.2–6).

4. Medicinal uses

According to Pliny (XXX.115) and dependent authors (Plinius Junior III.4; Marcellus IV.38), the *cossus*, either crushed and applied directly, or burnt and applied with anise, is effective as a treatment for all forms of ulcer.

Certain red larvae from trees are also prescribed as a remedy for ear ailments (Galen XII.649; Pliny XXIX.135; Marcellus IX.84; Q.Serenus 169–70).

5. Use as food

Pliny (XVII.220) informs us that the beetle larvae known as *cossi* were regarded as a delicacy in his day, being fattened in captivity by being fed on flour or meal. Fabre[218] discovered that the larva of *Ergates faber* could survive on a diet of flour and was indeed edible.

Jerome (1.2.i above) speaks of similar larvae being prized as a luxury item in Pontus and Phrygia.

6. Exotic species

(i) Aelian (XIV.13) describes certain *skolekes* produced in the date palm as being an article of diet in India. These are probably the larvae of the Palm weevil *Rhynchophorus ferrugineus* (Oliv.), which are known to be edible.[219]

(ii) According to Strabo (XII.7.3) and Dioscorides (*DMM.* I.66), the storax tree in Pisidia is infested by *skolekes* which are generated inside the trunk and bore through to the surface, pushing out sawdust which piles up at the foot of the tree. Pliny (XII.124) has a rather different account in which winged *vermiculi* fly around the trees in midsummer, gnawing the wood and producing sawdust. Bodenheimer[220] suggests that the references here are to larvae of *Zeuzera pyrina*.

---

[218] J. H. Fabre, op. cit., pp. 178–9, 181–3.

[219] Cf. Bodenheimer, *Insects as Human Food*, p. 63. The larva of *Oryctes rhinoceros* (Linn.) may be included also. Cf. Fröhlich and Rodewald, *Pests*, pp. 203–5.

[220] *Animal and Man*, p. 140.

VI

BEETLES

Insecta: Coleoptera (Beetles)

31. *Kantharos/Scarabaeus*

1. Identification: general

The Greek κάνθαρος[1] is not a precise equivalent of the Latin *scarabaeus*, since its use is normally restricted to the various species of dung-beetles, and is only rarely extended to cover other beetles of similar appearance.[2] Thus Aristotle does not employ it as a general term equivalent to our 'beetle', but instead supplies the need of such a term by coining κολεόπτερος of which *kantharoi* are a subdivision.[3] *Scarabaeus*, by contrast, though more often than not referring expressly to dung-beetles, does indeed cover much the same ground as our English 'beetle'. So that Pliny (XI.97–9) is able to use it as a group name encompassing stag-beetles, glow-worms, and other Coleoptera,[4] as well as the superficially similar cockroaches.[5]

The most important group of dung-beetles is the subfamily Scarabaeinae (family Scarabaeidae), to which may be added the family Geotrupidae. The former group includes the famous Egyptian scarab *Scarabaeus sacer* Linn., which is also found in Southern Europe, and which is the species most frequently in view among classical authors from Aristotle onwards.[6] This distinctive insect is noted for its habit of rolling

---

[1] On its etymology, cf. Fernandez, *Nombres*, pp. 226–7.

[2] It is used for a chafer by Theocritus (32.1 below), and for blister-beetles (33.1 below). In *Geoponica* XIII.10.10 it is a pest of vines, captured by the placing of ivy foliage in which it will congregate. In *C.Gl.L.* V.422.40 (12.1 above) it seems to be used for the cricket (cf. *C.Gl.L.* V.394.42 *Scarabaeus: genus locustae*). Its appearance in LXX *Hab.* 2.11 is due to a mistranslation, although some patristic commentators (for example, Cyril Alex. and Theodore of Mopsuestria ad loc. MPG 66.437, 71.882) take the Greek text at face value.

[3] In *HA* 522b 30 ff. (also referred to by Basil, *Hex.* MPG 29.172) Aristotle refers to a certain κολεόπτερος μικρός, lacking a name, which construct τρώγλαι from mud on walls or gravestones, in which they produce their σκωλήκια. These have not been satisfactorily identified.

[4] *Scarabaeus* is often used by Pliny to refer to chafers (Greek μηλολόνθη; cf. 32 below). The *scarabaei* mentioned as corn pests (XXVIII.78) are probably the *kantharides* mentioned in the next sentence (33.3 below). In Pliny XXIV.168 an unidentified herb *eriphia* is said to be so named because it has a *scarabaeus* inside its stem which runs up and down with a sound like that of a kid.

[5] 13 above. Columella (*RR* IX.7.5) lists *scarabaei* and *blattae* as if they are distinct pests of hives, but this is probably an error on his part (cf. 13.4 above). *Scarabaeus* may also be used by Pliny as a name for the cricket (cf. 12.1 above, and note 2 above), which itself was often confused with the cockroach.

[6] So identified by Sundevall, *Thierarten*, pp. 194–5, Aubert and Wimmer, *Thierkunde*, I. p. 165, Gossen, in PW 10.1484, and others.

balls of dung, which it transports to a suitable retreat for use as food. The females lay their eggs in pear shaped balls constructed in underground chambers from material brought there for the purpose.[7]

This dung-rolling habit is also found in many related species. Many of the Scarabaeinae are ornamented with varying numbers of horns on their head or thorax, and a number of these are described as distinct varieties by ancient authors. Horned species from Southern Europe include *Copris lunaris* (Linn.), *C. hispanus* (Linn.), and members of the genera *Onitis* and *Onthophagus*.

In Egypt, similarly, many species occur alongside the well known *S. sacer*. Although the latter, with its distinctive shape, is the most frequent model for man-made representations, there are examples which have been identified as representing other genera, including *Copris*, *Hypselogenia*, the horned *Catharsius*, and the brightly coloured metallic *Gymnopleurus*.[8] It has been suggested[9] that the giant *Heliocopris* spp., which roll correspondingly enormous balls of dung, of Upper Egypt had a part to play in impressing the habits of dung-beetles upon the imagination of the Egyptians.

The members of the Geotrupidae are similar in appearance to the scarabaeines, but none of them exhibit the habit of dung rolling. The family includes such species as the hornless *Geotrupes stercorarius* (Linn.) and the three horned *Typhaeus typhoeus* (Linn).[10]

1.2. Identification: alternative names

(i) *Κανθαρίς*. The diminutive of *κάνθαρος*, normally referring to the blister-beetle,[11] is used rather incongruously by Plato Comicus (1.3.v below) for the giant *Aitnaios kantharos*.

(ii) *Κάνθων*. The *Etymologicum Magnum* (489.1) says that 'the *kantharos* is so called because it is generated from the dung of *κάνθωνες*, i.e., of asses'. This is probably not a real insect name, but simply an attempt to explain Aristophanes' joke in *Pax* 82.[12]

(iii) *Σήραμβος*. According to Hesychius, 'a type of *kantharos*'. The word may be related to the beetle names *κάραβος* and *καράμβιος*.[13]

(iv) *Βυρρός*. An Etruscan synonym of *kantharos* listed by Hesychius.[14]

(v) *Σκορόβυλος*. Listed as an equivalent of *kantharos* by Hesychius.[15]

---

[7] Imms, *Gen. Textbook*, p. 862.

[8] Bodenheimer, *Geschichte*, p. 34 and pl. III.

[9] Bodenheimer, *Animal and Man*, p. 81.

[10] Other horned beetles such as *Oryctes* spp., though not dung feeding, may have been mistaken for scarabs in antiquity, as Gossen, in PW Supp.VIII. 241, suggests.

[11] 33 below.   [12] Fernandez, op. cit., p. 50.   [13] Ibid., p. 229.

[14] Fernandez, ibid., p. 236, includes it as a non-Greek name.

[15] Hesychius (s.v. *βρυτίνη*), with reference to Cratinus fr. 96, says that *βρύτον* is the name for a creature 'similar to a *kantharos*, from which comes the fabric *βρύτινον*, otherwise known as *βομβύκινον*'. LSJ recognise *βρύτον* as a type of beetle, but consider the adjective to have been simply invented by Cratinus '*παρὰ προσδοκίαν* for *βύσσινος*'. According to Fernandez, op. cit., p. 243, the etymology suggests a meaning 'covering', referring to Hesychius' fabric. Perhaps *βρύτον* was a name for the silk-moth, which Pausanias (27.2.iii above) compared to a *kantharos*.

(vi) Σίλφη. Elsewhere a name for the cockroach,[16] this seems to be used by Sextus Empiricus (*Pyrrh.* III.221) to refer to the Egyptian scarab.

1.3. Identification: named varieties

(i) Ἡλιοκάνθαρος, κάνθαρος ἡλιακός, αἰλουρόμορφος. These are clearly specific names for *Scarabaeus sacer*, the pre-eminent sacred scarab of the Egyptians, which is more often referred to simply as *kantharos* or *scarabaeus*.[17] The first is given by Alexander of Tralles (*Febr.* 7) and refers, like the second, to the fact that the insect was regarded as a symbol of the sun. The second is found in the magical papyri (IV.751, VII.973 ff.), the first of these references describing the beetle as 'having twelve rays (ἀκτῖνας)'. The same insect is referred to in *PMag* XII. 271 ff. as κάνθαρος ἀκτινωτὸς ἱερός. What these 'rays' are is uncertain: they may, as Keller[18] suggests, be the sharp processes on the front of the beetle's head.

Horapollo (*Hier.* I.10), whose account of scarabs, derived from Apion,[19] is the most detailed to have survived, also describes this insect as 'rayed' and as associated with the sun. It is said to have thirty δάκτυλοι symbolising the days of the month. The meaning of δάκτυλος here is as uncertain as that of ἀκτίς: possibly it refers to the strong spines on the insect's legs which it uses for digging and for manipulating its dung balls.[20] Horapollo gives no direct explanation for his name αἰλουρόμορφος. The cat was indeed, as he informs us, another animal sacred to the sun, but in what sense it was said to resemble the scarab is not stated.[21]

(ii) Κάνθαρος σεληνιακός, ταυρόμορφος, ταυροειδής. The *tauroeides* is the second of Horapollo's (loc. cit.) three species of scarab, characterised as having two horns and as being sacred to the moon, which is itself associated with the bull. It appears also in the magical papyri under the first (IV.2456,2688) and second (IV.65) of the above names.

Unlike the preceding, this variety would comprehend a number of present day species, including for example in Europe *Typhaeus typhoeus*, with its two long straight horns flanking a shorter central spine, and species of *Onthophagus*, and in Egypt species of *Catharsius* and *Heliocopris*.[22]

(iii) Ἰβιόμορφος.[23] The third of Horapollo's scarabs is described as having a single horn and as being associated with the god Hermes, as was the bird from which it took

---

[16] 13.1 above.

[17] So identified by M. Wellmann, 'Der Physiologus. Eine Religionsgeschichtlich-naturwissenschaftliche Untersuchung', *Philologus*, Supplementband XXII, Heft 1, 1930, p. 78, and Fernandez, op. cit., pp. 212–13. Misidentified by Gossen, in PW Supp. VIII.236.

[18] *Ant. Tierwelt*, II p. 411.

[19] M. Wellmann, art. cit., p. 67.

[20] LSJ Supp. says 'joints of a beetle's tarsi', but *S. sacer* lacks front tarsi.

[21] LSJ suggest the name means 'cat-faced', like an Egyptian zoanthropomorphic deity.

[22] Other suggestions are given by Wellman, art. cit., p. 79, and Gossen, in PW 10.1486.

[23] This emendation of the mss. ἰδιόμορφος is supported by Wellman, art. cit., p. 79 and Fernandez, op. cit., p. 53.

its name. It is not referred to explicitly in the magical papyri, although an undifferentiated *kantharos* is associated with Hermes in *PMag* V.213 ff. There are a considerable number of scarabaeines which bear a single horn, including for example members of the genera *Copris* and *Onitis*. The name *ibiomorphos* probably results from the similarity between the slender backward-curving horn of scarabs like *C. hispanus* and the long curved bill of the ibis.[24]

(iv) *Taurus, pediculus terrae.* According to Pliny (XXX.39) the insect so named is a form of *scarabaeus* resembling a tick[25] and possessing small horns, which digs up earth. It is presumably to be identified as one or other of the horned dung-beetles of Southern Europe, such as *Typhaeus* or *Copris* spp., the latter having smaller projections on the thorax as well as their single backward-curved horn.[26] The *taurus* is also listed by Isidorus (*Or.* XII.8.5). It seems to be identical to the *scarabaeus cui sunt cornicula reflexa* described elsewhere by Pliny (XXX.100).[27]

(v) Αἰτναῖος κάνθαρος. A popular belief that Etna in Sicily was inhabited by dung-beetles of enormous size is attested by a number of Greek dramatists. In Aristophanes *Pax* 73, the giant *kantharos* upon which Trygaeus flies to Olympus is described as a native of Etna. The scholia on this passage display some uncertainty as to the meaning of this line and give three possibilities, of which the most likely is that the region in question did indeed have a reputation for producing large *kantharoi*.[28] Quotations referring to the Aetnaean beetle are cited by the scholia from Epicharmus (fr. 76), who was a native of Sicily, Sophocles (fr. 162), Aeschylus (fr. 233), who compared Sisyphus rolling his boulder to an Aetnaean beetle pushing its ball of dung, and Plato Comicus (fr. 37), who writes of the insects being reputed to be as large as a man. The creature also appears in Sophocles' *Ichneutai* (300),[29] where it is described as 'horned'.[30]

2. Life history

According to Aristotle's account (*HA*552a 17 ff.), reproduced with slight alterations by Pliny (XI.98), the dung-beetle rolls balls of dung, lies hidden in them during the winter,

[24] Wellmann, art. cit., pp. 78–9, supported by Fernandez, op. cit., pp. 53–4, argues that the list of three *kantharoi* in Horapollo I.10 corresponds with the three *scarabaei* in Pliny XXX.99, both deriving independently from Apion, but in fact the two accounts are not parallel. Horapollo's ἰβιόμορφος cannot be equated with Pliny's third beetle *fullo*, which is without doubt the chafer *Polyphylla fullo* (32.1.2.vii below), since the latter is not a scarab and is not horned. Nor is there any similarity between the colouration of the ibis and that of *P. fullo*, the latter being dark speckled with white and the former particoloured.

[25] 8 above. Perhaps they seemed similar in shape to engorged ticks.

[26] Wellmann, art. cit., p. 79, equates this with Horapollo's ταυροειδής.

[27] Ibid.

[28] Cf. the scarab which appears on a coin of Aetna; B. V. Head, *Historia Numorum*, 1911, p. 131.

[29] A. C. Pearson, 'Αἰτναῖοι κάνθαροι', *CR*, XXVIII, 1914, pp. 223–4, argues that the Αἰτναῖος κάνθαρος was simply an invention of the comedians, but there is no reason why a paradoxographic tale of giant beetles should not have been associated with this region. Cf. Davies and Kathirithamby, *Greek Insects*, pp. 86–7.

[30] D'Arcy Thompson, *A Glossary of Greek Birds*, London, 1936, pp. 87–8, suggests that the δίκαιρον of Ctesias and Aelian (IV.41) was a scarab.

and in them gives birth (ἐντίκτουσι) to small larvae (σκωλήκια), out of which a new generation of beetles develop (Pliny . . . *parvosque in iis contra rigorem hiemis vermiculos fetus sui nidulantur*). D'Arcy Thompson[31] understands this as containing an allusion to the female's guarding of her offspring to maturity, a habit characteristic of *Copris* spp.

The Egyptian version of the insect's life history, based upon that of *Scarabaeus sacer*, is recorded, with certain variations, by Plutarch (*Mor.* 355a,381a), Aelian (X.15), Horapollo (*Hier.*I.10), Porphyry (*de Abst.* IV.9), Clement (*Strom.* V.4), Eusebius (*Praep.Ev.* MPG 21.173), Ps. Eustathius (*Hex.* MPG 18.748), Suidas (s.vv. *kantharos*, *onthos*), and Augustine (MPL 32.1372). As reported by Horapollo, the story was that all dung-beetles are male, and that therefore they are able to reproduce themselves without mating. They mould balls of dung, roll them along backwards, and bury them in the ground for 28 days. After this time, they dig up the balls and cast them into water, in which they open up, releasing new beetles. The other sources add that the insects inject their seed into the dung balls[32] and omit the casting into water. According to Porphyry and Eusebius, the former process precedes the rolling of the ball, here described as made from mud. A number of authors do not mention that the ball is buried in the ground: only Augustine, who cites the insect as an example of spontaneous generation, does so explicitly, while Aelian simply says that the ball is incubated for 28 days, following which the adult 'leads forth its young'. Clement and Ps. Eustathius say that the beetle spends six months of the year above ground and six months below.

Jerome's simple statement that the *kantharos* is spontaneously generated from dung (MPL 25.1297) probably presupposes the above life history (cf. Augustine loc. cit.). There are a few references to the ball rolling habit not explicitly linked with the insect's life history (Lucian *Pseud.* 3; Pliny XXX.99; cf. Aeschylus loc. cit. and 3 below).

Alongside the above essentially correct accounts, there also existed a folk belief to the effect that *kantharoi* were spontaneously generated from dead asses (Plutarch *Cleom.* 39; Sextus Empiricus *Pyrrh.* I.41; Origen *c. Celsum* IV.57; Isidorus *Or.* XI.4.3) or horses (Isiorus *Or.* XI.4.3, XII.8.4).[33]

3. Habits

Aristotle (*PA* 682b 21 ff.) includes *kantharoi* as examples of insects which protect themselves by becoming motionless when touched. They are also described as clumsy in flight (*PrA* 710a 10 ff.).

The beetles' unsavoury diet of dung is often commented upon: Dio Chrysostom (XXXII.97) says that they reject honey even if it is poured out for them. They are said to be especially fond of ox (Horapollo loc. cit.; *Aesopica* 84 Perry; Ps. Eustathius loc. cit.) or ass (*EM* s.v. κάνθων; Suidas s.v. κάνθαρος) dung. Hipponax (fr. 92 Knox) depicts them as being attracted in numbers to their food, guided by scent. In one Aesopic fable (84 Perry) dung-beetles are portrayed as feeding on dung all the year round, flying off in winter in

---

[31] *Hist. An.*, V.19 note.
[32] Aelian σπείρει ἐς; Plutarch τίκτουσι τὸν γόνον εἰς; Suidas ἀποσπερμαίνειν.
[33] Cf. similar beliefs about wasps, 40.2 below.

search of new feeding grounds and returning to their original location in spring. In another (*Aesopica* 112), however, we find a version of the more familiar ant and cicada story[34] in which the improvident beetle starves to death after the winter rains have washed away the dung upon which it subsisted.

Some writers at least display an awareness that *kantharoi* roll dung not only in order to reproduce their kind, but also as food for themselves (Plutarch *Mor.* 355a): in Aristophanes *Pax* 7–8 the beetle moulds the dung it has been given into a ball with its feet before devouring it.

4. Popular beliefs and attitudes

Because of their unpleasant feeding habits, dung-beetles are characterised in Artemidorus' work on dreams (*On.* II.22) as insects of ill omen (cf. Jerome MPL 25.1297). A proverb κανθάρου σοφώτερος is said (Zenobius IV.65) to derive from the Aesopic fable (*Aesopica* 3: Aristophanes *Pax* 129, *Vesp.* 1443, *Lys.* 694; Lucian *Icarom.* 10) in which the dung-beetle brings about the destruction of the eggs of the eagle.[35]

Because of their delight in so ill-smelling a substance as dung, *kantharoi* were believed to positively detest (Plutarch *Mor.* 710e, 1058a; Artemidorus *On.* II.22; Sextus Empiricus *Pyrrh.* I.55) or to be actually harmed by sweet scents (Theophrastus *de Odor.* 4). It was widely held that they could be killed by the odour of roses (Ps. Aristotle *de Mir.* 845b 2; *Geoponica* XIII.16.3) or perfume (Aelian 1.38, VI.46), or by scattering of rose petals (Aelian IV.18) or anointing with rose oil (Clement *Paed.* II.8; Ps. Eustathius *Hex.* MPG 18.735).

In the paradoxographical tradition, there is a frequently repeated story concerning a locality in Thrace known as Κανθαρώλεθρος. This was said to be a spot somewhat larger than a threshing floor, on entering which dung-beetles were unable to find their way out, but would wander miserably around in circles until overcome by starvation (Ps. Aristotle *de Mir.* 842a5 ff.; Plutarch *Mor.* 473e; Strabo VII. fr. 30; Antigonus Carystius *Mir.* 14; Pliny XI.99).

5. Scarabs in Egyptian religion

The importance of animals in Egyptian religion was a subject of considerable curiosity among the Greeks and Romans, and is often commented upon. The dung-beetle or scarab frequently appears in lists of the particular creatures involved, and the Egyptians are often described as worshipping, honouring or revering these insects (Apion *ap.* Pliny XXX.99; Manetho and Hecataeus *ap.* Diogenes L.I. 10; Plutarch *Mor.* 381a; Porphyry *de Abst.* IV.9; Ps. Eustathius *Hex.* MPG 18.478). The more accurate authorities, such as Plutarch, are aware that they did not strictly speaking 'worship' dung-beetles but held them sacred as symbols of deity, but other writers do not concern themselves with fine distinctions. Such 'worship' of animals is the subject of scorn and ridicule among the philosophers, and especially in Christian polemical writing, since it seemed as if the

[34] 16.5 above.
[35] On the Aesopic fable, cf. Davies and Kathirithamby, *Greek Insects*, pp. 3–4.

Egyptians had gone out of their way to select the most incongruous creatures as objects of reverence. Thus Arnobius (*Adv.Nat.* I.28; cf. Eusebius loc. cit.), detailing the follies of paganism, writes scathingly of the erection of temples to cats, bullocks, and dung-beetles.

The scarab's name in Egypt had the same consonants as the term for 'to be' or 'to become'. It was taken as a symbol of the self-generated deity Kheper, associated with the initial creation of the universe and with the rising sun, the daily recapitulation of creation.[36] The idea of self-generation and association with the sun, the creation of the cosmos, and the succession of days and months are prominent in classical accounts of Egyptian beliefs, and are reflected in the descriptions given of the insect's life history.

Horapollo (loc. cit.), Plutarch (*Mor.* 381a), Pliny (XXX.99, citing Apion), Porphyry (loc. cit.), Clement (*Strom.* V.4), and Ps. Eustathius (loc. cit.) make it clear that it was the beetle itself which was conceived of as representing the sun and its ball as representing the sphere of the heavens. Just as the insect pushes its ball along backwards with its hind legs, so, as Plutarch explains, 'the sun seems to turn the heavens in the direction opposite to its own course, which is from east to west'. Similarly, in the magical papyri (IV.943, VII.520, XII.44,57, XIII.128; cf. 1.3.i above) the sun god is commonly addressed as *kantharos*. Horapollo states that the scarab's development mirrors that of the cosmos, and he is careful to point out the calendrical patterns present in the account of its life history, for example the 28 day incubation period. Since scarabs were believed to be all of one sex and to reproduce without mating, they were held to represent the concept of self-generation (Horapollo I.10,13; *PMag* IV.943) and the male principle (Horapollo I.10). According to Aelian (X.15) and Plutarch (*Mor.* 355a), it was with the latter idea in mind that Egyptian warriors used the scarab image on rings and seals.

As has been mentioned (1.3.ii–iii above), particular varieties of scarab other than *S. sacer* were associated with deities besides the sun god. According to Sextus Empiricus (*Pyrrh.* III.221), scarabs are sacrificed to Thetis at Alexandria.

6. Scarabs in the magical papyri

Scarab beetles figure prominently in various charms and spells described in the magical papyri for the purpose of invoking the aid of and gaining power over the deities associated with the sun and moon, Helios/Mithras (*PMag* IV.751) and Selene/Hecate (IV.65,2456,2688). For each of these the appropriate variety of scarab, *heliakos* or *seleniakos* (1.3.i–ii above) is prescribed. The instructions given involve complex sequences of ceremonies and incantations, sometimes extending over several days, employing a variety of articles besides the scarabs. We also read of the scarab emblem being engraved upon rings and amulets credited with the power of bringing good fortune and of preservation from danger, and with various magical properties (*PMag* V.213 ff., XII.271 ff., XXXVI.178–85). Pliny (XXXII.124) records a Magian claim that emeralds engraved with a scarab had the power of averting natural disasters such as hail and locust

---

[36] Keller, *Ant. Tierwelt*, pp. 409–13; Bodenheimer, *Geschichte*, pp. 31 ff.; R. T. R. Clark, *Myth and Symbol in Ancient Egypt*, London, 1959, p. 40; J. G. Griffiths, *Plutarch's de Iside et Osiride*, notes ad. loc.

swarms, if used with appropriate incantations. He also refers (XXXVII.155) to a certain stone said by the Magi to have the power of raising storms if dropped into boiling water with a scarab.

7. Medicinal and other uses

Both *S. sacer* (Pliny **XXX**.99; Alexander of Tralles *Febr.* 7; cf. 1.3.i above) and one of the horned scarabs (Pliny **XXX**.100; cf. 1.3.iv above) are described as effective amulets against quartan fever (scarabs are also associated with fever in Horapollo II.41). Earth thrown up by the burrowing of the *taurus* (1.3.iv above) is prescribed as an application for scrofula, gout and similar conditions (Pliny **XXX**.39). Galen (XIV.243 K) says that dung-beetles in general may be boiled with oil and used for the treatment of ear complaints. In veterinary medicine, dung-beetles are prescribed for the treatment of animals suffering from lethargy (Chiron 376). Dionysius' *Ixeutikon* (*Epit.* III.11) describes the capture of cranes by means of a lure consisting of a dung-beetle enclosed in a hollow gourd smeared with birdlime.

32. Melolonthe

1. Identification: general

The Greek μηλολόνθη, which has no precise Latin equivalent, covers large, conspicuous, actively flying beetles known to us as chafers. Most frequently mentioned under this name are the brilliant metallic green rose chafers (Scarabaeidae: Cetoniinae), including the well known *Cetonia aurata* (Linn.) and a number of larger Mediterranean species, which were used by children as playthings. These are attractive day flying insects which feed at flowers: their larvae, unknown in antiquity, are found among roots, leaf mould or decaying wood.[37]

It is probable that the name also covered members of the related subfamilies Melolonthinae and Rutelinae, a number of which are important pests as adults and larvae. These are generally duller in colour and nocturnal. The former subfamily includes the well known cockchafer *Melolontha melolontha* (Linn.), which swarms around trees and bushes at dusk and may cause severe damage by devouring the foliage. Its larvae, which hatch from eggs laid in the ground, feed upon the roots of grasses and cereal crops, and are thus important as pests in their own right.[38] There are a number of references to such larvae in ancient authors, but no evidence that they were linked with their appropriate adult stage. A larger relative of the cockchafer, *Polyphylla fullo* (Linn.), would also have been well known in antiquity, and is clearly described by Pliny.

Chafers are often described as resembling the *kantharos* (e.g. Schol. Aristophanes *Nub.* 761; Eustathius *in Il.* 1243.30 ff.), which is not a term for beetles in general but normally restricted to the dung-beetle.[39] There is one apparent example of *kantharos* being used in

[37] Imms, *Gen. Textbook*, p. 861.
[38] Ibid., pp. 861–2.
[39] 31.1 above.

place of *melolonthe* (Theocritus V.114–15), but the usage here is determined by metrical considerations.[40] Latin has no distinct term for chafers and usually refers to them by the name *scarabaeus*, which, unlike *kantharos*, is a general term equivalent to our 'beetle'.[41]

1.2. Identification: distinct names and named varieties

(i) Μηλολόνθη; variant forms μηλολάνθη (Pollux), μηλόνθη (Eustathius 1329.25 ff.), μηλάνθη (Herodas). In Aristotle, the *melolonthe* appears as a form of beetle (*HA* 490a 15, *PA* 682b 13 ff.), clumsy in flight (*PrA* 710a 10 ff.), which produces a buzzing sound (*de Resp.* 475a 5). It is not possible to determine whether the reference here is to cetoniines or melolonthines or both.[42] It has been suggested, in view of Aristotle's erroneous statement that its larvae develop in dung, that he may have confused the true *melolonthe* with some beetle such as *Geotrupes*,[43] but the error may simply derive from a false etymology, noted by Eustathius (loc. cit.), which associated the word with ὄνθος.

Artemidorus (*On.* II.21) associates *melolonthai* with the *kantharos* as insects of ill omen, and might therefore be thought to have a pest species in mind, but since he also includes the glow-worm[44] it is perhaps more probable that he has simply reproduced a list of beetles from Aristotle or some other source. In Aristophanes (*Nub.* 762–3), Herodas (fr. 12) and Pollux (IX.124) the *melolonthe* is clearly the attractive *Cetonia* spp., as the scholia and Eustathius make clear. The lexicographers (for example Photius, Hesychius s.v.) commonly define the word as equivalent to χρυσοκάνθαρος.

The etymology of the word is uncertain, Strömberg[45] suggests that it is a compound of μῆλον and ὀλύνθος, with reference to those Melolonthinae and Rutelinae which defoliate trees, such as that mentioned by Theocritus (2 below) as attacking fig trees. In that case, the variant -ανθη would be the result of a popular etymology erroneously connecting the word with ἄνθος.

(ii) Χρυσοκάνθαρος. This term is given as a synonym of the preceding by the scholiast on Aristophanes loc. cit., Hesychius and Photius, with reference to the chafers played with by children. It is described as an insect resembling the *kantharos*, with golden wings. These sources clearly have in mind the beautiful golden green *Cetonia aurata* and related species such as *C. speciosissima* Scop.[46] The diminutive

---

[40] H. Gossen, 'Die Tiere bei den Griechischen Lyrikern', *AGM*, XXX, 1938, p. 334.

[41] 31.1 above.

[42] Identified as covering both groups by Sundevall, *Thierarten*, p. 194, Keller, op. cit. II p. 409, and D'Arcy Thompson, *Hist. An.*, V.19 note. Gossen, in PW 10.1488, takes μηλολόνθη to be *Melolontha* alone, restricting *Cetonia* to the χρυσο-compounds, but actual usage does not support this distinction.

[43] So Aubert and Wimmer, *Thierkunde*, I p. 167.   [44] 35 below.

[45] Gr. *Wortstudien*, pp. 5–10, supported by Fernandez, *Nombres*, pp. 231–3.

[46] So identified by Gossen, art. cit., 1488–9. Davies and Kathirithamby, *Greek Insects*, pp. 89–91, argue that the χρυσο-compounds, like κυνόμυια (52 below), are not genuine insect names at all, but that χρυσοκάνθαρος is a mere lexicographical invention to explain what was no more than a 'term of endearment' in *Vesp.* 1341. However, it seems unnecessary to suppose this, in view of the evident appropriateness of χρυσο-compounds as descriptive terms for the golden green cetoniines.

form χρυσοκανθαρίς (*Anecdota Bekkeri* III.1432) is probably a simple synonym rather than a term for metallic coloured blister-beetles.[47]

(iii) Χρυσομηλολόνθιον. This equivalent of the preceding is found only in the diminutive form, in Aristophanes *Vesp.* 1341. Fernandez[48] suggests that the word is simply a neologism coined by the playwright for the occasion, on the grounds that the addition of the prefix to a term already synonymous with *chrysokantharos* is strictly tautologous. However, this would only be the case if *melolonthe* was restricted entirely to cetoniines and never extended to cover other less colourful species.

(iv) Χρυσαλλίς. Another term referring to the colouration of *Cetonia*,[49] this is noted by Eustathius (1329.25 ff.) as a synonym of *melolonthe*. Hesychius' 'certain animal' seems also to refer to this meaning of the word.[50]

(v) Pliny's brief reference (XI.98) to certain *scarabaei* which *volitant magno cum murmure aut mugitu* would cover members of the three subfamilies considered here, as well as other heavy flying species such as *Geotrupes*.

(vi) *Scarabaeus viridis*. A Latin designation, given by Pliny (XXIX.132), for the metallic green cetoniines.[51] Marcellus (VIII.100) terms them 'emerald coloured beetles'.[52]

(vii) *Fullo*. Pliny (XXX.100) refers to a specific variety of *scarabaeus qui vocatur fullo, albis guttis*. There is no reason to doubt the traditional identification of this insect as the large and impressive chafer *Polyphylla fullo*, which is speckled and dusted with white markings, from which it obtains its name.[53]

2. Life history and habits

Aristotle (*HA* 552a 15 ff.) states erroneously that *melolonthai* develop from *skolekia* found in the dung of cattle. Since chafers were regarded as similar to dung-beetles, they may by association have been credited with a similar life history: such a belief would have been supported by the popular etymology (1.2.i above) which derived their name from ὄνθος.

Eustathius (*in Il.* 1329.25 ff.), Pollux (IX.124), Suidas, and the scholiast on

---

[47] There are other examples of *kantharis* where one would expect *kantharos* and vice versa (31.1.2.i above; 33.1 below).

[48] Op. cit., pp. 101–2.

[49] More usually the word refers to pupae of Lepidoptera (23.1.2 above).

[50] Χαλκῆ μυῖα, name of a children's game mentioned by Herodas (fr. 12) and Pollux (IX.122), is taken by Eustathius (1243.30 ff.; cf. Hesychius s.v) to be also a name for the rose chafer, but this is probably due to confusion on his part. There is, however, an insect χαλκομυῖα mentioned by Aetius (VII.100) as a treatment for eye ailments. It has been suggested (Davies and Kathirithamby op. cit., p. 160) that χαλκῆ μυῖα as a term for the game 'blind man's buff' contains a reference to the μύωψ or horsefly, which was believed to be blind or short-sighted (49.4 below) and was described as brassy in colour (for example Schol. *Odyssey* XXII.299).

[51] So identified by Leitner, *Zool. Terminologie*, p. 217.

[52] In *C.Gl.L.* II.31.38 *bulli* is glossed as χρυσοκάνθαροι. This may mean that *bullus* was a Latin term applied to *Cetonia* because of its similarity to the amulets so named, or that the beetles were used as amulets, as certain other beetles were (30.1.3.v above).

[53] So Keller, op. cit., II. p. 409; Gossen, art. cit., 1481; Leitner, op. cit., p. 123. It is misidentified by Wellmann (31 note 24 above) and A. Blanchet, *Bull. Arch. Comm. Trav. Hist.*, 1946–9, pp. 617–18.

Aristophanes *Nub.* 761 all associate the golden *Cetonia* spp. with the blossom of apple trees or with flowers in general, which is in keeping with the feeding habits of these insects. They are variously described as sitting upon flowers, being spontaneously generated from or simultaneously with apple blossom, and flying toward the latter as soon as it begins to appear.

Theocritus (V.114–15) alludes to certain flying *kantharoi* as pests of fig trees. These would probably be one or other of the defoliating chafers, for example *Amphimallon solstitialis* (Linn).[54]

Although the fact was apparently unknown in antiquity, the larvae of certain chafers, for example *Melolontha* and *Amphimallon*, feed upon roots and constitute important pests of grassland and cereal crops. Such root feeding larvae (σκώληκες, *vermes, vermiculi*) are described by Theophrastus (*HP* VIII.10.4), Pliny (XVIII.151,158), Columella (*RR* II.9.10), and Palladius (X.3.2) as attacking wheat, barley and other crops. According to Theophrastus (*CP* III.22.4), they are spontaneously generated when the roots begin to rot under the combined influence of warmth and moisture. Hesychius records the term ψώμηξ, related to the verb ψωμίζω[55] as a specific name for these pests.

Pliny, Columella and Palladius state that if the seeds are soaked prior to sowing in the juice of houseleek or wild cucumber, or mixed with crushed cypress leaves, they will be safe from attack prior to germination, and from attacks on the roots later. Alternatively the furrows may be sprinkled with the above herbal liquids or with *amurca*. Pliny also notes the practice of sowing before a full moon, and of carrying a toad round the field and burying it in the centre. Other pests causing damage to seeds, seedlings, stem bases or roots of cereal crops, which would not have been distinguished from the above in antiquity, include the larvae of ground beetles of the genus *Zabrus*,[56] of click-beetles such as *Agriotes* and of craneflies (Tipulidae).

3. Chafers as playthings

The attractive green rose chafers of the genus *Cetonia* were popular playthings among Greek children, as they are today,[57] who would capture them, tether them by a long thread, and then let them fly. It is this practice which is alluded to by Aristophanes in the couplet ἀλλ' ἀποχάλα τὴν φροντίδ' εἰς τὸν ἀέρα λινόδετον ὥσπερ μηλολόνθην τοῦ ποδός (*Nub.* 762–3), and more indirectly in *Vesp.* 1341,[58] and by Herodas (fr. 12), as is explained by the scholia on Aristophanes, Suidas, Eustathius (*in Il.* 1243.30 ff., 1329.25 ff.), and Pollux (IX.124) in his account of children's games. These accounts vary somewhat, and although the general picture is clear enough there are certain details where it is not entirely clear what the author is envisaging. Eustathius' second description probably means that the children run along after the insect as it flies on ahead of them, still

---

[54] The species named by Gossen, in PW Sup.VIII.241.
[55] Fernandez, op. cit., p. 118.
[56] Cited by Gossen, in PW Supp. VIII.236–7.
[57] Sundevall, op. cit., p. 194.
[58] A third reference has been claimed in *Ach.* 920 ff., but a different insect is in view here (cf. 18 above).

holding onto their end of the thread, clapping their hands in delight as they go. The 'small piece of wood which the chafers cannot lift' referred to by Suidas and the scholiast on *Nub.* 761 is presumably something tied onto the child's end of the thread, so that if it should lose hold the insect will not be able to escape: the 'piece of wax' mentioned in Eustathius' first account probably fulfilled a similar function.[59]

4. Medicinal and other uses

According to Pliny (**XXX**.100), the *fullo* was recommended by the Magi as an amulet against quartan fever. The green *Cetonia* spp. are described as having been used by engravers of gems, who would look at them from time to time in order to rest their eyes (Pliny XXIX.132; Isidorus *Or.* VI.3). The colour green was recognised in antiquity as being the most restful to the eyes, green marble being recommended for the paving of libraries.[60] Marcellus states in more general terms (VIII.100) that the insects are of use in promoting sharpness of vision (*scarabaeus coloris smaragdini tantum beneficii oculis praestare dicitur ut visionem ei acutissimam reddat qui eum contemplatus fuerit adsidue*).

33. Kantharis

1. Identification: general

Κανθαρίς, though in origin simply the diminutive of κάνθαρος,[61] had by the time of our earliest sources already acquired a more specialised meaning. It is mainly used to refer to certain beetles of the family Meloidae which, along with the closely related *bouprestis*,[62] played an important role in ancient pharmacology. Members of this family, most notably the genera *Lytta* and *Mylabris*, are sources of the blistering agent cantharidin, which has been used continuously in medicine up to modern times.[63]

In mediæval and more recent times the most important of these blister-beetles has been the so called 'Spanish fly' *Lytta vesicatoria* (Linn.), an attractive metallic green insect which feeds gregariously upon the foliage of ash, elder, privet and other plants.[64] In antiquity, however, the most widely used species were the yellow and black banded members of the genus *Mylabris*.[65] Aristotle lumps all *kantharides* together as a single 'species',[66] but later medical sources distinguish a number of varieties.

[59] It has been suggested (cf. 18 note 80 above) that 'lighted wax' is meant here, but in the absence of supporting evidence it seems best to understand this passage in line with the other descriptions.

[60] Cf. Seneca *de Ira* III.9.2; Isidorus *Or.* VI.2–3.

[61] Fernandez, *Nombres*, p. 65, suggests, as does Keller, op. cit., II, pp. 414–15, that in for example Theophrastus, Hesychius and the *Geoponica* κανθαρίς means nothing more specific than 'small beetle'.

[62] 34 below.

[63] J. Scarborough, 'Nicander's Toxicology II', *Pharmacy in History*, XXI, 1979, pp. 13–14; J. L. Cloudsley-Thompson, *Insects and History*, p. 205; Imms, *Gen. Textbook*, p. 891.

[64] J. L. Cloudsley-Thompson, op. cit., p. 205.

[65] Keller, op. cit., II p. 414, names *Mylabris cichorii* (Linn.), and Leitner, *Zool. Terminologie*, p. 71, *M. variabilis* Pall, and *M. floralis* Pall.

[66] Sundevall, *Thierarten*, p. 195, and Aubert and Wimmer, op. cit., I p. 165, identify the Aristotelian *kantharis* as *Lytta vesicatoria*, and D'Arcy Thompson, *Hist. An.*, V.19 note, as *Lytta* plus *Mylabris*. But cf. 33.2.

The Meloidae have very complex life histories which were unknown in antiquity, involving in the case of *Lytta* and other species parasitism in the nests of solitary Hymenoptera.[67]

It is quite probable that besides the genuine producers of cantharidin other small colourful beetles were also credited with similar medicinal properties. This may explain some of the references to *kantharides* as pests of crops. Latin has no equivalent term of its own for these medicinal insects, but took over the Greek name in the form *cantharis* (late Latin forms *cantaris, cantharida, cantareda,* etc.). There are rare examples in Greek (Diogenes L. VI.44; *Hippiatrica Ber.* 15.9) of the use of κάνθαρος to refer to blister-beetles.[68] A single synonym of *kantharis*, ἐαρίς, is recorded by Hesychius.[69]

1.2. Identification: named varieties

(i) Ποικίλη κανθαρίς. Dioscorides (*DMM.* II.61), Galen (XII.363 K), and Pliny (XXIX.94) agree in declaring that the most powerful of the blister-beetles,which they grade according to the strength of the active principle contained in them, are those which are coloured with yellow bands running across their wings. The former describes them as longish in body, stout, of a fatty substance (ἐμπίμελοι; Pliny *pinguis*) like cockroaches.[70] They are also referred to in the *Hippiatrica* (*Ber.* 15.9). This variety is clearly to be identified with the genus *Mylabris*.[71]

(ii) The *cantharides variae et oblongae* from Alexandria described by Scribonius Largus (231) are probably the same as the preceding.

(iii) The fourth in Pliny's list (XXIX.94) of five varieties of *cantharis* are described as much less powerful than the yellow banded form, *minutae latae pilosae*. The description suggests that these may not be true blister-beetles at all, but rather soldier-beetles of the family Cantharidae and genera *Cantharis* and *Rhagonycha*. These are conspicuous diurnal insects, generally bright orange or yellow in colour, their wing cases covered with fine hairs (hence *pilosae*). Soldier-beetles are similar to meloids in general form, and are similarly soft-bodied; it is therefore not unlikely that they should have been thought of as possessing similar properties.[72]

(iv) The least potent *cantharides* on Pliny's list (XXIX.94) are described as *unius coloris macrae*. Dioscorides (*DDM.* II.61) similarly refers to an inferior unicolorous variety. These are perhaps soldier-beetles like the preceding.[73]

---

[67] Imms, op. cit., pp. 890–1.

[68] There are similar examples of the use of the diminutive where *kantharos* might be expected (31.1.2.i, 32.1.2.ii above).

[69] Fernandez, op. cit., p. 165, suggests a derivation from ἔαρ, 'spring'.

[70] 13 above. Both cockroaches and meloids are soft-bodied.

[71] So Keller, op. cit., II p. 414; Gossen in PW 10.1483; Bodenheimer, *Geschichte*, p. 82; Leitner, op. cit., pp. 70–1. *Mylabris* is recognisably illustrated in the illuminated mss of Dioscorides discussed by Kadar, *Zoological Illuminations*, p. 60 and pl. 68, 88, 93. Cf. also 7b1.3.ix above.

[72] Identified by Gossen, art. cit., 1483, and Leitner, op. cit., p. 70 as *Lytta lutea* Waltl.

[73] Identified by Gossen and Leitner, ibid., as *Lytta nuticollis* Muls.

(v) Pliny's first variety of *cantharis* (XXIX.94) is characterised as developing from a *vermiculus* found in the robin's pincushion gall[74] on the wild rose and *fecundissime* on ash, but nothing is said of its physical appearance. Although the association with rose galls is erroneous, the mention of ash is in keeping with the most characteristic foodplant of *Lytta vesicatoria*, so it is probable that it is this species which Pliny has in mind.[75] A second variety, associated with the white rose, is said (loc. cit.) to be less potent.

Aristotle (*HA* 552b 1 ff.) lists the wild rose as one of the foodplants of the *kampai* from which blister-beetles arise. Palladius (I.35.4) speaks of *cantharides quas in rosis invenire consuevimus* as an effective preventive agent against pests of vines,[76] and the *Geoponica* (V.30) refers to these same insects in a parallel passage as *kampai*. Similarly, Plutarch (*Mor.* 874b) characterises *kantharides* as harmful creatures lurking in roses. It is not impossible that in the last three passages there is some confusion with rose chafers,[77] which have a similar green colouration to that of *Lytta*. Alternatively, those passages which associate *kantharides* with roses may refer to ladybirds (vi below), which frequently occur on these plants as predators of aphids.

Unmistakable references to *Lytta vesicatoria* are found only in the Latin glossaries, where *cantaridae* are defined as *muscae virides in fraxino* (III.559.12) or *vermes ex fraxino* (III.588.14, 610.2).

(vi) Χελωνίας. This term is defined as ἡ ποικίλη κανθαρίς by Hesychius. Although it is possible that this is simply another name for the ποικίλη κανθαρίς of Dioscorides (i above), it is more probable that, as Gossen[78] suggests, the word refers to the various species of ladybird (Coccinellidae), which, being red or yellow with black spots, could be described with equal appropriateness by the adjective ποικίλος. Furthermore, the name suggests an insect shaped like a tortoise,[79] which would fit the rounded Coccinellidae very well. There is evidence from mediæval sources[80] that ladybirds were regarded as a form of *kantharis* with medicinal properties similar to those of the Meloidae.

2. Life history

According to Aristotle (*GA* 721a 6 ff.), *kantharides* belong to that class of insects which, although they mate and produce offspring, produce not creatures like themselves, but only *skolekes*, and which are themselves generated spontaneously from decaying

---

[74] 47.3 below.

[75] So Leitner, op. cit., p. 71. Gossen, art. cit., 1482, names *L. dives* Brull.

[76] Cf. 25.3 above.

[77] 32.1, 1.2.ii above.

[78] *Zool. Glossen* p. 124; in PW Supp. VIII.238; though no particular species of ladybird is in view.

[79] Cf. Fernandez, op. cit., p. 52.

[80] Bodenheimer, *Geschichte*, p. 137. It has been suggested too (Davies and Kathirithamby, op. cit. p. 92) that the saying φεύγετε κανθαρίδες, λύκος ἄγριος αἷμα διώκει, preserved by Pliny (XXVII.100), is parallel to our 'Ladybird, ladybird . . .'. Although it is difficult to understand, even bearing in mind the medicinal uses of κανθαρίδες, why Pliny should introduce a rhyme about ladybirds as an incantation to cure skin complaints.

substances. They are described (*HA* 542a) as having the characteristic mating posture of beetles. Elsewhere, however, we read (*HA* 552b 1 ff.) that they develop from *kampai* or *skolekes* found on fig, pear, fir, and wild rose.[81] The same list of plants, with the addition of the garden rose, is given by Pliny (XI.118), who elsewhere (1.2.v above) says that the beetles derive from *vermiculi* found in rose galls and on ash. Aelian (IX.39) simply states that they are generated among wheat, poplar, and fig. Spontaneous generation is also implied by Theophrastus (*HP* VIII.10) and Isidorus (*Or.* XII.5.5). If the *kantharis* of Aristotle and Theophrastus covers other brightly coloured beetles besides Meloidae, e.g., ladybirds and leaf-beetles (Chrysomelidae), it is possible that some genuine observations lie behind these accounts.[82]

3. Kantharides as pests

A number of authors (Theophrastus *HP* VIII.10; Nicander *Alex.* 115; Dioscorides *DMM.* II.65; Galen XII.363 K; Hesychius; Pliny XVIII.152) characterise the *kantharis* as a pest of wheat. Elsewhere Pliny (XXVIII.78) states that these creatures will perish if the affected fields are circled by a menstruating woman: he records on the authority of Metrodorus of Scepsis that a severe infestation in Cappadocia was dealt with in this way.

The particular insects in view here have been identified as belonging to the genus *Lytta*,[83] although the term *kantharis* could also cover chrysomelids such as the metallic blue and red *Lema melanopa* (Linn.).

Kantharides are also spoken of as pests of vines (Palladius I.35.6; *Geoponica* V.49.1, 22.3, XIII.16; Hesychius) and of garden vegetables (Hesychius).[84]

4. Predation

It was popularly believed that vipers (Galen XIV.364 K; Aristotle fr. 334) replenished their resources of venom by feeding upon toxic creatures such as blister-beetles, *boupresteis*, and *pityokampai*.[85] Sextus Empiricus (*Pyrrh.* I.57) states that blister-beetles, though toxic to most animals, can be eaten by swallows without harm.

5. Medicinal and other uses

Blister-beetles were widely used in ancient pharmacology, although it was recognised that they were dangerous if taken internally. Particular varieties were graded according

---

[81] Similar contradictory life history details are given for flies (48.2 below).

[82] Unlike meloids, whose parasitic larvae were unknown until modern times, ladybirds and pest species of Chrysomelidae have conspicuous larvae which the ancients are likely to have observed, those of the former feeding on aphids and those of the latter on foliage.

[83] *Lytta segetum* (Fab.), according to Gossen, art. cit., 1483, Leitner, op. cit., p. 70, and J. Scarborough, 'Some Beetles in Pliny's Natural History', *The Coleopterist's Bulletin*, XXXI. 1977, pp. 294–5. The identification as the grain weevil *Sitophilus* (Gossen, art. cit., 1483; Leitner, op. cit., pp. 69–70), which was known in antiquity as *kis/curculio* (36.1 below), is impossible.

[84] The identity of these vine pests is impossible to determine. In *Geoponica* XIII.16 they are confused with dung-beetles or *kantharoi* (cf. also 31 note 2 above). Elsewhere *kantharides* are said to be effective against vine pests (33.1.2.v above).

[85] 28 above.

to their potency (1.2 above). They are described (Pliny XXIX.95) as being prepared for use by being placed under a covering of roses in an earthenware vessel closed with a linen cloth, the vessel being suspended over boiling vinegar and salt: the same process, omitting mention of roses and salt, is given by Dioscorides (*DMM.* II.61) and Galen (12.363 K).

There was some dispute as to where the active ingredient was concentrated. Pliny (XI.118, XXIX.94) records that some were of the opinion that it was contained in the legs and head, but that others disagreed with this. Hippocrates (passim) and the Cyranides (p. 40) recommend that the insects be used with the wings, legs and head removed, but Galen (loc. cit) says that, although this is the practice of some, while others do the exact opposite, he is in the habit of using the entire insect. In fact, more cantharidin is contained in the wing cases than in the rest of the body.[86]

Since cantharidin is a blistering agent, *kantharides* are commonly prescribed, sometimes mixed with other substances, as an external application for the treatment of various skin diseases (Pliny XXIX.93,95, XXX.120; Dioscorides *DMM.* II.61; Galen XI.681, XII.363 K; Celsus V.8, 22.2, 28.12; Cyranides p. 40; Marcellus XIX.1; Cassius Felix p. 19; Isidorus *Or.* XII.5.5), *alopeciae* (Pliny XXIX.110), warts (Pliny XXX.81; Galen XII.363), sores and abscesses (Hippocrates VI.420 L; Pliny XXX.75), and conditions of the nails (Pliny XXX.111; Galen XII.363), for the removal of brand marks (Cassius Felix p. 21), and for the extraction of objects embedded in the flesh (Pliny XXX.122). Pliny (XXIX.93) records a fatality that occurred as a result of taking blister-beetles internally for a skin complaint.

Hippocrates frequently prescribes *kantharides* in his gynaecological works, to promote menstruation, to bring away the afterbirth, and to cause ejection of a dead foetus (VII.314,338,348,414,428, VIII.119,158,176,180,182,208,220,400,426 L). Such uses are referred to briefly by Pliny (XXIX.95) and Dioscorides (*DMM.* II.61), but are rejected out of hand by Soranus (*Gyn.* I.71), as an archaic practice that could only cause harm to the patient.

Blister-beetles are noted as an effective diuretic if taken internally in small quantities, together with other substances to counteract harmful effects (Pliny XXIX.95; Dioscorides *DMM.* II.61; Galen XI.368,609, XII.363, XIV.248, XV.913). They were also taken internally as a remedy for jaundice (Hippocrates VIII.258) and dropsy (Hippocrates II.512 L; Pliny XXIX.95; Dioscorides *DMM.* II.61; Galen I.667). Galen, however, records the unfortunate experience of a 'bold doctor' who tried blister-beetles both with and without removal of their appendages as a treatment for dropsy and had a fatality on both occasions. It is not surprising, therefore, that Caelius Aurelianus (*Chron.* III.140–1, V.54), who belonged to the same school of medicine as Soranus, dismisses these uses as too dangerous to be recommended.

Being toxic if taken internally, blister-beetles could also be used as a poison, and prosecutions involving them are recorded (Pliny XXIX.96; Cicero *Tusc.* V.117, *ad Fam.* IX.21.3; Valerius Maximus VI.2.3; Marcian *Dig.* 48.8.3.3; cf. Ovid *Ib.* 306, Diodorus

---

[86] J. Scarborough, 'Nicander's Toxicology', p. 14.

XXII.1, Tertullian *adv. Marc.* I.14). Some authors write as if the insects themselves are actively venomous in the manner of scorpions and *phalangia* (Plutarch *Mor.* 525 f., 537a, 874b; Diogenes L. VI.44; Jerome *adv. Ruf.* III.42).[87] The symptoms of *kantharis* poisoning are described by Nicander (*Alex.* 115 ff.), Ps. Dioscorides (*Peri Del.* 1), Paulus Aegineta (V.31), and Scribonius (189).[88] Suggested antidotes include a wide variety of herbal preparations (Nicander *Alex.* 128 ff.; Dioscorides *DMM.* I.45; *Eup.* I.156; Galen XIV.141; Pliny XX.133,220, XXIII.80,87; Celsus V.27.12), milk (Nicander; Dioscorides *DMM.* II.70, *Eup.* 1.156; Galen XII.269; Pliny XXVIII.128,160–1, XXIX.105; Celsus loc. cit.; Sextus Placitus V.b.16), fat or broth from various animals (Nicander; Dioscorides *DMM.* II.76, *Eup.* I.156; Galen XIV.141; Pliny XXVIII.160, XXIX.105), goose flesh (Dioscorides *Eup.* I.156), must (Pliny XXIII.29,62), and Samian or Lemnian earth (Nicander; Galen XII.174–5; Dioscorides *Eup.* I.156).[89] It was also believed that the wings or legs of *kantharides* were themselves effective as an antidote (Plutarch *Mor.* 22b, 554a; Galen XII.363, XIV.141; Dioscorides *DMM.* II.61; Pliny XI.118, XXIX.94): similarly, Pliny (XXIX.76) says that they provide an antidote to the poison of the salamander. Blister-beetles were employed in veterinary medicine for birth problems (*Hipp.Ber.* 15.9) and as a treatment for skin complaints (*Hipp.Ber.* 76.2; Chiron 922).

34. *Bouprestis*

1. Identification

The βούπρηστις, which acquired its name as a result of its reputation for being poisonous to cattle,[90] is an insect closely associated with the *kantharis* and described as having similar medicinal properties.[91] It is characterised by Dioscorides (*DMM.* II.61) as a 'type of *kantharis*' and by Galen (XII.364 K) as 'a kind of animal resembling the *kantharis* in appearance and potency'. The scholiast on Nicander (*Alex.* 335), who does not himself describe the insect, states that it resembles the *kantharis* or the *phalangion*, and that it is the same as the *staphylinos* of Ps. Aristotle.[92] According to Pliny (XXX.30), it is rare in Italy and very similar to a long-legged *scarabaeus*.[93] The Latin veterinary writers (Vegetius *Mul.* II.79.10; Chiron 453; cf. *C.Gl.L.* V.493.16) describe the insect as resembling a spider (*araneus*).

These details fit very well the oil-beetles of the genus *Meloe*, for example, the large blue-black *M. proscarabaeus* Linn., and the smaller metallic green *M. variegatus* Don.[94]

---

[87] There appears to have been some confusion between *phalangia* and *kantharides* (7b.1.3.ix above).

[88] Scarborough, art. cit., p. 13, says that Nicander's description is quite accurate.

[89] Nicander's antidotes are discussed by Scarborough, ibid., pp. 73–9, who concludes that only the Samian earth would have been of genuine benefit.

[90] Fernandez, *Nombres*, pp. 136–7.

[91] 33 above.

[92] 38 below.

[93] 31.1 above.

[94] Gossen, in PW 10.1480–1; Bodenheimer, *Geschichte*, p. 83; Leitner, *Zool. Terminologie*, pp. 63–4; J. Scarborough, 'Nicander's Toxicology', pp. 20–1, Davies and Kathirithamby, op. cit., pp. 91–2.

These beetles belong to the same family as the blister-beetles *Lytta* and *Mylabris*, and are sources of the same pharmaceutical product cantharidin.[95] They are, however, very distinctive in appearance and thus merited a separate name. Oil-beetles are so named as a result of their defensive habit of exuding an oily liquid. They are ungainly soft-bodied beetles, long-legged (hence the comparison with spiders), with short wing-cases which leave most of their abdomen exposed. They inhabit cattle pastures,[96] and may indeed cause poisoning to animals which ingest them with their food.[97] There is evidence of the survival of the name in dialectical Greek with clear reference to *Meloe*.[98] The life history of oil-beetles is complex and was unknown in antiquity: the larvae live as parasites in the nests of solitary bees.[99]

There is no native Latin equivalent of βούπρηστις, the Greek term being adopted in the form *buprestis* (late Latin *bubrostis*, *bubestris*). An alternative Greek name βουδάκη is recorded by Hesychius. A small variety is distinguished by Hippocrates (VII.360 L).[100]

2. Oil-beetles and domestic animals

Oil-beetles are frequently mentioned in veterinary and other sources (Nicander *Alex.* 344 ff.; Aelian VI.35; Pliny XXX.30; Vegetius *Mul.* II.79.10, II.142; Chiron 453,506; *Hippiatrica Ber.* 86.14, *Cant.* 71.1; *Geoponica* XVII.18; Isidorus *Or.* XII.8.5) as insects poisonous to cattle, which were liable to inadvertently swallow them while grazing, with potentially fatal results. The symptoms are said to be swelling of the belly and refusal to eat (Vegetius; *Hippiatrica*). Pliny and Aelian exaggerate somewhat by stating that the animals continue to swell until they burst open. It is recommended that affected animals should be bled, or given raisin wine or an infusion of cabbage with *garum* and oil to drink, or fed with wheat mixed with raisin wine and leeks, or that oil or fig juice be poured through their nostrils.

3. Medicinal and other uses

Oil-beetles are described as being prepared in a similar way to *kantharides*[101] for medicinal use (Pliny XXIX.95; Dioscorides *DMM.* II.61; Galen XII.364). They are said to have identical properties, being effective for gynaecological purposes (Hippocrates VII.361, VIII.158–60) and for the treatment of skin diseases (Pliny XXX.30).

Like *kantharides*, they had a reputation as a poison (Marcian *Dig.* 48.8.3.3), and their effects are similarly described (Nicander *Alex.* 335 ff.; Ps. Dioscorides *Peri Del.* 3; Paulus V.32; Scribonius 190). Suggested antidotes correspond to those listed for *kantharides* (Nicander *Alex.* 347 ff.; Pliny XXIII.29,62,80,87, XXVIII.74,128, XXIX.105, XXXI.119; Dioscorides *DMM.* I.45, V.113, *Eup.* II.157; Galen XIV.141 K; Cyranides p. 55).[102]

---

[95] J. Scarborough, ibid., p. 21.   [96] Bodenheimer, *Geschichte*, p. 83.   [97] Soulsby, *Helminths*, p. 521.
[98] Gow and Scholfield, *Nicander*, pp. 195–6.   [99] Imms, *Gen. Textbook*, p. 891.
[100] Perhaps *M. variegatus* contrasted with the larger *M. proscarabaeus*.
[101] 33.5 above.
[102] Cyranides p. 55 erroneously suggests that the insects actively bite or sting (cf. 33.5 and note 87 above).

4. Exotic species

Lucian (*Dips.* 3) includes the *bouprestis* in a list of venomous creatures occurring in Libya in large numbers and growing there to an unusual size.

35. *Lampyris/Cicindela*

1. Identification

The above names and their several synonyms cover the very distinctive beetles of the family Lampyridae, well known for their ability to produce light. The family includes both the glow-worms, for example *Lampyris noctiluca* Linn. and *L. splendidula* Linn., the females of which are wingless and resemble the larvae while the males are of normal coleopteran form,[103] and the fireflies of the genus *Luciola*, of which both sexes are fully winged.[104]

Both larvae and adults possess luminous organs which are situated on the hind segments of the abdomen, the emission of light being under the control of the insect. In the case of adults, the light serves as a signal to bring the two sexes together. The wingless females of *Lampyris* are sought out by the actively flying males, while in the gregarious genus *Luciola* both sexes are active. Adult Lampyridae take little or no food, but their larvae are carnivorous, preying in the case of *Lampyris* upon slugs and snails.[105]

1.2. Nomenclature

(i) *Λαμπυρίς* (v.l. *λάμπουρις*). The commonest Greek term for glow-worms and fireflies is used by Aristotle (*PA* 642b 34), who unusually employs two names for the same insect, Artemidorus (*On.* II.21), the Cyranides (p. 91), Hesychius, and Suidas. In the Latin glossaries it is the standard equivalent for *cicindela*. It is used by Pliny (XI.98, XVIII.250) in a latinised form *lampyris*.[106]

Λαμπυρίς and the following four synonyms are etymologically interrelated, and Strömberg and Fernandez[107] have suggested processes of development by which all five might have been derived from a single original.

(ii) *Πυγολαμπίς*. Aristotle's alternative name for the Lampyridae, used in *HA* 523b 19 ff. and 551b 24 ff.

---

[103] Sundevall, *Thierarten*, pp. 195–6, and Aubert and Wimmer, op. cit., I, p. 162–3, identify the Aristotelian *λαμπυρίς* as covering members of the genera *Lampyris* and *Luciola*.

[104] For example *L. italica*, cited by Sundevall, ibid., and *L. graeca*, cited by Gossen, in PW Supp.VIII.237. Gossen's identifications, in PW 10.1480, 1487–8 and Supp.VIII.237, of the various Greek synonyms as referring to distinct species are rightly rejected by Fernandez, *Nombres*, p. 84. The genus *Drilus* is not luminescent.

[105] Imms, *Gen. Textbook*, pp. 871–2.

[106] *Laparis*, which occurs in Polemius Silvius' list of insect names, is probably a corruption of *lampyris*: Ernout and Meillet, *Dict. Ét.*, p. 340.

[107] Strömberg, *Gr. Wortstudien*, pp. 13–14, suggest *λαμπυρίς* is the earliest form, giving rise to an inverted form *πυριλαμπίς*, and that in turn by a folk etymology to *πυγολαμπίς*, of which *κυσολαμπίς* would be a variant with the same meaning, with *χρυσολαμπίς* as a derivative. Fernandez, op. cit., pp. 83–6, argues that *πυγο-* and *κυσολαμπίς* are the older names, being folk appellations referring to the light at the end of the insect's 'tail': *πυρο-* and *χρυσολαμπίς* would then be more literary variants, with *λαμπυρίς* as an inversion of the former.

(iii) Πυρολαμπίς, πυριλαμπίς. Given as a synonym of *lampyris* by Photius and Suidas, and defined by them and Hesychius as 'a winged creature which shines in the dark', this also occurs as a v.l. for the preceding in mss of Aristotle's *HA*.

(iv) Κυσολαμπίς. Recorded only by Hesychius, who defines it as 'the *kantharis*[108] which shines by night'.

(v) Χρυσολαμπίς. Recorded only by Phrynichus Atticista (*Praep.Soph.* 126b).[109]

(vi) *Cicindela*. The native Latin equivalent of λαμπυρίς, derived from *cando*,[110] is used by Pliny (XVIII.250–3) and recorded by the grammarian Festus (p. 37 L) and by Isidorus (*Or.* XII.8.6), as well as occurring in the Latin glossaries (*C.Gl.L.* III.319.38, V.54.20, etc.). Festus describes the creatures as a *genus muscarum, quod noctu lucet*, while Isidorus says that *cicindela scarabaeorum genus est, eo quod gradiens vel volans lucet*.

(vii) *Luciculia*. A second Latin name, probably a compound of *lux* and *culus*,[111] which has been preserved only by *C.Gl.L.* III.319.40 and III.529.22, where it is given as equivalent to the Greek *lampyris*.

2. Appearance and habits

Glow-worms and fireflies are commonly described correctly as a form of beetle (Artemidorus *On.* II.21; Pliny XI.98; Hesychius s.v. *kysolampis*; Isidorus loc. cit.), although Festus (loc. cit.) describes them vaguely as a *genus muscarum*. The Cyranides (p. 91) describes the *lampyris* as a winged σκώληξ[112] which flies in summer and shines like a star during the night. Pliny is aware that the emission of light is under the control of the insect, but is under the impression that this is achieved by opening and closing the wing cases which he conceives of as covering the luminous parts of the body (XI.98, XVIII.252): he says that they *lucent ignium modo noctu laterum et clunium colore*.

Although there are no surviving references to glow-worms and fireflies in imaginative literature, Pliny provides (XVIII.250–3) an uncharacteristically poetic description of their relationship with Roman farmers, which makes it clear that, like cicadas,[113] these were among the few insects for which the ancients felt any affection.[114] Here we are told that the appearance of *cicindelae*, shining in the evening like stars scattered across the fields, is noted as a sure sign of the correct time to sow millet and harvest barley. This, Pliny

---

[108] Cf. 33 above, esp. note 68.

[109] Fernandez, op. cit., p. 235, includes κανδῆλα as a name for lampyrids on the strength of *C.Gl.L.* II.338.24, but *cicindela* probably has its secondary meaning of 'lamp' here.

[110] Ernout and Meillet, op. cit., p. 119.

[111] Th.L.L. s.v. Cf. Greek πυγολαμπίς.

[112] An unusual usage. Cf. 30.6.ii above.

[113] Cf. 16.1 above.

[114] Artemidorus' inclusion (*On.* II.21) of lampyrids as insects of ill omen would seem to be a mistake. In this section he has *kantharoi* chiefly in mind (cf. 31.4 above), and the presence of *lampyrides* and *melolonthai* (cf. 32.1.2.i above) is probably to be explained by his having copied out a list of 'types of beetle' from some other source. Cf. 11.1.2.iv above and 51.3 below.

continues, is an example of the *incredibilis benignitas naturae*; for she has not only provided the stars in the heavens as weather signs, but terrestrial stars as well (*non . . . his contenta terrestres fecit alias veluti vociferans, Cur caelum intuearis, agricola? . . . Ecce tibi inter herbas tuas spargo peculiaris stellas*).

3. Life history

The only surviving account of the development of *lampyrides* is provided by Aristotle (*HA* 551b 24 ff.). He notes elsewhere (*HA* 523b 19 ff., *PA* 642b 34) that these insects occur, like ants,[115] in both winged and wingless forms, but he displays some confusion as to their respective status. The wingless form, he relates, develops from a certain small hairy *kampe*,[116] and this in turn undergoes a further transformation giving rise to a winged creature known as βόστρυχος.[117] Sundevall[118] suggests that these three stages are to be identified respectively as the larvae (though these are not 'hairy'), wingless females, and winged males of *Lampyris noctiluca* and its relatives: in this case, βόστρυχος would also cover both sexes of adult *Luciola*. According to Hesychius (s.v. *lampyris*), these insects are generated ἐκ φρυγάνων (dry twigs or undershrubs).

4. Uses to man

According to the Cyranides (p. 91), a glow-worm or firefly placed in a leather amulet will prevent conception. The same work states that if one of these insects is thrown down in a place which is infested with fleas, the pests will be driven away.

36. Kis/Curculio

1. Identification and nomenclature

The Greek κίς belongs to a small group of mainly monosyllabic names referring to small insect pests.[119] Its use is normally[120] restricted to insects infesting stored grain or pulses, pests of the latter being sometimes denoted by distinct names (4 below).

The term therefore comprehends a considerable number of modern day species, including certain moths with grain-feeding larvae, most notably *Nemapogon granellus* (Linn.) and *Sitotroga cerealella* (Oliv.),[121] and members of a number of families of Coleoptera. The most significant of these beetles are the grain weevil *Sitophilus granarius* (Linn.), the saw-toothed grain beetle *Oryzaephilus surinamensis* (Linn.),[122] *Tenebroides mauritanicus* (Linn.), the anobiid *Stegobium paniceum* (Linn.) and members of the

---

[115] 44.2.1 below.

[116] Cf. 23.1.2 above.

[117] Cf. Fernandez, op. cit., p. 35.

[118] Op. cit., pp. 195–6.

[119] Κορίς, φθείρ, σής, ἴψ, σκνίψ, θρίψ. Cf. 17, 22, 25–6 above.

[120] It is equivalent to θρίψ (37 below) in Hesychius, and probably in Pindar fr. 222.

[121] The former referred to by Keller, op. cit., II p. 413, and Leitner, *Zool. Terminologie*, p. 107.

[122] *S. granarius* is named by Keller, op. cit., II p. 413, Gossen, in PW 10.1483–4, and Leitner, op. cit., p. 107. From this species derives the traditional translation of κίς and *curculio* as 'weevil'.

Tenebrionidae such as *Gnathocerus cornutus* (Fab.), *Tribolium confusum* Duv., *T. castaneum* Herbst, and the mealworm beetles *Tenebrio molitor* Linn. and *T. obscurus* Fab.[123] Some of these are more particularly associated with grain products such as flour. Specimens of a number of the above species have been identified from excavations of Roman granaries and other sites.[124]

The usage of the Latin *curculio* (later Latin *gurgulio*, Palladius and Isidorus *Or.* XII.8.17; *conculio*, *C.Gl.L.* III.281.19; *curcurio*, *C.Gl.L.* V.282.46) corresponds exactly with that of κίς.[125]

Grain pests are commonly spoken of as a type of *skolex* (Theophrastus *HP* VIII.11.2; Anaxilas fr. 33; Suidas s.v. *kis*; Eustathius in *Il.* 113.45) or *vermis* (Augustine MPL 38.412; *C.Gl.L.* IV.501.32 *vermis frugibus inimicus*, V.282.46, VI.297).[126] They are also referred to by the louse names φθείρ (Lucian *Ep. Sat.* II.6; *C.Gl.L.* II.470.58) and *pediculus* (*C.Gl.L.* II.576.6, V.281.19, 448.47 *peduculus frumenti*),[127] and by the Greek σής, normally referring to the clothes moth,[128] (*C.Gl.L.* III.188.54) and its Latin equivalent *tinea* (*C.Gl.L.* V.281.19).[129] An apparent Greek synonym σκωλοβάτης[130] is preserved only by Hesychius, who defines it as an ὄνομα ζῳοῦ μικροῦ παραπλησίου ἐρισύβῃ[131] which causes damage to grain.[132]

2. Life history

The true life histories of the various species of grain pests were unknown in antiquity. Instead, as with most noxious insects, a general belief prevailed that they were spontaneously generated within the grain store itself (for example Theophrastus *HP* VIII.11.1–3, *CP* IV.15.4; Pliny XVIII.304; Varro *RR* 1.63). According to Theophrastus, they develop under the influence of moisture as the grain upon which they feed begins to rot.

3. Importance as pests

Κίες or *curculiones* are generally acknowledged as a major threat to stores of grain, agricultural writers devoting considerable space to means of preventing or combating

---

[123] Cf. Imms, *Gen. Textbook*, pp. 885–6.

[124] Including *Sitophilus, Oryzaephilus, Stegobium, Tenebrio, Cryptolestes*, and *Palorus*. Cf. P. J. Osborne, 'An Insect Fauna from the Roman Site at Alcester', *Britannia*, II, 1971, pp. 162–4; R. A. Hall and H. K. Kenwood, 'Biological Evidence for the Usage of Roman Riverside Warehouses at York', *Britannia*, VII. 1976, pp. 274–6.

[125] A. Thomas, 'Le Laterculus de Polemius Silvius', *Romania*, XXXV, pp. 171–2, suggests that *corgus* in Polemius Silvius' list of insect names is equivalent to *curculio*.

[126] On *skolex* and *vermis* cf. 23.1.2 above.

[127] Cf. 22 above.

[128] Cf. 26.1 above.

[129] Cf. 26.1 above, 37.1.3.i below.

[130] Rendered 'weevil' in LSJ, but not included by Fernandez, *Nombres*.

[131] On ἐρυσίβη cf. 11.1.2.xxiv above.

[132] The confused gloss *C.Gl.L.* II.433.46 defines *curculio* as 'the σκνίψ (cf. 58 below) which some call ἀγριομύρμηξ (Fernandez, *Nombres*, p. 160, and LSJ take the gloss at face value and give this word as a synonym of *kis*, but it is more likely to refer to those ants known as pests of grain. Cf. 44.1.2 below), and according to others a type of κώνωψ (cf. 50 below)'.

their depredations. As such, they are mentioned by a number of non-technical writers (for example Aristophanes fr. 141; Anaxilas fr. 33, who describes them as boring into wheat grains one by one; Plautus *Curc.* 586–7, *Rud.* 1325; Virgil *G.* I.185–6 *populatque ingentem farris acervum curculio*; Symphosius *Aen.* 24 in *Anth.Lat.* 440; Arnobius *adv.Nat.* II.47; Prudentius *C.Or.Symm.* II.1054).

The most effective countermeasures proposed by classical agriculturalists were concerned with the actual siting and construction of granary buildings. It is generally recommended (Varro *RR.* I.57; Columella *RR* 1.6.15–6; Vitruvius *Arch.* VI.6.4) that granaries should be raised somewhat off the ground, and that they should be open to the air on the east and north sides. This advice displays an awareness that keeping the grain cool and dry is the most effective precaution against infestation by pests.[133] In some regions outside Italy, as Varro and Columella (loc. cit.) inform us, underground storage was practised: the effectiveness of this method would have lain in keeping the grain free of oxygen.[134] It is also generally recommended (Cato *Ag.* 92; Varro loc. cit.; Columella *RR* 1.6.14; Pliny XV.33; Palladius I.19) that the interior of the granary be painted with a mixture of *amurca*, clay and chaff or dried olive leaves, or with marble cement. The ancients would not have been aware of precisely why this would be effective, but in fact it would serve to cover over cracks and crevices in which insect eggs might otherwise survive periods during which conditions would be inimical to their development.[135] Varro (loc. cit.) and Pliny (XVIII.305) recommend sprinkling the stored grain with *amurca*, wormwood, chalk, or Olynthian or Cerinthian earth; and Palladius (loc. cit.) mentions a similar use of coriander or fleabane foliage. The latter advises that grain should not be put into storage immediately after removal from the threshing floor, but should be aired for a few days. Pliny (XVIII.303) states that the time for storage should be carefully chosen to minimise the risk of pests developing. He also records a folk belief involving the hanging of a toad at the threshold of the store before taking the grain inside.

Columella (*RR* I.6.16), followed by Pliny (XVIII.302), states that some were of the opinion that if an infestation of insect pests had already begun it was useful to winnow the grain in situ in order to air it; this, however, is in his view an error, since such a practice only serves to distribute the pests through the whole mass, whereas if it is left undisturbed they do not penetrate deeper than a palm's breadth. Varro (I.63) suggests that infested grain be brought out into the open and vessels of water placed around it: according to him the pests will congregate around the latter and drown themselves.

4. Pests of stored pulses

The pests of stored pulses[136] are closely associated by ancient authors with those of cereals. The modern day species particularly in view here would be the pea and bean weevils of

---

[133] G. Rickman, *The Corn Supply of Ancient Rome*, Oxford, 1980, p. 123.

[134] Ibid.

[135] Ibid.

[136] They are not always clearly distinguished, especially by Theophrastus, from the pests of growing pulses, on which cf. 23.1.4.ii above.

the family Bruchidae.[137] Some bruchids, for example *Bruchus pisorum* (Linn.), have larvae which feed on the living plants, their eggs being laid on the young pods and the resultant larvae boring inside, while others attack dried seeds in store, for example *Acanthoscelides obtectus* Say.[138]

Theophrastus (CP IV.15.4) restricts the term κίς to pests of wheat and barley, denoting those which occur in beans by the name μίδας,[139] used elsewhere only in Hesychius, who defines it similarly as 'a certain small creature which eats through beans'. He adds that similar pests are found in other pulses such as peas, lentils, and tares, but gives these no specific name. In *HP* VIII.11.2, however, all such insects, whether associated with cereals or pulses, are described as *skolekes*. Two apparent synonyms of Theophrastus' μίδας, τρώξ and κοίελος, are preserved only in the lexicographical tradition. The first, derived from τρώγω, appropriately enough for an insect pest,[140] is defined (Photius = Strattis fr. 80; *EM* 770.43; Hesychius; Tzetzes *ad* Hesiod *Op*. 418) as 'the small creature in pulses'; while the second is defined (Hesychius) as 'the small creature in beans'.

In Latin sources pests of beans, lentils and other pulses are generally termed *curculiones* (Columella *RR* II.10.11, 11.16; Palladius VII.3.2). There are examples of their being loosely termed *vermes* (Augustine MPL 38.412) or even *musca* (Augustine MPL 32.1372 *musca quae in faba gignitur*).

According to Columella (*RR* II.10.11), beans are less liable to be infested if before the seeds are sown they are soaked in *amurca* or soda, or if (also Palladius VII.3.2) they are gathered and safely transferred to the store house before the new moon. Discussing lentils, the same author (*RR* II.16) states that before storage any infested seed should be separated out by immersing the whole crop in water, since that which is infested will float, and that the remainder should be dried in the sun and treated with vinegar and silphium root (cf. Pliny XVIII.308).

Pliny (XVIII.308) and Palladius (VII.3.2) record that some were of the opinion that leguminous seeds could be protected by piling them over jars of vinegar placed on a bed of ashes and coated with pitch, or by storing them in plastered casks that have contained salted fish.

5. Medicinal uses

Pliny (XXII.121) refers to a certain *in farre vermiculus teredini*[141] *similis* which was said to be effective for the removal of decayed teeth.

---

[137] Gossen, in PW 10.1488 and Supp.VIII.239–40, names various species.

[138] Cf. Imms, op. cit., p. 894.

[139] Gossen, in PW 10.1488, supported by Fernandez, op. cit., p. 193, puts forward an unlikely explanation for the origin of this name.

[140] Fernandez, op. cit., pp. 103–4.

[141] On the name *teredo* cf. 30.1.2.iv above.

37. Thrips/Tinea

1. Identification: general

The above names, along with their several synonyms, cover various species of small wood boring pests. They are particularly associated with cut timber, but are also referred to as attacking living trees. Normally they are clearly distinguished from the larger wood boring larvae known by various names,[142] but due to the close link between these two groups of insects the nomenclatural boundary between them is not always precisely drawn. This is particularly evident in connection with the Greek term σκώληξ, which Theophrastus uses as his distinctive name for the larger larvae,[143] but which at one point is also used by him to refer to the smaller θρῖπες. Such usage results from the ambiguity of σκώληξ, which, although it literally means a 'worm' or 'larva',[144] etymologically a creature which twists or curves its body,[145] can also be applied to the small nondescript pests of the *ips–knips* complex, as with our English expression 'wood*worm*'.[146]

The insects involved here belong to a number of families of Coleoptera. Of those which infest cut timber, the most important are members of the Anobiidae, which includes the well known domestic furniture-beetle, *Anobium punctatum* (deG.), and the more dangerous death-watch, *Xestobium rufovillosum* (DeG.), attacking furniture, roof beams, and flooring.[147] The females of these insects lay their eggs on wood surfaces, and the larvae tunnel inwards on hatching. The adults of the death-watch are noted for the tapping sound which they produce while still in their larval tunnels prior to emergence.[148] Related to the Anobiidae are the powder-post beetles of the family Lyctidae, for example *Lyctus fuscus* (Linn.) and *L. brunneus* (Steph.), and members of the Bostrychidae.[149]

Those attacking living trees belong chiefly to the family Scolytidae. Most of these beetles tunnel directly beneath the bark, constructing entrance tunnels through the bark as far as the wood surface, and then elongated galleries between the two. The eggs are laid along the walls of these galleries, and the resultant larvae bore their own tunnels at right angles. The pattern of borings which results is often very complex, and is recognisably distinct for different species.[150]

1.2. Identification: Greek nomenclature

(i) Θρίψ. This, the most usual Greek term for the smaller timber pests, belongs to a complex of names referring to small insect pests[151] and is occasionally used to refer

---

[142] 30 above.   [143] 30.1.2.i above.   [144] Cf. 1.1 and 23.1.2 above.
[145] Fernandez, *Nombres*, p. 147.   [146] Cf. 26.1.2 and 36.1 above.
[147] Gossen's various identifications, in PW 10.1488 and Supp.VIII.238–41, which have been largely accepted by Leitner, *Zool. Terminologie*, pp. 231–2, 235–8, are over-precise and mostly dubious. He identifies *thrips* chiefly with the bark feeding pine-weevils of the genus *Hylobius*.
[148] Imms, *Gen. Textbook*, p. 874.
[149] Specimens of *Lyctus* and *Anobium* have been recovered from archaeological contexts. Cf. P. J. Osborne, 'An Insect Fauna . . .', *Britannia*, II, 1971, p. 162.
[150] Imms, op. cit., pp. 901–2.   [151] Cf. 36 note 119 above.

to other species.[152] It is etymologically related to τερηδών and the Latin *tarmes*, which all suggest a creature which bores, drills or perforates.[153]

Theophrastus (*HP* V.4.5) divides the pests of timber into three groups: the marine τερηδών,[154] the σκώληξ,[155] and the θρίψ. He characterises the latter as an insect resembling the *skolex* which bores its way little by little through timber. There is no explicit mention of the *thrips* in the *Historia Animalium*, though at 616b 29 a species of bird is described as θριποφάγος.

The *thrips* is referred to by Menander (fr. 540), Polybius (VI.10.3, linked with τερηδών; cf. *AP* XII.190), and Plutarch (*Mor.* 49b) as a characteristic pest of timbers.[156] Elsewhere, Plutarch (*Mor.* 924a) alludes to them in passing by saying that the inhabitants of the other side of the world 'do not hang upside down like *thripes* or geckoes'. Here the *thrips* is characterised as a domestic insect like the gecko, a creature well known for its ability to move over ceilings: perhaps the reference is to the insects living suspended in roof beams. Lexicographical sources (Ammonius *Diff.* 244; *EM*, Suidas, Photius s.v. θριπηδέστατον) commonly define the insect as one associated with timber.

The associated adjectives θριπήδεστος and θριπόβρωτος tend to be used as general terms for 'worm eaten' timber (for example Theophrastus *HP* III.8.5, V.1.2; Hesychius and Suidas svv. θριπήδεστον, θριπηδέστατον, θριπόβρωτος; IG 2.1628–9). Their meaning is occasionally extended further to cover damage to, for example, stored medicinal roots (Theophrastus *HP* IX.14.3).[157]

(ii) Σκώληξ. In *HP* V.1.2, Theophrastus clearly uses this term for what, according to his later definition (i above), should more properly be termed *thripes*.[158] Similarly, Suidas (s.v. θριπηδέστατον) states that '*thripes* are a type of *skolex*' and *EM* (iv below) speaks of 'the *skolex* in timber'.

(iii) Σής. More usually referring to the clothes moth,[159] this term is used to describe insects destroying wooden sticks (Hermas *Sim.* VIII.1) and roof timbers (Cyril Alexandrinus *in Hab.* 2.11 MPG 71.882). Defining the meaning of *thrips*, Hesychius (s.v. θριπήδεστον) states that 'the σῆτες in timber are so named'.

(iv) Κίς. Normally a term for grain pests,[160] κίς is applied to small wood borers as early as Hesiod (*Op.* 435), who describes the type of wood most resistant to pest damage

---

[152] It is used for clothes moths (26.1 above), for pests attacking horn (*EM*, Suidas and Photius s.v. θριπηδέστατον), which would also be clothes moths (cf. 26.1, 4 above), and for the fabulous 'snow worm' (59 below).

[153] Fernandez, op. cit., pp. 114–15.

[154] Cf. 30 note 177 above.

[155] 30.1.2.i above.

[156] Cf. 26.3, 30.1.2.i above.

[157] Cf. 30.1.2.ii above.

[158] Some of the other references to σκώληκες cited under 30.1.2.i above, especially Apollodorus I.9.12 (cf. 30.3), may in fact belong here.

[159] 26 above.

[160] 36 above.

as ἀκιώτατον. The *Etymologicum Magnum* (s.v. ἀκιώτατον) explains this passage by saying that 'the *skolex* in timber is called κίς'. Similarly, Hesychius defines *kis* as 'a small creature generated in timber and in grain'. *Kis* probably has this meaning in Pindar fr. 222, where gold is characterised as that which οὐ σὴς οὐδὲ κίς δάπτει.

(v) *Κνύψ*. This term[161] is probably synonymous with *thrips* in two passages of the *Historia Animalium* (593a 3, 614b 1), where it is applied to small insects living under the bark of trees and hunted by woodpeckers along with *skolekes*.[162] Theophrastus (*HP* IV.14.10) uses it once to refer to insects generated in wood. Plutarch (*Mor.* 636d), discussing the subject of spontaneous generation, associates *sknipes* and *teredones* as creatures both produced in decaying wood: he seems to link the first with trees and the second with cut timber. Both Hesychius (s.v. *knipes*) and Suidas (s.v. *sknips*) have 'small timber feeding creature' among their several definitions of this word, as does Zenobius (V.35).

(vi) *Κάνθαρος*. Cyril of Alexandria (loc. cit.) speaks of small κάνθαροι otherwise known as σῆτες infesting old roof timbers, but, since he is here attempting to explain a mistranslation in the LXX which had incongruously introduced *kantharoi* into the text,[163] this is not necessarily evidence for the regular use of *kantharos* in this sense.

(vii) *Ἀχραδίνης*, defined as 'certain timber feeding creatures' by Hesychius, seems to be derived from ἀχράς 'wild pear', and so could be taken to refer to a wood boring pest associated with this tree.[164]

1.3. Identification: Latin nomenclature

(i) *Tinea*, commonly used for clothes moths and booklice,[165] seems to be the standard Latin equivalent of θρίψ, being used in this sense by Pliny (XVI.197,220,223) and Vitruvius (*Arch.* II.9.12–14, V.12.7). In the latter passage *tineae* are linked with *teredines*, *teredo* probably referring here to the shipworm, since shipyards are under discussion, rather than to another variety of terrestrial wood borer.[166] There are no clear examples of the Latin term being used to refer to pests of living trees.[167]

(ii) *Thrips*. The latinised form of the Greek name is used only by Pliny, in his rendering (XVI.220) of Theophrastus' passage (1.2.i above) on the various types of wood boring pests. Pliny extends Theophrastus' list to four, *teredo*, *vermiculus*, *tinea*, and *thrips*, presumably because of uncertainty as to whether the latter two names were synonymous. Theophrastus' statement that *thripes* are similar to the *skolex* is here mistranslated as *culicibus similes*.

---

[161] Cf. 58 below.
[162] Cf. the designation of another bird as θριποφάγος in *HA* 616b 29.
[163] Cf. 31 note 2 above.   [164] Fernandez, op. cit., p. 166.   [165] 26 above.
[166] Cf. 30.1.2.iv above.
[167] The identity of Columella's (*RR* V.10.9; cf. Pliny XVII.256) *tineae* attacking fig seedlings is unclear.

(iii) Pliny's *scarabaei . . . excavantes nocturno stridore vocales* (XI.98) are probably to be identified as crickets rather than, as has been suggested by some, Coleoptera such as the death-watch.[168]

2. Life history

Thripes were generally held to be spontaneously generated within the wood upon which they feed, as a consequence of its decay (Theophrastus *HP* V.4.4; Polybius VI.10.3; Hesychius s.v. κίς).

3. Thripes as pests

These insects are commonly referred to as causing damage to the structural timbers of buildings, it being noted that some forms of wood are more liable to be attacked than others (Theophrastus *HP* V.4.1–6; Plutarch *Mor.* 49b; Pliny XVI.223; Vitruvius *Arch.* II.9.12–14). As a preventive measure, Pliny (XVI.197) recommends the treatment of timbers with cedar oil.

4. Uses to man

According to Theophrastus (*HP* V.1.2) and Hesychius (s.v. θριπόβρωτος), wood which had been marked by the tunnelling of *thripes* beneath the bark of trees was used for the making of seals. The latter says that this was particularly a Spartan practice, said to have been invented by Herakles. Such θριπήδεστα σφραγίδια are also alluded to by Aristophanes (*Thesm.* 726), Lucian (*Lex.* 13), and Lycophron (*Alex.* 508). As Theophrastus' description makes clear, the reference here is clearly to the labyrinthine patterns formed by the tunnelling of Scolytidae, which would have provided an infinite variety of distinctive imprints.

38. Sphondyle and Staphylinos

Although probably referring to unrelated species, this pair of beetle names are closely associated in ancient authors. Aristotle uses the former once (with the variant spelling σπονδύλη) alongside flies and *kantharides* as an example of insects which mate, but he fails to provide a description of the insect in question. Theophrastus (*HP* IX.14.3) uses it to refer to an undescribed pest attacking the roots of all kinds of plants, and in a latinised form *sphondyle* it appears in Pliny XXVII.143 in the same sense, characterised there as a *genus serpentis*.[169] In the inauthentic book IX of the *Historia Animalium* (619b 22) the creature is included in a list of the prey of nocturnal birds, along with mice and lizards, and similarly in Aelian (VIII.13) it figures in the diet of an Ethiopian variety of scorpion. The latter author also includes the *sphondyle*, together with mice, centipedes, and others, in a list of creatures observed to depart from the town of Helike en masse shortly before it

---

[168] Cf. 12.1 above.

[169] *Serpens* is very unusual for an invertebrate, but cf. 3.1.4.i above.

was devastated by an earthquake (XI.19). Aristophanes (*Pax* 1078) refers to the *sphondyle* as a creature which emits an unpleasant smell as it flees,[170] the scholiast on this passage defining it as a type of *silphe* or cockroach.[171] Similarly, Hesychius describes it as a creature ὁμοίον σαλαφίῳ[172] which produces an ill odour if it is touched.

The σταφυλῖνος is mentioned in *HA* 604b 19 as an insect of the same size as a *sphondyle* which is poisonous to horses if swallowed. The scholiast on Nicander *Alex.* 335 equates it with the *bouprestis*[173] which was noted as similarly dangerous to oxen. The *Hippiatrica* (*Ber.* 119) contains a full chapter detailing its effect on horses and prescribing possible treatments; here the insect is described as resembling the *sphondylai* which live in houses, but as larger in size; it is common in pasture land, and walks around with its tail raised in the air. Hesychius calls it σταφυλῖνος ἄγριος, 'an animal the same size as the *sphondyle*'.

Since it is generally mentioned in passing, the *sphondyle* was clearly a familiar insect which ancient authors expected their readers to recognise without the need for explanation. The *Hippiatrica* explicitly describes it as a domestic species, and of the other references Aelian XI.19 implies the same habitat. In view of this, and the comparison with cockroaches, it is most probably to be equated with Pliny's *blatta exacuta clune*,[174] which is similarly described as ill smelling and can be identified as the tenebrionid beetles of the genus *Blaps*. These have a similar domestic habitat to cockroaches, and do indeed produce a distasteful odour.[175]

The *staphylinos*, however, judging from the description in the *Hippiatrica*, would appear to be one of the large rove beetles such as *Ocypus olens* (Muell.).[176] These are suitably sinister looking insects, noted for their habit of curving the tip of their abdomen over their back in a defensive posture.[177] Their association with the *bouprestis* or oil-beetle is readily explicable as due to similarity of appearance, both groups of beetles being elongated and soft-bodied, with very short wing-cases which leave much of their abdomen exposed.[178]

---

[170] E. K. Borthwick, 'Beetle Bell Goldfinch and Weasel in Aristophanes' Peace', *CR* NS XVIII, 1968, p. 138, suggests that in this passage Aristophanes is referring to the mammal γαλῆ, said by Hesychius to be also called σφονδύλη, probably because like the insect it was noted for producing an odour.

[171] 13 above.

[172] Σαλαφίον, not recognised by LSJ, is accepted as a genuine insect name by Gossen, *Zool. Glossen*, p. 112, though his suggested identification is purely speculative.

[173] 34 above.

[174] 13.1.2.iv above.

[175] So identified by Sundevall, *Thierarten*, pp. 237–8. Gossen's suggested identifications, in PW 10.1488, PW Supp.VIII.237,239, and *Zool. Glossen*, p. 112, all refer to insects which are neither ill smelling nor domestic; the bombardier beetles, *Brachinus* spp., mentioned also by Davies and Kathirithamby, op. cit., p. 95 and various commentators on Aristophanes, produce a volatile liquid rather than an offensive odour. Theophrastus' reference to root feeding must be due to some error.

[176] So identified by Gossen, in PW VIII.237. Sundevall, op. cit., pp. 237–8, suggests the insect may be a staphylinid, but prefers to identify it as identical to the *bouprestis*. The etymology of σταφυλῖνος is uncertain: cf. Fernandez, *Nombres*, pp. 42–3.

[177] Cf. Imms, *Gen. Textbook*, p. 855.

[178] Cf. 34.1 above.

39. *Kleros* or *Skleros*

In book IX of the *Historia Animalium*, in a passage which is followed by later writers on the subject (605b 9 ff.), three distinct types of pest infesting beehives are listed: an insect resembling the moths which fly to light,[179] a type of caterpillar known as τερηδών,[180] and a *skolekion* known as κλῆρος, which causes damage to the combs and produces cobwebs like those of a spider.[181] Later in the same book (626b 15 ff.), the *kleros* is described as a νοσήμα in which small larvae are generated on the floor of the hive: as they grow, they produce webs which take over the whole hive and cause the combs to decay.

The variation between these two passages is reflected in the lexicographical tradition, Hesychius referring s.v. κλῆρος to 'a certain small larva generated in beehives', and s.v. σκλῆρος to a νοσήμα involving the spread of cobwebs and decay of the combs.[182] Hesychius also lists a related term σκῆνος, not found elsewhere, which he defines as a πάθος affecting bees and consisting in the spontaneous generation of *skolekes* in the hive.[183]

The three types of hive pest distinguished by Ps. Aristotle recur in Latin sources (Pliny XI.65, XXI.81; Columella *RR* IX.14.1; Palladius IV.15.4), where they are termed respectively *papilio, teredo* or *tinea*, and *araneus*.[184] The latter are described, as in the *Historia*, as damaging the combs and covering them with webs (Columella *aranei qui favos corrumpunt*; Pliny *cum praevaluere ut intexant enecant alvos*). The same methods are proposed for combating these pests as are prescribed for dealing with wax moths.[185]

Since the *kleros* or *araneus* is so inextricably confused with real spiders and with wax moths and their larvae, which are also described as producing webs,[186] it is not possible to determine with certainty what kind of insect is being referred to. As the cobwebs described by our sources would have been solely the result of the activities of moth larvae, we are clearly dealing with some other pest which was observed to infest beehives and was erroneously credited with the spinning of some or all of these conspicuous webs. It has been suggested[187] that the insect in question is the clerid beetle *Trichodes*, whose larvae infest beehives and prey upon the bee larvae in their cells.

---

[179] Cf. 24.1,2 above.

[180] Cf. 24.2 above.

[181] The sentence which follows is plainly corrupt and is best left out of account. The text as we have it would make κλῆρος synonymous with the moth name πυραύστης (24.1.2.ii above), which is clearly wrong, as has been recognised by editors from Schneider onwards (cf. D'Arcy Thompson, *Hist. An.*, VIII.28 note), although Fernandez, *Nombres*, pp. 90–1, argues against this view. The phrase ἐντίκτει . . . οἶον ἀράχνιον is probably simply a repetition of the statement that the insect produces webs, if we omit ὅμοιον ἑαυτῷ as Dittmeyer and D'Arcy Thompson propose.

[182] Cf. Pliny XI.64, where *claros* is simply described as a disease.

[183] Cf. also the term σκήν (23.1.3.iii above). Fernandez's suggested etymology for these names, op. cit., pp. 90–1, depends upon his erroneous identification of them as referring to the wax moth.

[184] There is confusion here with genuine spiders, which could also constitute a threat to hive bees: cf. 7a.7 above.

[185] Cf. 24.4 above.

[186] Cf. 24.2, 4 above.

[187] Sundevall, *Thierarten*, p. 196; Aubert and Wimmer, op. cit., I p. 166; D'Arcy Thompson, op. cit., VIII.28 note.

VII

BEES, WASPS AND ANTS

Insecta: Hymenoptera (Bees, Wasps and Ants)

40. *Sphex* and *Anthrene*; *Vespa* and *Crabro*

1. Identification: general

These two pairs of names cover the various species of colonial wasps along with their larger relative the hornet, plus certain solitary wasps which were sometimes distinguished by name from their allies.[1]

The colonial wasps, which are far more conspicuous than the solitary forms, belong to the family Vespidae. They include species which nest underground, for example *Vespula vulgaris* (Linn.), *V. germanica* (Fab.) and *V. rufa* (Linn.), and others which suspend their nests from the branches of trees and bushes, for example the tree wasp *Vespula sylvestris* (Scop.). The closely related hornet, *Vespa crabro* Linn., prefers to nest in hollow trees. Colonies of these species last only for a single season, the workers and males dying at the end of the year, and the females or queens hibernating. On emerging in spring, the females set about constructing their nests out of wood rasped from weather worn surfaces and mixed with saliva. The cells are constructed in tiers from the top downwards, each tier attached by a single pedicel. The larvae which hatch from eggs laid in these cells are fed upon portions of animal material, usually other insects. Near the end of the summer, especially large cells are built in which new queens are reared.[2]

Belonging to the same family is *Polistes gallicus* Linn., also a social species but with colonies which are much smaller: its nest consists of a single tier of cells without an external envelope.[3]

The solitary wasps of the families Eumenidae, Pompilidae, and Sphecidae construct burrows or mud nests in which to lay their eggs, providing food for their offspring in the form of whole insects or spiders which they kill or paralyse by stinging.[4] In Greek terminology, σφήξ (Doric σφάξ, Theocritus V.28) is the more common and generally used term, with ἀνθρήνη as a subdivision. Traditionally,[5] *anthrene* has been seen as the

---

[1] Cf. Keller, *Ant. Tierwelt*, II pp. 431–5; W. Richter, in PW Supp. XV.902–8. Sundevall, *Thierarten*, p. 218, and D'Arcy Thompson, *Hist. An.*, V.22 note, identify Aristotle's ground nesting species as for example *Vespula rufa*, *V. germanica*, and *V. vulgaris*.

[2] Imms, *Gen. Textbook*, pp. 1247–9.   [3] Ibid., p. 1249.   [4] Ibid., pp. 1245–7, 1251–3.

[5] So Richter, art. cit., 902; although Sundevall, op. cit., pp. 220–1, and Aubert and Wimmer, *Thierkunde*, I pp. 159–60, express uncertainty as to its identity. Latin *crabro* is identified as the hornet by Keller, op. cit., II p. 435, and Leitner, *Zool. Terminologie*, p. 102.

larger hornet, but actual usage tends to be imprecise, and the two words are to some extent interchangeable. In the authentic portion of Aristotle's *HA*, *anthrene* may well be the hornet, but in the pseudonymous book IX what is clearly the hornet is termed ἄγριος σφήξ, while *anthrene* is applied to a non-colonial species. In non-technical literature, a corresponding imprecision is found. Aristophanes (*Vesp.* 1080, 1107) uses ἀνθρήνιον instead of σφηκείον for the nests of what he otherwise refers to as σφῆκες, and Philostratus (*Im.* 13) apparently uses the same word to mean 'honeycomb'. A similar confusion between wasps and bees occurs in *Vesp.* 1114, where mention is made of drones feeding upon the honey stores of their colony (cf. Herodotus II.92), and perhaps in Aelian IV.39.

Among the lexicographers (Hesychius; Suidas), *anthrene* is defined both as 'a kind of *sphex*' and as 'a kind of bee'. Suidas explains that 'some use the word for bees, as the *anthrene* is related to the bee'. In the Latin glossaries, *vespa* is defined on four occasions as a 'wild bee' (III.319.45, 507.5 ἀγρία μέλισσα; III.436.13 ἀγριομέλιττα; V.335.53 *apis rustica*). A number of authors view wasps as degenerate versions of the honey bee (Plutarch *Mor.* 96b; Lucian *Char.* 15; Tertullian *adv.Marc.* IV.5).

Latin nomenclature is similarly variable. Traditionally *sphex* has been equated with *vespa*, and *anthrene* with *crabro* (late Latin *cabro, crabo, scabro,* etc.).[6] But Pliny, rendering Aristotle, translates the Greek in the opposite way, so that the ἄγριος σφήξ of book IX becomes *crabro silvestris*. The same equation of *sphex* with *crabro* is found in Avienus' version of Aratus, the Vulgate as compared with the LXX, and the Latin glossaries, where *sphex* is defined only once as *vespa* but eight or nine times as *crabro*, while *vespa* is correspondingly rendered in four places by a variant of *anthrene*. In addition to these bilingual glosses, *crabro* is defined on five occasions as a particular type of *vespa* (IV.314, etc., *vespa longa*; IV.167.62, V.243.4 *vespae longiora crura habentes*).[7]

1.2. Alternative names and distinct varieties

(i) Ἀνθηδών. This term is clearly a simple synonym of ἀνθρήνη as it is used in the Latin glossaries (*C.Gl.L.* II.207.28, 227.21, III.319.37, 474.15), where it is equated with *vespa*, and by Galen (XIV.91 K), where it figures in a list of stinging invertebrates. The lexicographers (Hesychius; *EM*; *Et. Gud.*) define it as a bee, doubtless as a result of the general confusion between these and wasps. Aelian uses the word (XV.1) as one of a trio of Hymenoptera, alongside the honey bee and the *sphex*.[8] Fernandez[9] takes the word to be a variant form of ἀνθρηδών, itself etymologically a derivative of ἀνθρήνη,[10] resulting from the false etymology (given in *EM*) which linked it with ἄνθος.

---

[6] The *sc-* forms are influenced by *scarabaeus*; hence the confusion in *C.Gl.L.* II.338.25, III.258.29; cf. Ernout and Meillet, *Dict. Ét.*, p. 147.

[7] The erroneous gloss II.374.40 apparently equating *vespa* with the Greek μύωψ is considered by Fernandez, *Nombres*, p. 82.

[8] Cf. 53 below.

[9] Op. cit., p. 43.

[10] Cf. 41 below.

(ii) Δέλλις. A local synonym of *sphex*, which survives in some modern South European dialects.[11] It is defined by Hesychius as 'sphekes or a creature like the honey bee'. The second of these definitions is repeated by the grammarian Herodian (*Tech.* II.709,741,761), who also gives the meaning τὸ μικρὸν σφηκίον. Hesychius also records the word δελλίθιον, 'wasp's nest', which he says is variously explained as equivalent to ἀνθρήνιον or κηρίον 'honeycomb' (cf. 1 above).

(iii) Πάρνοψ. According to Suidas, this locust name[12] was also applied to wasps.

(iv) Ἀλακάται. Defined by Hesychius as 'earthworms or, according to some, sphekes'.[13]

(v) Φερέοικος. According to Hesychius, Photius, and Suidas, this term could refer either to the snail, or to a small mammal, or to 'a creature larger than the *sphex*'.

(vi) In HA 554b 22 ff., *sphekes* and *anthrenai* (Pliny XI.71 *crabrones* and *vespae*) are two forms of wasp, but the difference between them is not stated. The latter may be the hornet, which is clearly paired with smaller *Vespula* spp. in book IX, or, as D'Arcy Thompson[14] suggests on the basis of what is said of its nest, *Polistes*. A different nomenclature appears in 627b 22 ff., where ἡμερωτέροι and ἄγριοι σφῆκες (Pliny XI.73 *crabrones silvestres*) are distinguished. The latter, probably to be identified with the hornet[15] are characterised as scarcer, larger, longer, darker in colour, and more aggressive than their relatives, inhabiting hilly country and nesting in hollow trees.

(vii) Ἰχνεύμων. The *ichneumon* or 'tracker'[16] wasp is described by Aristotle (HA 552b 26 ff.; cf. 609a 5), whose account is reproduced by Pliny (XI.72), as a form smaller than the common variety which is notable for hunting *phalangia*.[17] The spiders they kill are carried off to a wall or anywhere where there is a convenient hole and there deposited: the hole is coated with mud, and there the wasps produce their offspring.

The insects described here are clearly hunting wasps of the families Pompilidae and Sphecidae. The pompilids all prey upon spiders, which the females pursue and paralyse by stinging: these are then used to stock the burrows, or in some species earth nests attached to walls, etc., in which the wasp's eggs are laid, providing food for the resultant larvae. A number of sphecids also hunt and store spiders, nesting in mud cells.[18] D'Arcy Thompson[19] suggests that Aristotle's account is based upon observation of the sphecid *Sceliphron*, since this is especially fond of building its nests in human habitations.

---

[11] G. P. Shipp, *Modern Greek Evidence for the Ancient Greek Vocabulary*, p. 209; Fernandez, op. cit., p. 73.
[12] 11.1.2.ii above.   [13] Cf. 1.1. above.   [14] Op. cit., V.23 note.
[15] So Sundevall, op. cit., p. 218; Aubert and Wimmer, op. cit., I. p. 171; D'Arcy Thompson, op. cit., IX.41 note.
[16] Cf. Fernandez, op. cit., p. 176.   [17] 7b above.
[18] Imms, op. cit., pp. 1247–6, 1251–3.   [19] Op. cit., V.20 note.

(viii) The *anthrene* of *HA* 628b 32 ff. is clearly not the same insect as that which goes under this name in book V of the *Historia* (vi above). Here it is described as a flesh-eating insect which, although it will take sweet fruit, prefers to hover around patches of dung waiting to capture the large flies which frequent them.[20] Having taken their prey, they remove the insect's head and fly off with the rest. These *anthrenai* nest underground in cavities which they excavate themselves, the nests being progressively enlarged so as to contain sometimes as much as three or four basketfuls of 'combs'. They are social, possessing leaders and workers. If individuals stray from their home nest, they may construct layers of cells in some tree and there produce a new leader who will lead them off to found a fresh underground colony.

This exceedingly confused account appears to result from confusion between two or more distinct species of Hymenoptera. As D'Arcy Thompson[21] points out, the references to fly hunting seem to be based upon observation of the sphecid wasp *Bembix rostrata* Linn., the females of which make underground nests for their young and feed them from day to day on large Diptera of various species.[22] However, the rest of the account displays confusion with the large underground nests of social wasps (*Bembix*, although gregarious, is solitary), as well as with the swarming habits of bees. The reference to fruit feeding also represents confusion with social wasps, which, like *Bembix*, may also hunt flies.

(ix) *Pseudosphex*. The name is preserved only in a latinised form by Pliny (XXX.98), as that of an insect used for medicinal purposes. It is said to be a *vespa quae singularis volitat*. Leitner[23] suggests that it is to be identified as one or other of the solitary wasps, and that it is the same insect as the *seiren*[24] of Ps. Aristotle.

(x) Theophrastus (*HP* VII.13.3) makes mention of a type of *skolex* found inside the stem of the asphodel which turns into a ζῷον ἀνθρηνοειδές. When the stem finally withers, the latter flies off. This may perhaps be a reference to one of the stem sawflies of the family Cephidae, whose larvae bore into the stems and shoots of various plants and pupate there: some of these possess a black and yellow wasp-like colouration.[25]

2.1. Life history: colony founding and development

Most of the classical data on the life history of the social wasps is contained in two sections of the *Historia Animalium*. The first of these, the work of Aristotle himself (544b 22–555a 12), consists of a unified account of the natural history of *sphekes* and *anthrenai*, the two not being clearly distinguished. The second account, which is independent of the preceding, is found in the non-Aristotelian book IX, and consists of a description of the life cycle of the common and 'wild' (cf. 1.2.vi above) *sphekes* (627b 22; 628 b), followed by a series of disconnected notes on wasps in general (628b ff.). Pliny (XI.71–4) gives a composite account conflating material from both sections of the *Historia*.

---

[20] 48.1.2.i below.   [21] Op. cit., IX.49 note.   [22] Imms, op. cit., p. 1259.
[23] Op. cit., p. 206.   [24] 43 below.   [25] Cf. Imms, op. cit., p. 1204.

As regards the distinct castes within the colony, Aristotle explicitly names only the queens, which he calls 'leaders' (ἡγεμόνες): he states rather cryptically that the two forms of wasp may operate either with or without such a leader, speaking of them as in some cases wandering in search of one.[26] Ps. Aristotle, followed by Pliny (XI.73), says that both forms of *sphex* possess leaders, otherwise known as 'mothers' (μῆτραι), thus correctly identifying their sex,[27] and workers (ἐργάται), the leaders being larger but more mild tempered. A little later, it is said that the leaders are less strong on the wing and that, although they probably do have stings, they do not use them. Aelian (V.16) states similarly that wasps have a 'king' for whom they work, who is stingless and docile.[28] Plato (*Phaedo* 82b) compares the social life of wasps with that of ants.

Ps. Aristotle and Pliny (XI.74) note the existence of male wasps, which they describe as 'drones', on the analogy of honey bees. The former correctly describes these as stingless, and notes that some believed them to be female and all other wasps male. Both sources record a popular belief that all wasps shed their stings in winter.

The hibernation of the queens of the common wasp is mentioned by Ps. Aristotle and Pliny (XI.73), who correctly state that only they survive the winter, the workers all dying. They are said to hide in hollow trees or in clefts in the ground. However, among the disconnected notes which follow the main life history account in *HA* IX, the erroneous statement is made that both stinged and stingless wasps have been discovered together in holes in winter. The *agrios sphex* is said to differ from its relatives in surviving the winter in all its castes, hibernating in hollow trees.

Describing the preferred nesting sites of wasps, Aristotle states rather confusingly that, when they have no leader, *anthrenai* nest in elevated situations and *sphekes* in holes, but that, when they do, both nest underground. Pliny (XI.71) simplifies this by writing that *vespae* nest high up, while *crabrones* do so in cavities or below ground. Ps. Aristotle also refers to underground nesting, and says that the *agrios sphex* (cf. Pliny XI.73) nests in hollow trees: this is characteristic of the hornet.

Wasp nests in hollows below ground, by the roadside, are mentioned as early as the *Iliad* (XII.167 ff.; XVI.259 ff.), and by Callimachus (fr. 191 Pf). Ovid (*Fast.* III.747–54) speaks of *crabrones* nesting in a hollow tree, and Quintus of Smyrna (VIII.41–4, XIII.54.7) similarly of *sphekes*.

There is some uncertainty as to what the nests were actually made of. Aristotle describes them as consisting of 'a bark-like and cobweb-like substance', which is rendered by Pliny (XI.71) as *e cortice . . . araneosae*. The arrangement and construction of the cells is compared with that of honey bees, the same terms κηρία and *cerae* being used as for honeycombs. Ps. Aristotle says that the cells are made from wood chippings or shavings and earth; hence Pliny's *e luto* in XI.71. The same author accurately describes how they hang from a single pedicel 'as from a root'. Aristotle says that the nest of the *anthrene* is

[26] D'Arcy Thompson, op. cit., V.23 note, understands this as a confused reference to the distinction between solitary and social wasps.

[27] Queen honey bees were commonly regarded as male and termed 'kings': cf. *HA* 553b 6.

[28] It is Ps. Aristotle who is correct about their stings.

neater in appearance (γλαφυρώτερον) than that of the *sphex*: D'Arcy Thompson[29] suggests that this is a reference to the small nests of *Polistes*. The development of the colony is described by Ps. Aristotle (= Pliny XI.74) as beginning with a small nest of four cells[30] or, as Pliny translates it, with four entrances, constructed by the leader at the approach of summer. In this the workers are produced. As these grow up, they add on extra 'combs' of cells until the end of autumn, the queen meanwhile no longer working but being fed by her workers. Finally, at the end of the season especially large cells (Pliny incorrectly translates *alios maiores nidos*) are constructed in which a new generation of leaders is reared.

2.2. Life history: larval development and rearing

In *GA* 761a 2 ff., Aristotle states that the manner of reproduction in *sphekes* and *anthrenai* is the same as that in honey bees inasmuch as the leaders give rise both to the workers and to their own kind, although he is under the misapprehension that there is a difference in that the former mate with each other, while bees are parthenogenetic. The first life history stage, the egg in our terms, is said by Aristotle (*HA* loc. cit.) to resemble a drop of liquid adhering to the side of the cell. This γόνον develops into a larva or *skolex*, and that in turn into a motionless pupa or νύμφη from which the adult emerges, breaking out of the cell which has meanwhile been sealed up (cf. 551a 29 ff.). The leaders do not produce their offspring all at one time, since at one time there may be found various stages in different cells: this is the only part of the Aristotelian account reproduced by Pliny (XI.71).[31]

There is no evidence that feeding of the larvae by the workers had been observed in antiquity. Aristotle (*HA* loc. cit.; *GA* 759a 1 ff.) says that they have no external source of nourishment but must possess some internal food reserve, since the fact that they do feed in some way is indicated by their production of excrement. In the case of *anthrenai*, however, it is said (*HA* loc. cit.) that a drop of honey is found in the cell next to the larvae.

Alongside this scientific account of the matter, there coexisted a popular belief to the effect that wasps are spontaneously generated from the dead bodies of horses (Archelaus *ap.* Varro *RR* III.16.4; Nicander *Ther.* 741; Plutarch *Cleom.* 39; Sextus Empiricus *Pyrrh.* I.41; Aelian I.28; Pliny XI.70; Ovid *Met.* XV.368; Phaedrus *App.* 31; Origen *c.Celsum* IV.57; Isidorus *Or.* XII.8.4).

3. Feeding habits

Although wasps are carnivorous, they also feed upon nectar, ripe fruit, and honeydew: their attraction to sweet substances is especially noticeable at the end of summer after the rearing of young in the colony has ceased.

---

[29] Op. cit., V.23 note.

[30] So translated by D'Arcy Thompson, ibid., IX.41 note.

[31] The sentence at 555a 8 ff. would seem to contradict this and may perhaps have been interpolated: it uses σχαδών, a term normally used only for the early stages of the honey bee.

Ps. Aristotle records that they live for the most part on animal food but also on certain flowers and fruits. Pliny (XI.72) omits mention of the latter and makes them exclusively carnivorous, though elsewhere (XV.67) he does describe them as attacking stored fruit. He adds that *vespae muscas grandiores venantur*, a statement which in Ps. Aristotle belongs to a distinct variety of hymenopteron (1.2.viii above). Ps. Aristotle also notes that queen wasps have been observed collecting a sweet exudation from elm trees.

Wasps are described as attacking flies, though not necessarily for predatory purposes, in Ps. Callisthenes II.16.

4. Popular attitudes and beliefs

Wasps were not popular insects in antiquity. Though they were not regarded as positively malignant like the scorpion[32] or deliberately troublesome like the mosquito,[33] they had a reputation for being easily roused and provoked, for being liable to attack and sting at the slightest provocation. As Aristophanes (*Vesp.* 1104–5) puts it, οὐδὲν ζῷον ἠρεθισμένον μᾶλλον ὀξύθυμον ἐστιν οὐδὲ δυσκολώτερον. They therefore became proverbial as typical of quick tempered, irritable and unpredictable human beings, dangerous when roused, such as the cantankerous old jurymen of Aristophanes' *Wasps* (*Vesp.* 430, 1104–5; cf. Callimachus fr. 380; Plutarch *Mor.* 96b, 461a; Lucian *Char.* 15, *Bis Acc.* 13; *AP* VII.71,405,408; Ps. Plato *Eryx.* 392b; Dio Chrysostom VIII.3; Arrian *Epict.Diss.* II.4.6; Clement *Strom.* II.20; Artemidorus *On.* II.22; Plautus *Amph.* 707). The *Iliad* (XII.167 ff., XVI.259 ff.) compares warriors to wasps emerging en masse to attack those who disturb their nests, although the author here is more sympathetic towards them and commends their courage in defence of their home and offspring. Quintus Smyrnaeus (VIII.41–4, XIII.54–7; cf. Ps. Ausonius *App.* V.51–6) uses similar imagery, describing how the insects pour forth in anger when their hollow tree is assaulted by the woodcutter's axe.

There is a story in Ps. Aristotle *de Mir.* 844b 32 ff. concerning wasps in Naxos which feed upon the flesh of adders and thus become endowed with venomous stings. It appears also in Pliny (XI.281) and Aelian (IX.15), though without the local reference. The latter author (V.16) states elsewhere that whenever wasps spot a dead adder they swoop down on it and treat their stings with venom.

Theophrastus (*Sign.* 47) and Aratus (1064 ff.), in their works on weather signs, note that the appearance of swarms of wasps in autumn is an indication of the onset of stormy winter weather.

5. Importance as pests

Most of the classical writers who deal with the subject of bee keeping include wasps or hornets (Greek sources always *sphex*; Latin variously *vespa* or *crabro*) on their lists of creatures which prey upon honey bees and are therefore to be regarded as pests (Ps. Aristotle *HA* 626a 8; Aelian I.58, V.11; Lucian *Char.* 15; Varro *RR* III.16.19; Pliny XI.61;

---

[32] 5.4 above.
[33] 50.5 below.

Virgil G. IV.245; Palladius IX.7; *Geoponica* XV.2). Aelian (I.58) says that they may be trapped by means of a cage containing a small fish hung up in front of the hive. Quintus Smyrnaeus (XI.146–9) gives a vivid depiction of wasps swooping down upon bees as they feed at ripe fruit, or as they emerge from their hive.

Of the social wasps, the hornet in particular is known as causing problems to the beekeeper, as is a species of solitary wasp, *Philanthus triangulum* (Fab.) which excavates burrows in the ground for its offspring and stocks them with captured bees.[34]

Nicander, Pliny, and other medical sources deal with the subject of wasp stings and how to treat them. Nicander (*Ther.* 811 ff.) associates them with truly venomous invertebrates like the scorpion, and gives a composite list of remedies for the stings or bites of all such creatures.[35] Particular remedies prescribed include various herbs (Pliny XX.29,133,173,223, XXI.149, XXIII.118,152; Dioscorides *DMM.* II.118, V.109, *Eup.* II.126–7,134), wine (Pliny XXIII.43), salt in vinegar (Pliny XXXI.99), and animal dung (Galen XII.300; Cyranides p. 64). The wearing of the bill of a woodpecker or of part of an owl was reputed to protect the wearer from stings (Pliny XXIX.92). The use of a verbal spell to relieve the pain of a sting is mentioned by Achilles Tatius (II.7).

Ps. Aristotle (*HA* 627b) and Pliny (XI.73) state that the sting of the *agrios sphex* is more severe than that of the ordinary kind. According to the latter, its stings are usually followed by fever, and twenty-seven of them are reputed to be fatal. As is the case with a number of invertebrates,[36] stories are reported of whole populations being driven out under attack from wasps or hornets (Aelian XI.28). Certain Biblical references to wasps or hornets (LXX σφηκία in *Ex.* 23.28, *Deut.* 7.20, *Josh.* 24.12, though this term is elsewhere only used in the sense of 'wasp's nest'; σφῆκες in *Wis.* 12.8), though originally intended in a metaphorical sense,[37] were understood by Philo (*Praem.* 96) and patristic commentators (Theodoret MPG 80.277; Augustine MPL 34.511,630) as referring to some literal hymenopterous invasion of the kind described by Aelian.

Domestic animals were also liable to be stung, and applications to prevent attack and treatments when attacks have taken place are described in the *Hippiatrica* (*Cant.* 71.14). In Ovid *Met.* XI.334–5, a *crabro* takes the place of the usual gadfly[38] in the familiar description of an animal driven to headlong flight by the stinging of an insect.

Wasps are also mentioned as pests of grapes and other fruit, either when still on the vine or when in store (Pliny XV.67; *Geoponica* IV.10). They are said by these authors to be deterred by sprinkling with oil, which was reputed (Sextus Empiricus *Pyrrh.* I.55) to be fatal to them. Colonies of troublesome wasps could be disposed of by smoking the insects out (Euripides *Cyc.* 475; Aristophanes *Vesp.* 457; Lycophron *Al.* 181–2).

6. Medicinal uses

Galen (XIX.740 K) says that wasps rotted in cedar oil have the same properties as *pityokampai*.[39]

---

[34] Imms, op. cit., p. 1254.   [35] Cf. 7b.3 above.   [36] Cf. 3.2, 7b.1.3.vi above.
[37] Bodenheimer, *Animal and Man*, p. 74. He suggests that the allusion is to the eastern hornet *Vespa orientalis*.
[38] 49.5 below.   [39] 28 above.

7. Wasps as portents

A swarm of wasps flying into the forum at Capua and settling on the temple of Mars is recorded as a portent by Livy (XXXV.9.4).

41. Tenthredon, Anthredon, Pemphredon, Bembix

1. In the Historia Animalium

The τενθρηδών is listed in book IX of the *Historia Animalium* (623b 8 ff.) as a form of wasp or bee alongside the *sphex*, the *anthrene*, the *seiren*, and the *bombylios*.[40] It is described (629a 31 ff.) as being similar to the *anthrene* but speckled, and similar in width to a honey bee. It flies into food shops and alights on slices of fish and other foods. It breeds underground as *sphekes* do, constructing a nest which is larger and longer than theirs.[41]

It has been suggested[42] that *tenthredon* is simply an alternative name for the ground nesting social wasps normally comprehended under the name *sphex*, and that Ps. Aristotle is in error in listing it as a distinct species.

2. In medical and lexicographical sources

Nicander (*Ther.* 805–12) associates the honey bee, the *sphex*, the πεμφρηδών[43] and the βέμβιξ[44] as insects for whose stings he suggests treatments. The same four names recur in *Alex.* 182–4, where they are all depicted as attacking ripe grapes.[45] The *bembix* is characterised in both passages as an insect of the hill country, and in the former *pemphredon* is described as an insect of small size. In the absence of further descriptive details, it cannot be determined which Hymenoptera the poet has in mind here.[46] The description of the *pemphredon* given by the scholiast on *Alex.* 183 suggests that he regarded the word as equivalent to *tenthredon* and *anthredon*, since it is clearly parallel to descriptions we have of Cleitarchus' exotic *tenthredon* (3 below). According to this, the insect is wasp-like and intermediate in size between an ant and a honey bee, patterned with black and white. It feeds from flowers in its upland habitat and flies into hollow trees.

In the same scholion, the *bembix* is uninformatively described as a type of bee which resembles a wasp and otherwise known as βεμβίς. In the scholiast on *Ther.* 805, however, we are told that it is black in colour, and that it was referred to by the poet Parmeno

---

[40] Cf. 40.1.2.viii above; 42–3 below.

[41] The term τενθρήνιον, parallel to ἀνθρήνιον etc. (cf. 40.1 above), is used. A corresponding adjective τενθρηνιώδης is used in the sense of 'honeycombed' by for example Democritus *ap.* Aelian XII.30, Hippocrates *Anat.* 1, Plutarch *Mor.* 721 f.

[42] Sundevall, *Thierarten*, p. 219; Aubert and Wimmer, op. cit., I p. 171. There is certainly strong evidence for regarding ἀνθρήνη, -ηδών, τενθρήνη, -ηδών, and πεμφρηδών as essentially synonyms.

[43] A name derived, like τενθρηδών, from the insect's buzzing. Cf. Fernandez, *Nombres*, p. 129.

[44] An onomatopoeic word belonging to the same complex as βόμβυξ, -ύκιον, etc. Cf. Fernandez, ibid., p. 133 ff., 27.1.2.iii above, and 42 below.

[45] Cf. 40.5 above.

[46] Gow and Scholfield, *Nicander*, p. 193, equate βέμβιξ with the Ps. Aristotelian βομβύλιος, therefore identifying it as a bumble bee.

(fr. 4). Dioscorides (*DMM.* V.109) uses the term *tenthredon* alongside *sphex* in a chapter on stinging insects. It appears in a similar context in the *Hippiatrica* (*Cant.* 71.14), where it is listed along with honey bees, *sphekes*, *anthrenai*, *bombylioi*, and *bembikes*.

Hesychius equates *tenthredon* with *anthredon* (s.v.) and defines the former (s.v.) as 'one of the sting-bearing insects similar to a wasp, or, according to others, an ἀγρία μέλισσα'.[47] Τενθρήνη,[48] a form not found elsewhere, is used by Nicander (*Alex.* 547) apparently to refer to the honey bee.

3. In sources dependent on Cleitarchus

We learn from Demetrius (*Eloc.* 304) that the Alexander-historian Cleitarchus,[49] among other exciting stories of strange sights encountered by the Macedonian army on its journey eastward, included an account of the ferocious variety of bee found in the region of Hyrcania by the Caspian Sea. In this passage Demetrius criticises the historian severely for employing grotesquely exaggerated language concerning the *tenthredon*, 'an animal resembling the honey bee' which 'roams over the mountains and flies into hollow oaks', language out of all proportion to the insect's real nature and more appropriate for 'a wild bull or the Erymanthian boar'.

Demetrius does not, unfortunately, reproduce the offending statements of Cleitarchus, which presumably portrayed the insect as engaging in fierce attacks on man and stinging viciously. Neither are they reproduced by Diodorus (XVII.75) in his passage derived from Cleitarchus. Here the creature in question is named ἀνθρηδών[50] and characterised as smaller than the honey bee, but μεγίστην ἔχει τὴν ἐπιφάνειαν.[51] It roams the mountains gathering nectar from flowers and nests in rocky hollows and lightning blasted trees, constructing combs in which it manufactures a sweet substance not much inferior to honey.

A brief reference to Hyrcanian bees nesting among rocks is found in Strabo II.1.14.

It is probable that this insect is to be identified as *Apis dorsata*, an eastern relative of the honey bee, which constructs a single large comb, sometimes three or four feet in length, suspended from rocks, branches or buildings. Whatever Cleitarchus wrote about its ferocity may not have been so extravagant as Demetrius believed, since *A. dorsata* is a bee which is easily irritated and readily attacks man or domestic animals, sometimes even with fatal results.[52]

---

[47] Ἀγρία μέλισσα is simply a term for a wasp (40.1 above).

[48] Etymologically it is only a variant of τενθρηδών. Cf. Fernandez, op. cit., pp. 129–30.

[49] Cf. W. W. Tarn, *Alexander the Great*, Cambridge, 1948, II pp. 89–90; L. Pearson, *Lost Histories of Alexander the Great*, 1960, p. 220.

[50] This latter phrase would appear to contradict what is said about the insect being smaller than the honey bee. Pearson, loc. cit., suggests the reading λειπόμενον δὲ μεγέθει μελίττης, μελίττης ἔχει τὴν ἐπιφάνειαν. Tarn, loc. cit., prefers to leave the text as it stands, on the grounds that there would otherwise be nothing to justify Demetrius' criticism, but it seems that Demetrius was referring to words about the bee's ferocity which have not been preserved.

[51] Tarn and Pearson disagree as to whether Cleitarchus originally wrote τενθρηδών or ἀνθρηδών. The latter is simply a variant of ἀνθρήνη: cf. Fernandez, op. cit., pp. 73–5.

[52] Imms, *Gen. Textbook*, p. 1263.

42. Bombylios and Bombykion

The βομβύκιον is described by Aristotle (*HA* 555a 13 ff.) as a relative of bees and wasps which constructs a pointed (ὀξύ) nest of clay against a stone or some other such surface and smears it with something like saliva.[53] Inside this nest, which is so thick and hard that it can scarcely be broken with a spear, are produced white larvae in a black membrane; separate from the membrane there is a certain amount of wax, yellower than that of honey-bees. The insect described here would appear to be a mason-bee of the genus *Chalcidoma*, belonging to the Megachilidae, one of several families of solitary bees.[54] The females construct very hard cells out of soil and salivary secretion, with small pebbles mixed in, which they affix to walls or large stones. When eight or nine cells have been built, the nest is roofed over to give a domed appearance.[55] The yellow wax referred to by Aristotle is no doubt the food store of honey provided by the bee to maintain its larvae.

Bombykion and the word of which it is the diminutive, *bombyx*, were also used respectively for the cocoon and adult of the silkmoth, though originally the names of the two types of insect would have been of separate origin.[56] Pliny was misled by this into classifying the silkmoth as a kind of wasp or bee and crediting it with two alternative life histories, one of which applies to the moth, the other of which is a version of Aristotle's passage above. His account therefore reads (XI.75): *Quartum inter haec genus est bombycum, in Assyria proveniens, maius quam supra dicta* (i.e., the honey-bee, the *vespa* and the *crabro*). *Nidos luto fingunt salis specie adplicatos lapidi, tanta duritia ut spiculis perforari vix possint. In his ceras largius quam apes faciunt, dein maiorem vermiculum.* Hesychius defines *bombyx* (s.v. *bombykes*) as a winged creature like a wasp, and (s.v. *bombyx*) as a silkmoth, adding that some call the *bombylios* by this name. The use of *bombykion* in Schol. Aristophanes *Nub.* 158 as a general term for buzzing insects would seem to be in error.

The βουβύλιος is one of the Hymenoptera catalogued in *Historia Animalium* IX, and the brief account of its life history (629a 29 ff.) states that it nests under stones on the ground, constructing two cells or a few more, in which may be found a certain amount of rather poor honey. The insect has been identified,[57] reasonably enough, as a kind of bumblebee, genus *Bombus*, a group of bees which form small underground colonies. The females hibernate and found new nests in spring, constructing them from vegetable material. Within the nest are constructed cells in which the larvae feed upon a food store of pollen, and also waxen receptacles containing honey. Workers produced from the first

---

[53] Accepting with D'Arcy Thompson, op. cit., V.24 note, Aubert and Wimmer's emendation ὥσπερ σίαλῳ of the corrupt mss text at this point. Other suggestions have been made, for example Schneider's ὡσπερεὶ ἁλὶ which would explain Pliny's *salis specie adplicatos* in XI.75.

[54] So identified by Sundevall, *Thierarten*, p. 216; Aubert and Wimmer, op. cit., I, p. 162; D'Arcy Thompson, op. cit., V.24 note; naming *Chalcidoma muraria* Linn.

[55] Imms, *Gen. Textbook*, pp. 1256–7.

[56] Cf. Fernandez, *Nombres*, pp. 132–6, and 27.1.2.iii above.

[57] Sundevall, op. cit., pp. 215–16; Aubert and Wimmer, op. cit., I p. 162; Keller, *Ant. Tierwelt*, II p. 435; D'Arcy Thompson, op. cit., IX.43 note. Sundevall mentions the Xylocopidae alongside *Bombus*, but these, though not dissimilar in appearance, nest in timber or large plant stems.

eggs laid by the female soon take over the maintenance of the nest, and the colony expands up to the end of the season.[58] Ps. Aristotle's description, with its limited number of cells, would refer to the early stages of the colony.

Isocrates (*Helen* 12) mentions the *bombylios*[59] along with the honey bee, as does Aristophanes (*Vesp.* 107), who implies that they both produce wax. The scholiast on the latter passage, along with Suidas and the *Etym. Magnum*, notes that it is an insect similar to the bee. Hesychius says that it is an insect related to the *sphex* or a large bee, or a fly (*myia*). The word is also mentioned by Athenaeus (XI.784d) and Eustathius (*Comm.in Il.* 945.23), who calls it a beetle.[60]

43. Seiren

The term σειρήν[61] is used by Ps. Aristotle (*HA* 623b 10 ff.) to refer to two varieties of solitary Hymenoptera. Unlike the other Hymenoptera listed in this passage,[62] no details of their life history are given. Of the two forms of *seiren*, one is characterised as small and greyish (φαίος) in colour, the other as larger in size, black and speckled (ποικίλος). In view of the association of the word elsewhere with the honey bee, these insects should perhaps be identified, with Sundevall,[63] as solitary bees rather than, as other authors[64] have suggested, as solitary wasps.

Eustathius refers to the *seiren* twice (*Comm. in Od.* 1709.52, 1914.40) as an insect mentioned by Aristotle which produces a faint wailing sound. Suidas reproduces a popular expression Σειρὴν μὲν φίλον ἀγγέλλει, ξεῖνον δὲ μέλισσα and explains that 'the *seiren* is a creature which builds a comb and resembles the honey bee'.

44. Myrmex/Formica

1. Identification: general

The various species of ants (family Formicidae), Greek μύρμηξ and Latin *formica*, held a great fascination for the ancients, and they were noted for the qualities of intelligence, hard work, good organisation, and harmonious social life.[65] Certain particular motifs

---

[58] Imms, op. cit., pp. 1259–60.

[59] The word is onomatopoeic and related to βόμβυξ, -ύκιον, and like them it has a silkmoth name of parallel form. Cf. 27.1.2.iii above.

[60] Perhaps he has in mind the heavy flying μηλολόνθη (cf. 32.1.2.i,v above).

[61] The word is also occasionally used (Aelian V.42; Pliny XI.48) as a name for the drone caste of the honey bee. Fernandez, *Nombres*, pp. 214–18, suggests that it does have a genuine connection with the Sirens of mythology, either because it alludes to the sound made by the insect, or because it was originally an ancient name for a deity associated with the honey bee.

[62] Cf. 41.1 above.

[63] *Thierarten*, p. 221. He identifies the small σειρήν with the bees of such families as Andrenidae, Megachilidae, and Colletidae but the larger as the wasp *Bembix*.

[64] Aubert and Wimmer, op. cit., p. 70, who, followed by D'Arcy Thompson, op. cit., IX.40 note, identify the two as belonging to the genera *Eumenes*, *Synagris*, etc.; Leitner, *Zool. Terminologie*, p. 221, who equates with Pliny's *pseudosphex* (40.1.2.ix above).

[65] Cf. Keller, op. cit., II pp. 416–21.

with regard to their habits are constantly repeated in literature, making them one of the most frequently mentioned of all insects. Though there are many distinct species of ant native to Europe, differing in size, colour, and more especially in habits, these were generally not distinguished in antiquity, most authors considering the whole family as constituting a single 'species'.

1.2. Identification: alternative names and distinct varieties

(i) The Greek μύρμηξ occurs in a number of dialectical forms.[66] Of these, the Doric μύρμαξ is used by Theocritus (IX.31, XV.45, XVII.107), and μύρμος by Callimachus (fr. 753 Pf) and Lycophron (*Al.* 176), while others are preserved only in lexicographical sources: Doric μυρμηδών (Hesychius; *C.Gl.L.* III.361.68);[67] βύρμαξ, βόρμαξ (Hesychius); ὅρμικας (Hesychius).

(ii) There is a curious reference in Pausanias (III.26.3) to a place on the coast of Laconia called Pephnus, where the ants are whiter in colour (λευκότερον) than elsewhere. They are also mentioned by Aelian (IV.5). Keller[68] identifies these as termites, a view rejected by Venmans[69] who understands λευκός as meaning 'shining'. There are in fact two species of termite (Isoptera) native to Southern Europe, *Kalotermes flavicollis* and *Reticulitermes lucifugus*, which are colonial like ants and excavate galleries in trees and in building timbers.[70]

(iii) The only surviving unequivocal reference to termites,[71] although not from Europe, is found in Aelian (XVI.15), who cites Juba as his source. He gives an accurate description of the earth built mounds (οἰκίσκοι, χώματα) with a single entrance and containing a maze of passages, which are constructed by termites in India such as *Odontotermes obesus* (Ramb.).[72] He calls them Ἰνδοὶ μύρμηκες and depicts them as storing seeds in the way that true ants were noted for.

(iv) The ἱππεύς μύρμηξ is given a single brief mention by Ps. Aristotle (*HA* 606a), giving an example of anomalous distribution in animals: καὶ ἐν μὲν Σικελίᾳ ἱππεῖς μύρμηκες οὐκ εἰσίν. For the manuscript reading Bekker and other editors have read ἱππομύρμηκες, but Fernandez rejects[73] this alteration as unnecessary, particularly in view of the entry in Hesychius under ἱππῆς, which notes λέγονται καὶ μύρμηκες οὕτως. Dittmeyer reads οἱ πτερωτοὶ μύρμηκες to bring the text in line with the obviously parallel statement in Pliny (XI.110) *Non sunt in Sicilia pennatae* (sc. *formicae*).

---

[66] Fernandez, *Nombres*, pp. 23–4.

[67] The racial name *Myrmidones* apparently has no real connection with ants, contrary to the ancients' general belief: cf. Fernandez, ibid., p. 23.

[68] Op. cit., II p. 421.

[69] L. A. Venmans, '*Leukoi Myrmekes*', *Mnemosyne*, NS 58, 1930, pp. 318–22.

[70] Imms, *Gen. Textbook*, pp. 606 ff.

[71] So identified by Gossen, in PW XII.1039.

[72] Cf. Imms, op. cit., p. 632.

[73] Op. cit., p. 55.

Sundevall and later authors[74] agree in understanding the *hippeus* as a distinct species of ant, the large *Campanotus herculeanus* (Linn.). Fernandez,[75] however, suggests a harmonisation of Pliny and Aristotle by interpreting 'horseman ant' as a name for the winged male and female ants which were taken by Aristotle and others to be distinct from the apterous workers.

(v) A type of ant called *formica Herculanea* is referred to by Pliny (XXX.29) and Marcellus Empiricus (XIX.14) as being of medicinal use for skin diseases when ground up with a little salt.[76]

(vi) Aelian (X.42), quoting Telephus of Pergamum, mentions the existence of certain venomous (θανατηφόρος) ants called λαέρτης, a name which he says is also applied to certain wasps. Fernandez[77] comments on the etymology of the word and suggests its appropriateness for colonial insects.

(vii) Μεταλλεύς. Although Hesychius says 'Certain ants are so called', 'miner' is probably a descriptive name for ants in general, referring to their excavation of tunnels in the ground. Aelian (XV.15) calls ant workings μεταλλεῖαι.

(viii) Ἀγριομύρμηξ. In *C.Gl.L.* II.433.46, *curculio* is defined as 'the σκνίψ which some call *agriomyrmex*'.[78] Although Fernandez[79] and LSJ accept this gloss at face value, it is more probable that the word *agriomyrmex* refers not to the corn pests *kis* or *curculio*, but to the corn gathering ants (3.3 below) commonly associated with the former as pests of grain.

(ix) The Cyranides (pp. 68, 261, 294), in marked contrast to all other surviving sources, divides ants into seven distinct varieties. We have (*a*) the common (κοινοί) form known to everyone; (*b*) a long-headed (ἀδροκέφαλοι) form, black in colour, which Fernandez[80] suggests is the 'soldier' caste with large head and mandibles possessed by some species of ant; (*c*) a small slender form, red or reddish yellow (ξανθός) in colour, known as σκνίψ (pl. σκνῖπαι; Latin version *zinyphes*);[81] (*d*) the large ants with wings, which would be the queens described by Aristotle (2.1 below); (*e*) the medium sized field (ἀρουραῖοι) ants, which carry grain (cf. 3.3 below), whose name is parallel to *agriomyrmex* (viii above); (*f*) the small roadside (ἐνόδιοι) ants; and (*g*) the 'ant-lion'.[82]

(x) Μυρμήκιον. In the Cyranides (pp. 40, 44, 246), we read of certain small insects found in the centre of the flowers of the plant *chrysanthemos* (*Helichrysum siculum*).

[74] Sundevall, *Thierarten*, pp. 222–3; Aubert and Wimmer, op. cit., I p. 168; Keller, op. cit., p. 421; Gossen, *Zool. Glossen*, p. 40.
[75] Op. cit., p. 55.
[76] Identified by Leitner, *Zool. Terminologie*, p. 122, as one of the large ants of the genus *Campanotus*.
[77] Op. cit., p. 193.
[78] 36 note 132 above.
[79] Op. cit., p. 160.
[80] Ibid., p. 81.
[81] Cf. 58 below.
[82] 61 below.

They are variously described as 'resembling small black *myrmekia* with short wings' or as 'small black creatures, very tiny, like winged *myrmekia*', and are said to be used as ingredients in a magic ointment to bring success in personal relationships and good fortune in one's various dealings. These insects are unlikely to be real ants, but rather Thysanoptera or other minute black insects which feed inside flowers.[83]

2.1. Life history

Ants are social insects, and their colonies, like those of bees and wasps, are each founded by a single fertilised female or queen. This founding female begins laying eggs and feeds the resultant larvae until they are ready to pupate. The adults that emerge from these pupae are the first members of the worker caste, infertile females. The workers now take over the care of the young, and from this point the queen's activity is restricted to egg laying. Winged males and females are also produced, and those of the same species from the same locality emerge from their nests synchronously on a particular day for their nuptial flight. After mating the males die, and the females shed their wings and excavate a chamber in the earth to begin a fresh colony.[84]

The distinction between queens and workers and their respective functions are seldom mentioned, though the ancient world was well aware of caste differences in bee and wasp colonies. Aristotle (*PA* 642b 30 ff. and 643b 2) notes the existence of winged and wingless ants as alternative forms of one species, as with the glow-worm.[85] Similarly, Artemidorus (*On.* III.6) distinguishes winged from wingless ants, terming the latter workers (ἐργάται). According to Hesychius (s.v. *nymphai*), Photius (s.v. *serphoi*) and Eustathius (*in Od.* 1736.5), winged ants were known as νύμφαι,[86] and, according to Photius (s.v.) as σέρφοι.[87]

Aristotle (*HA* 555a 19 ff.), in a description which Pliny (XI.108) reduces by half, credits ants with a two stage life history, involving small larvae (σκωλήκια; Pliny states only that a small larva 'like an egg' is produced; accordingly Aubert and Wimmer insert ὠοειδῆ into Ar.'s text) which are at first rounded but become jointed as they grow. The pupal stage is not recognised here, but it is not improbable that what Aristotle takes to be stage one is in fact the pupa, in view of the fact that Theophrastus (*De Sign.* 22) and others evidently mistake pupae for eggs (which would be considerably less conspicuous than the former).

2.2. Nest construction

Ant nests have an irregular and complex structure, consisting of galleries with chambers leading off them, these chambers being used for various purposes such as food storage and

---

[83] Cf. Imms, op. cit., pp. 782 ff.
[84] Ibid., pp. 1236 ff.
[85] 35.3 above.
[86] For other uses of the term νύμφη cf. 40.2.2 above and Fernandez, op. cit., pp. 208 ff.
[87] Cf. 62 below.

the housing of larvae and pupae. There is a tendency for classical writers to ascribe to them more architectural regularity than they in fact possess. Aelian (VI.43) compares ant nests to the Cretan labyrinth, and explains the winding nature of the passages as a device to protect against attack by enemies. He describes the nest as being divided into three distinct parts: the ἀνδρών for the males 'and any females that are with them', the γυναικών where the females give birth (this would be his interpretation of a chamber containing young being cared for by workers), and the θησαυρός or σιρός where grain is stored for food. Plutarch (*Mor.* 967d–68d) also describes three sections, in his case three cavities which terminate the main passage leading from the entrance: a common dwelling place, a store-room, and a chamber where 'they deposit the dying'. Galen (III.7 K) speaks of labyrinths and θησαυροί, and Aeschylus (*Pr.* 453) compares the dwellings of primitive man with the subterranean homes of ants (cf. Philostratus *Im.* II.22). In Aeschylus and others (Theocritus XVII.105; Babrius *Fab.* 108; Marcus Aurelius VII.3), it is implied that their life in such dark underground habitations is not an entirely pleasant one. Certain authors (Pliny X.206; Apuleius *Met.* VIII.22) also refer to ants nesting in hollow trees.

3.1. Care for the dead

The idea of ants caring for their dead is mentioned by Aelian, who says (V.49) that they are careful to remove dead bodies from the nest and that (VI.43) their relatives bury them in coffins consisting of the outer capsules of wheat grains. Pliny (XI.110) notes that ants are the only living creatures besides man that bury their dead, and Celsus (*ap.* Origen IV.84) goes so far as to say that they set aside a particular spot as their 'ancestral graveyard'. Ants carrying away their dead are also referred to in the *Geoponica* (XIII.10.4).

3.2. Foraging tracks

Many species of ant, for example the large wood-ants *Lasius* spp, have persistent trackways leading out from their nests which they use for foraging expeditions. These tracks are noticeable on account of the fact that the vegetation on them is short and sparse.[88] This is alluded to by Aristotle (*HA* 622b 24 ff.), who describes how ants 'always walk along one track' and Pliny, who says with some exaggeration that 'we see rocks worn by their passage and a track produced by their labour'. A number of authors refer in passing to the narrow tracks along which ants travel (Plautus *Men.* 888; Virgil *Aen.* IV.401 ff., *G.* I.380; Ovid *Met.* VII.624 ff., *Tr.* V.6.39–40; Aelian II.25).

There is a metaphorical reference to these narrow winding runways[89] in Aristophanes *Thesm.* 100 (cf. also Pherecrates fr. 145, where μυρμήκια is used in a similar reference), on which passage the Scholiast defines μύρμηκος ἀτραποί as αἱ τῶν μυρμήκων ὁδοί. Hesychius and Photius (s.v. μύρμηκος ἀτραπούς, ὁδοί) note that there was a certain track in Attica which went by the name of 'Ant's Path'.

[88] M. V. Brian, *Ants*, London, 1977, p. 62.

[89] Or perhaps to the passages inside ant nests, as E. K. Borthwick, 'Notes on the Plutarch *De Musica* and the *Cheiron* of Pherecrates', *Hermes*, 96, 1968, pp. 69–70, suggests. Although Aristotle uses ἀτραπός of foraging tracks, Aelian (VI.43) uses it of tunnels within the nest.

3.3. Food collecting

Worker ants spend much of their time foraging for the colony, gathering both for adults and larvae. Diet varies according to species but ancient authors pay almost exclusive attention to those which specialise in collecting grain. The species particularly in view here, both in Europe and in the Middle East, would be the members of the genus *Messor*.[90] These ants gather grain both from the ground and from the growing plants.[91] It was this constant activity that gave ants their reputation for tireless industry. Aristotle (*HA* 622b 24 ff.), Pliny (XI.109) and Aelian (IV.43) all report that they carry on working even on moonlit nights (cf. Antigonus *Mir.* 126), the latter drawing morals about not making excuses for inactivity. Incessant activity is pointed to in Virgil *Aen.* IV.401 ff. (407 *opere omnis semita fervet*) and by Artemidorus (*On.* III.6) and Babrius (*Fab.* 108).

It would appear to have been believed that ants' staple, if not their only, food was grain. All references to foraging which actually specify what is being collected mention grain, or occasionally other plant seeds such as beans (Lucian *Icar.* 19) or cypress (Pliny XVII.73) or poppy (Plautus *Tri.* 410) seeds (if we except the fanciful account of an expedition to collect honey in *AP* IX.438 and the gruesome story told in Apuleius *Met.* VIII.22). With regard to their method of food gathering, ants were credited with a far higher degree of organisation than is in fact the case. Whereas foraging ants normally operate alone, gathering assistance when loads too heavy for a single ant are found, classical writers believed that they functioned like a well-organised and well-led army. Aelian (II.25) depicts them in summer when corn is being threshed assembling in bands and setting off in single file or two or three abreast, arriving at the threshing floor and methodically carrying away their supply grain by grain (cf. Babrius *Fab.* 117); and in VI.43 he relates how they go on expeditions into cornfields led by the older specimens, who act as στρατηγοί and who ascend the wheat stalks to cut out the seeds and drop them to the younger ones waiting below to strip them of their outer coating. Both Virgil (*Aen.* IV.401 ff.) and Ovid (*Met.* VII.642 ff.; *A.A.* I.93) present the picture of an *agmen* or organised column of ants gathering grain, the former adding a mention of individuals with the particular task of keeping the army moving in good order (*pars agmina cogunt castigantque moras*). Ambrose (*Hex.* VI.16) depicts expeditions to granaries in similar terms, although he says, following LXX *Prov*.6.7, that they act in concert without any orders or compulsion.

The fact is often remarked upon that ants are able to handle very large loads in proportion to their size, carrying them in their mouth where possible (Pliny XI.108; Ovid *Met.* loc. cit., *A.A.* loc. cit.; Horace *Sat.* I.1.33 ff.), but if not pushing them along with their legs (Pliny loc. cit.; Virgil *Aen.* loc. cit.). Plutarch (loc. cit.) notes that they will divide up large food items into more easily manageable portions.

According to Plutarch (*Mor.* 915 f.), ants prefer wheat to barley because the grains of the latter are more difficult to carry, although Aelian (II.25) notes the collection of both

---

[90] Bodenheimer, *Animal and Man*, p. 78.
[91] Imms, op. cit., pp. 1239–40.

types of grain from threshing floors. In later sources (Physiologus pp. 48–9; Isidorus *Or.*
XII.3.9), this belief is strengthened into an assertion that ants will not touch barley at all.
Physiologus says that when gathering from standing plants in the field they are able to
discern by scent which variety of grain they are dealing with. This concept of an ability to
distinguish grain according to species appears in Apuleius' tale of Cupid and Psyche (*Met.*
VI.10), where friendly ants assist in the humanly impossible task of separating out a heap
of mixed grain.

3.4. Social life

In connection with their food gathering activities, Pliny (XI.109) attributes to ants a
degree of social life, telling how when they come together they converse with each other
(. . . *quam diligens cum obviis quaedam conlocutio atque percontatio*). Lucian (*Icar.* 19) paints a
picture of activity going on at the nest entrance, with some ants huddled together
discussing affairs of state (πολιτευομένους), while others carry out refuse or convey food
inside. Celsus (*ap.* Origen IV.84) argues that ants have a language by which they can
discuss things with one another and relate their experiences. And Plutarch (*Mor.*
967d ff.), who states that ants display in miniature every virtue, describes them as having
affection for one another. What particularly struck the classical writers was the fact that,
even when there were vast numbers of ants milling about, they never seemed to quarrel
with each other or get in each others' way, and that they seemed always ready to help
with each others' burdens (Celsus *ap.* Origen IV.83). Dio Chrysostom (40.32,40;48.16)
draws unfavourable comparisons between this exemplary behaviour on the part of ants
and quarrelsome uncooperative human beings. He notes how contentedly they live
together, how helpful they are, and how politely they give way to one another when
they meet on the path. Plutarch (loc. cit.) and Aelian (II.25) state that in particular those
that are not loaded give way to those that are (there is a muddled relation of this in
Physiologus p. 45 Sb.). In at least one species of ant it has been noted that 'a slight
tendency exists for incoming workers to move on the outside and outgoing ones in the
centre of the track',[92] but nevertheless confusion and collisions do occur, so that again the
classical picture is a somewhat idealised one.

   In a number of places great crowds on the roads or elsewhere are compared to ants
(Theocritus XV.45; Aeschrion fr. 2; *Com.Adesp.* fr. 823; Apollonius *Arg.* IV.1452–3;
Hesychius s.v. μυρμήκια). There is an uncharacteristically negative view of ants on the
part of Seneca (*de Tran.* 12.3), who draws a comparison with ants wandering aimlessly
(*inconsultus . . . vanusque cursus*) about bushes.

   The life of an ant colony is often said to mirror in miniature that of a Greek *polis* (Plato
Phaed. 82b; Plutarch *Mor.* 601c; Lucian *Icar.* 19; Dio Chrysostom IX.19; Celsus *ap.*
Origen IV.23 ff.; Basil MPG 29.140). Their communities are said to possess a system of
government (*reipublicae ratio* Pliny XI.108), to have positions of leadership and authority
(Celsus IV.81), and generally to reflect, proportionately to the life of ants, all the offices
and occupations which are found in a human city (Lucian loc. cit.).

   [92] M. V. Brian, op. cit., p. 52.

3.5. Food storage and preservation

So far as classical sources are concerned, food storage means specifically grain storage, and there is no awareness of the fact that only certain types of ant in fact display this habit. Aelian (II.25, V.43) and Plutarch (loc. cit.) describe particular parts of the nest as being set aside as storerooms or granaries, as does Ovid (*Tr.* V.6.39). It was especially noted that ants laid up stores of grain for use during the winter, and this led to their frequent mention as examples of forethought and preparedness (for example *Aesopica* 112,373,521 Perry; Hesiod *Op.* 778; Lucilius 561; Titinius *com.* 34; Virgil *Aen.* loc. cit., *G.* I.186; Horace *Sat.* loc. cit.; Juvenal VI.361; Hes. *Op.* 778; Aelian *VH* I.12; Celsus *ap.* Origen IV.83; *Anth.Lat.* 292; Symphosius *Aen.* 22b; Basil *Hex.* MPG 29.193; Ps. Eustathius *Hex.* MPG 18.748; Isidorus *Or.* loc. cit.; Physiologus p. 46). We are frequently informed that ants treat their collected grain in such a way as to prevent it germinating, the process being variously described as eating out the germ from which the shoot would emerge (Plutarch loc. cit.; Celsus *ap.* Origen IV.83), nibbling in general (Pliny XI.109), boring through the middle (Aelian II.25; Basil *Hex.* loc. cit; Ps. Eustathius *Hex.* loc. cit.) and cutting the grain in half (Philo *de Animal.* 42; Physiologus pp. 46–7 Sb.). Plutarch (loc. cit). and Pliny (XI.109) state that if the seeds become damp the ants will bring them up to the surface to dry them (cf. Isidorus loc. cit.; Babrius *Fab.* 140). This latter observation is accurate: transporting damp seeds to the surface and laying them in the sun for a while is effective in arresting germination. As for the initial treatment of seeds, the biting off of the radicle has been observed, but it has also been discovered that grain collecting ants secrete a herbicide which prevents germination.[93]

3.6. Relations betweeen colonies

Many ants possess territories around their nests, and workers from other colonies are liable to be attacked: they can distinguish strangers from occupants of their home nest by scent. Intercolonial battles can develop, for example over food, involving numerous specimens and resulting in large numbers of fatalities. Ants may also invade and take over nests belonging to other colonies.[94]

However, there are only two examples from the classical world of any awareness of these various forms of aggression: otherwise all the emphasis is upon cooperation. Seneca (*Qu. Nat.* I. *Praef.* 10) hints at the existence of territories when he compares ants with human empire-builders and says that if someone could give a human intellect to ants they would no doubt divide a single threshing floor into many provinces. Warlike relations between colonies are also implied in the story, related by Plutarch and Aelian (Plutarch loc. cit.; Aelian VI.50), of how the Stoic philosopher Cleanthes noted some ants arriving at the nest of another colony carrying a dead ant. They stopped while individuals emerged from the nest and seemed to consult with them before returning within. This was repeated two or three times until finally the ants in the nest brought up a *skolex* and

---

[93] Imms, op. cit., p. 1240.
[94] M. V. Brian, op. cit., pp. 83, 94–5, 152 ff., 162–3.

delivered it to the visitors 'as if as a ransom for the dead'. The latter accepted it and handed over the body, 'and the ants from the nest received it gladly as though recovering a son or brother'.[95]

4. Popular attitudes

Because of their remarkable habits, ants were a source of great fascination in antiquity. Though sometimes regarded as pests, they were nonetheless greatly admired for their skill and industry (Tertullian *adv.Marc.* I.14; Augustine *Civ.* XXII.24), and for the quality of their social organisation, particularly in view of their diminutive size. Moral lessons for mankind were frequently drawn from various aspects of their behaviour. Since they appeared constantly busy at their various tasks, they were regarded as epitomes of hard work and industriousness (LXX *Prov.*6.6; Clement *Strom.* I.6; Basil *Hex.* MPG 29.192 ff.; Ambrose *Hex.* VI.16); while their well known preparations for the winter made them popular examples of careful provision and thought for the future (Babrius *Fab.* 140; Phaedrus *Fab.* IV.25; *Aesopica* 112,166,235 Perry; LXX *Prov.* 6.6–8, 30.25; John Chrysostom MPG 64.672; Jerome *in Ps.* 91; Augustine MPL 36.475,805, 38.238, 42.168; Ambrose *Hex.* VI.16). However, men's accumulation of wealth is sometimes compared unfavourably with the storing habit of ants (Theocritus XVII.105; Crates Thebanus 10.7; Plutarch *Mor.* 525e; Julian *Or.* VI.199D).

The efficient organisation of labour within the nest, plus the apparent fact of so many individuals being able to coexist so harmoniously, led to ants being credited with all the personal qualities desirable in the members of a human community. Plutarch (loc. cit.) goes so far as to say that they display in miniature every virtue known to man. They are viewed as examples of altruism, and as a challenge to mankind in the way that they work together for the good of the community as a whole, each looking not to his own interests but motivated entirely by a spirit of service to the state (Cicero *Fin.* III.63; Quintilian V.11.24; Plutarch *Mor.* 738 ff.; Porphyry *de Abst.* III.11).

5. Views on the intelligence of ants

It was a matter of considerable debate whether all of this remarkable activity was a product of genuine intelligence and reason such as is possessed by man, or whether it was after all merely an exceptional example of instinct. Plutarch (*Mor.* 967d ff.) and Aelian (VI.50) inform us that it was through his observation of ants that the Stoic philosopher Cleanthes was forced to the conclusion that animals are endowed with reason. Cicero (*Nat.D.* III.21) declares that ants possess *mens, ratio, memoria*; but Galen on the other hand asserts (III.7.K) that they act 'by nature rather than by reason'.

There is a long debate on this subject in Origen's tractate against Celsus (IV.81 ff.). Here Celsus contends that animals are as much rational beings, and therefore of no less significance, as man, since everything in human civilisation which might appear to make man unique can in fact be paralleled from the life of ants. Origen, however, argues that

---

[95] Cf. Davies and Kathirithamby, *Greek Insects*, p. 41.

ants should not be praised for all they do and be set on the same level as man, since they are irrational creatures performing all their activities purely by instinct (ἀπ' ἀλόγου φύσεως καὶ κατασκευῆς ψιλῆς).

6. Ant behaviour as a weather sign

Classical meteorological sources generally include the observation that it is a sign of the approach of rain if ants are seen carrying their pupae—which are always mistakenly described as eggs—up from the interior of their nests (Theophrastus *Sign*. 22; Aratus 956–7; Plutarch loc. cit.; Pliny XVIII.364; Virgil *G*. 1.379). The scholiast on Aratus discusses suggested explanations of this behaviour seeing it as a response to changes in temperature below ground. Physiologus (p. 47 Sb) declares that it is a sign of rain if ants are seen carrying seeds up to the surface. Basil (*Hex*. MPG.29.196) and Ambrose (*Hex*. VI.20), on the other hand, say that this is a sign of good weather, which will hold so long as the seed stays in the open, as the ants' purpose is to dry it. Aelian (I.22) credits ants with the ability to measure the passage of time, as well as to predict the weather, stating that they always remain in their nests on the first day of the month.

7. Ants as pests

The grain storing ants are noted as pests of threshing floors (Aelian II.25), stored grain in general (Virgil *G*. I.186, *Aen*. loc. cit.; Cicero *Nat.D*. III.157; Ovid *Tr*. I.9.9; Prudentius *C.Or.Symm*. II 1054) and sown grain (Ovid *Fasti* I.685; *Geoponica* II.18.1, cf.X.66). The precaution generally advised to prevent infestation of threshing floors was to treat them with a coating of *amurca* (Cato *de Ag*. 91,129; Varro *RR*. I.51.1; Pliny XV.33; Palladius VII.1; *Geoponica* II.26.5): Varro says that the intention is to provide a surface which will not crack to produce crevices for pests to hide in, and also that the *amurca* itself acts as an insecticide. Pliny (VII.65) states that ants will not touch grain that tastes of birumen. According to the *Geoponica* (XIII.10.3, 11.29), grain stores may be protected by a circle of chalk, or sprinkling with the herb *origanon*. The same source (XIII.10.8) records that honey jars will not be touched by ants if they are banded with white wool, chalk, or red ochre—*AP* IX.438 describes the placing of a honey pot in water. A general deterrent prescribed by the Cyranides (p. 56) consists of cow dung boiled in oil.

   Pliny, Palladius and the *Geoponica* refer to ants as infesting various kinds of tree. The preventative measures they recommend are smearing the trunk with red earth and pitch (Pliny XVII.266; Palladius IV.10.29; *Geoponica* XIII.10.15), red earth with vinegar and ash (Palladius IV.10.21), juice of *portulaca* with vinegar and wine lees (Palladius XI.12.8) or *amurca* (Pliny loc. cit.; *Geoponica* XIII.10.7), bitumen and oil (*Geoponica* loc. cit.), or bull gall with pitch and *amurca* (*Geoponica* XIII.10.15); smearing the roots with lupin pounded with oil (Pliny loc. cit.); and hanging up a fish (Pliny loc. cit.; Palladius IV.10.29; *Geoponica* XIII.10.16): Pliny says this is to collect the ants in one spot. Ants infesting vine stocks may be captured by hanging up ivy foliage in which they will congregate (*Geoponica* XIII.10.10).

   With regard to ants which cause trouble in gardens (referred to by Martial XI.18), suggested formicides are *heliotropium* (Pliny XIX.178), water in which an unbaked brick has been soaked (Pliny loc. cit.), and white chalk and ash (Palladius I.34.2). It is also recommended to block up the entrances to their burrows with *limus marinus* (Pliny), ash (Pliny), or the remains of burnt snail shells (Palladius I.34.8; *Geoponica* XIII.10.4); or to paint around them with cedar oil (*Geoponica* XIII.10.2); or to pour in silphium juice (*Geoponica* XIII.10.6). It is reported that ants could be induced to vacate their nests by sprinkling the entrances with powdered *origanon* and sulphur (Aristotle *HA* 534b 22; Palladius I.34.8; Antigonus *Mir.* 85; *Geoponica* XIII.10.5), or by placing by the entrance the heart of an owl (Palladius I.34.2) or bat (Dionysius *Ixeutikon* I.16; Pliny XXIX.92) (Dionysius states that owls keep bat's hearts in their nests to keep ants away from their young). The burning of captured ants was believed to drive off others of their kind (*Geoponica* XIII.10.1,13), and fumigation with wild fig root or the fish *silouros* (*Geoponica* XIII.10.11) is said to be similarly effective.

Ants biting man are referred to by Babrius (*Fab.* 117; cf. *Aesopica* 235 P), Plutarch (*Mor.* 458c) and Seneca (*de Ira* II.34.1).

8. Medicinal uses of ants

Adult ants appear in medical contexts as treatments for styes and warts (Cyranides p. 261), as a cure for fevers (Pliny XXVIII.86), and as an antidote to *phalangion* bites (Pliny XXIX.88). Their pupae, which ancient writers mistook for eggs, are mentioned as a depilatory (Pliny XXX.41,134) and as a treatment for ear ailments (Pliny XXIX.133; Marcellus IX.120). The earth thrown up by ants from their excavations is given as a possible application for *strumae* and similar conditions and for gout (Pliny XXX.39). We read in the *Geoponica* (XIII.10.12) of the use of an ant-gathered wheat grain wrapped in hide as an amulet to prevent conception. According to Pliny (XI.196, XXII.50), part of the toad's liver is poisonous but part an antidote to all poisons: which is which may be discovered by throwing the whole to ants and observing which part they avoid.

Earth dug by ants is mentioned by Pelagonius (283; cf. Vegetius *Mul.* III.79; *Hippiatrica Ber.* 86.3) as a treatment, with wine, for horses bitten by venomous creatures. An odd story constantly repeated related that sick bears treat themselves by eating either ants or ant pupae (Aristotle *HA* 594b 9; Pliny VIII.101, XXIX.133; Aelian VI.3; Plutarch *Mor.* 918c, 947b; Sextus Empiricus *Pyrrh.* I.57).

9. Ants as portents

Artemidorus Daldianus states (III.6) that to dream about winged ants is always a bad omen, but that to dream about ordinary ones can be good for farmers—because they indicate a good harvest—and for those who are sick unless they appear to actually crawl around the person of the observer, in which case they foretell death. An ominous dream involving winged ants is described by Suetonius (*Nero* 46.1), who tells how Tiberius' pet snake being devoured by ants was taken for a portent (*Tib.* 72.2). Plutarch (*Cimon* XVIII.4) relates how Cimon's death was portended by ants carrying off pieces of congealed blood from a sacrificial victim and depositing them on his toe. Cicero (*Div.*

I.78) and Valerius Maximus (I.6. ext. 2) say that when the legendary Midas was a boy ants filled his mouth with grain as he slept, which was taken as a token of his future wealth.

45. *Myrmex Chrysorychos/Formica Indica*

The gold-digging ant is a creature which figures prominently in stories of the origin of gold imported from India. It appears for the first time in Herodotus (III.102–5), who locates it as an inhabitant of the Thar desert and describes it as being intermediate in size between a fox and a dog, although otherwise very similar to an ordinary ant. He adds that the Persian king possessed some specimens. These enormous insects excavate burrows for themselves, shovelling out quantities of gold-bearing earth. The Indians set out on camels to collect the gold, planning to arrive at the hottest time of day when the ants will be sheltering underground. They fill their sacks with the excavated material and make off as quickly as possible with the ants, which do not take long to scent them, in hot pursuit. The ants are of unparalleled swiftness, and if they managed to overtake the gold collectors none would escape.

Further accounts of the ants appeared in Nearchus' account of his observations during the campaigns of Alexander in the East, and in the historical and geographical survey of India written by Megasthenes. Material from these writers is preserved by Strabo (XV.1.44) and, in a more reduced form, by Arrian (*Ind.* XV.4–6). Neither of them claimed to have seen the ants themselves, but Nearchus did assert that their skins had been brought into the Macedonian camp in some numbers (Arrian, loc. cit.) and that they resembled those of leopards. We do not know what Nearchus had to say about their geographical location, but Megasthenes (in Strabo) put them in a more remote area than Herodotus, among a mountain tribe called the Derdai, where the ants conduct their mining operations at the base of a certain plateau. The account that follows in Strabo adds to Herodotus' information that the ants hunt living prey, and that they do their digging specifically in the winter, and also gives a different version of how the gold is obtained: here the ants are lured away from their burrows by pieces of meat being set out at various spots. In XV.I.69 of Strabo it is noted that some of the ants are winged.

Of later authors dependent on the above, Pliny (XI.111) follows Megasthenes for geographical location and Herodotus for the mode of obtaining the gold; he suggests that the ants remain below ground all summer and excavate in winter, and that it is therefore in summer that the gold can be collected. He also adds two extra details: that the ants are the colour of cats, and that a pair of their horns was on display in the temple of Hercules at Erythrai.

Dio Chrysostom (XXXV.23–4) essentially follows Herodotus, replacing camels with horses and chariots and suggesting that the ants might be killed themselves when they overtake the departing natives, but adds picturesque details about the excavated gold. He suggests that what the ants throw out from their burrows is all pure gold, καθαρώτατον καὶ στιλπνότατον, and that the whole plain so shines with it that to observe it with the sun shining is to risk being blinded.

Apart from these more detailed descriptions, the ants are alluded to as inhabitants of India by Callimachus (fr. 202), Lucian (*Gall.* 16, *Sat.* 24), Propertius (III.13.5) and Mela (III.62). Theocritus XVII.106–7 may also be an allusion to these insects (cf. Scholiast ad. loc.). Fierce ants of giant size figure in Ps. Callisthenes' fabulous account of Alexander's explorations. They are encountered both in the vicinity of the Ganges (III.10) and in a desert region at the far limits of India (II.29), where they launch an attack, seizing upon men and horses and carrying them off.

The tale of the μύρμηκες χρυσωρύχοι, as Strabo calls them, is closely associated with a parallel report concerning gold which emanated from the far North of Europe. Herodotus (III.116, IV.13,27) records, citing as his source the lost epic attributed to Aristeas of Proconnesus, the existence of a tribe of one-eyed men called the Arimaspoi who obtain gold by stealing it from the griffins which guard it, and this report is reproduced by many later authors (for example Pliny VII.10; Pausanias I.24.6; Dionysius *Ix.* I.2; Philostratus *Vit.Ap.* III.48; Solinus XV.22–3). Because of the similarity of the two stories, there was a tendency to confuse them. Thus Ctesias (*Ind.* 12; *ap.* Aelian IV.27) transfers the griffins to India in place of the ants, giving a highly coloured description of their appearance and the means employed to deal with them. Aelian, not surprisingly, has something to say about the ants as well (though an unknown amount of his chapter on the subject has dropped out of the text) but, since he has followed Ctesias in situating the griffins in India, he has to locate them somewhere else, namely in the territory of the Issedones (he incongruously still calls them Ἰνδικοί). This is in the same part of the world as the Arimaspoi were said to be, so that he has effectively transposed the two populations of animals.

Gold-digging ants are also described as inhabitants of Ethiopia. Philostratus (*Vit.Ap.* VI.1) states that this region and India are very similar in their geography, flora and fauna, and cites as an example the fact that 'the griffins of the Indians and the ants of the Ethiopians, though they are dissimilar in form, yet play similar parts, for in each country they are sung of as guardians of the gold' (cf. Clement *Paed.* II.12, where both creatures are mentioned together, though without any statement of their respective locations). A further variation is found in Helidorus' *Ethiopica* (X.26), where the African tribe known as Troglodytes bring to the Ethiopian king tribute consisting of χρυσός μυρμηκίας and a pair of griffins. Solinus (XXX.23) and Isidorus (*Or.* XII.2.24) both locate the ants *iuxta Nigrim* in Africa and describe them as possessing lion's feet, a motif which properly belongs to the griffin. Fernandez[96] suggests that the placing of the ants in Africa is connected with the originally independent reports of ant-like creatures of unusual appearance[97] found in this region.

The comedian Eubulus (fr. 20) has a joke about the Athenians going out to fight gold-guarding ants on Hymettus, a passage which the lexicographer Harpocration, in his article on the expression χρυσοχοεῖν,[98] attempts to explain by the unlikely story that once

[96] *Nombres*, pp. 59–60.   [97] On which cf. 61 below.

[98] LSJ 'Smelt ore in order to extract gold from it: hence proverbially of those who fail in any speculation, as the Athenians in their attempts to extract gold from their silver ores'; J. M. Edmonds on Eubulus fr. 20 ' "Gold smelting" seems to have been the Greek for "wild goose chase" '.

upon a time 'a rumour ran through all Athens that a quantity of gold dust had appeared on Hymettus and was guarded by fighting (μαχίμοι) ants: people seized weapons and took the field against them . . . (but) came back with all their trouble for nothing'.

The twin stories, before they began to be confused, of the Indian ants and the Arimaspian griffins resulted from the fact that gold was coming into the Greek world by a long route the start of which was unknown to those at the end; whether they developed spontaneously or whether, as Tarn[99] suggests, they were deliberately propagated by middlemen on the trade route anxious to keep the source of the supply as obscure as possible. In fact the gold that emanated from India was not a native product but came from the same source as the 'Arimaspian' gold, only by a different route. The ultimate source was Siberia, from which gold travelled down to southern Russia and Greek colonies like Panticapaeum in the Crimea, whose inhabitants put the griffin on their coinage, and also to Bactria and India (probably via Bactria).[100] It has been suggested that the origin of the griffin story was the use of this motif in south Siberian art.[101] As for the ants, they were already present in Indian folklore (*Mahabharata* II.2860 speaks of gold excavated by ants being brought as tribute).

Herodotus obtained his version of the story from the Persians, who would have had it from the Indians. As regards Nearchus, Pearson[102] suggests that he probably did not hear the story on the spot, but that he simply deduced the identity of the animal skins coming into the camp on the basis of what he had read in Herodotus. However, the idea of ants having pelts like leopards is a curious one not deducible from Herodotus, so that it is unlikely that Nearchus would have made such an identification of his own accord. And it is evident from Strabo that there were more versions of the tale in circulation than merely Herodotus'. There has been considerable speculation concerning actual members of the known Indian fauna which could be seen as having given rise to these stories, the two chief suggestions being the pangolin[103] and the marmot.[104] The first may safely be discounted, but if we decline to follow the opinion of Fernandez[105] and others that the creatures in question are entirely fabulous, the marmot is certainly the most likely candidate. As Hennig[106] points out, this is a plain dwelling mammal which lives in extensive colonies, burrowing into the ground and throwing up heaps of earth like so many ant hills. This would explain the persistent idea that the ants possessed furry pelts and the comparisons with various mammals, which are quite incongruous as applied to an insect and therefore unlikely to have arisen spontaneously. Possibly the Indians called these animals 'ants' metaphorically, and their usage was misunderstood or deliberately embellished by Greek traders and explorers.

[99] W. W. Tarn, *The Greeks in Bactria and India*, Cambridge, 1938, pp. 106 ff.

[100] Ibid., pp. 105–7.   [101] Ibid., p. 105.

[102] L. Pearson, *The Lost Histories of Alexander the Great*, American Philological Society Monographs no. XX, 1960, p. 125.

[103] G. Rawlinson, *History of Herodotus*, London, 1875, II p. 494.

[104] Keller, op. cit., I pp. 184–5; R. Hennig, 'Herodots Goldhütende Greifen und Goldgrabende Ameisen', *RhM*, 1930, pp. 326–32; W. W. How and J. Wells, *A Commentary on Herodotus*, Oxford, 1912, I p. 289.

[105] Op. cit., p. 59.   [106] Art. cit., pp. 331–2.

46. Psen/Culex ficarius

1. Identification

The ψήν was one of the smallest insects known to classical writers, but it was one which was of great interest to them because of its importance in the cultivation of figs. It is known today as the fig-wasp *Blastophaga psenes* (Linn.),[107] which belongs to a family of highly specialised hymenoptera, the Agaonidae, all of which are the exclusive pollinators of various species of fig throughout the world. The classical accounts of the insect are unusually accurate, demonstrating the particular attention and careful observation that it received. However, in the absence of true understanding of pollination, no one, despite much speculation, was able to explain the precise mechanism by which the insect was able to achieve an effect upon the figs.

The European fig (*Ficus carica*) occurs in two different forms: the wild or caprifig (Gk. ἐρινεός), which is the natural host of the fig-wasp, and the cultivated variety (Gk. συκῆ) which is deficient in that it produces female flowers only. The caprifig, in which the full life cycle of the fig-wasp can take place, produces at different times of the year three distinct types of receptacles. These receptacles are complete inflorescences: they are hollow with a small opening and contain on the inside tiny unisexual flowers. The receptacles of type I contain neuter flowers and a smaller number of male ones situated by the entrance. Female fig-wasps enter these, lay eggs in the neuter flowers, and die. When ovipositing, it introduces a secretion into the flower causing the ovule to develop into a gall that provides food for the resultant larva. The adult males, which are wingless and have reduced legs, emerge from the flowers first, whereupon they bore into those which are occupied by females and fertilise them. The winged females then emerge and make their escape from the receptacle, collecting pollen from the male flowers by the entrance as they leave.

By this time it is about June, and the females find their way to receptacles of type II, containing neuter and female flowers. Here they lay their eggs as before but only those in neuter flowers develop. Pollen carried by the wasps fertilises the female flowers, and these then set seed. In the autumn, a new generation of fertilised females emerges and enters receptacles of type III, which have neuter flowers only: here the wasps develop that will hatch in the following year and restart the cycle.

Female fig-wasps will also enter and fertilise the inflorescences of cultivated figs, but since these contain only female flowers they cannot lay eggs in them. The main commercial crop is produced from the receptacles of type II, which ripen in the late summer and autumn.[108]

2. Caprification and the fig-wasp's life history

Classical discussions of the *psen* centre upon the practice known as caprification, which

---

[107] Correctly identified by authors from Sundevall, *Thierarten*, p. 222, onwards.

[108] McLean and Ivimey-Cook, *Textbook of Theoretical Botany*, pp. 1327–9; Proctor and Yeo, *The Pollination of Flowers*, London, pp. 312 ff.; Imms, *Gen. Textbook*, pp. 1220–1.

has continued up to the present day and which consists in hanging fruits of the caprifig upon trees of the domestic fig so that wasps emerging from the former may enter and fertilise the fruits of the latter. This has always been done in the belief that the cultivated fruit will not otherwise come to maturity; but it would appear that in fact normal fruit can be produced in the absence of fertilisation. It has been suggested that it is simply a traditional practice dating from a time when fertile seeds, rather than cuttings, were required to propagate the plants.[109]

Suspending fruits of the caprifig on cultivated trees is alluded to by Theophrastus (*HP*, II.8.1–3, *CP* IX.9.5), Aristotle (*HA* 557b 25 ff.; *GA* 715b 22 ff.), Pliny (XV.81), Columella (XI.2.56), Plutarch (*Mor.* 700 f.), Ps. Eustathius (*Hex.* MPG 18.716) and Palladius (IV.10.28). Theophrastus' discussion of the subject is by far the most detailed, and his observations were not added to by any subsequent writer. The general reason given by the above authors was that without caprification the fruit would fall off before coming to maturity, but it was also known that the practice was not universally adopted. Theophrastus (*CP* II.9.12) states that some growers do not caprify, and that their figs are considered to be superior. He also lists various localities and situations where caprification was not practised: in S. Italy; in places with a N. facing aspect or with light soils (= Pliny XV.81), because (*CP* II.9.7) they are dried naturally and Theophrastus thinks it is moisture that causes the fruit to drop off; in places where the wind does not blow heavily from the North (explanation in *CP* II.9.7); where certain varieties of tree are planted which are resistant to fruit dropping (explanation in *CP* II.9.8; list in *HP* loc. cit.) Theophrastus (*HP* loc. cit. = Pliny XVII.256) notes that there is a particular type of caprifig, with dark fruit and coming from rocky country, that is especially useful because of its large quantity of seeds (which would produce a large quantity of fig-wasps), and that it is possible to recognise caprified figs by their appearance. The best time to hang the caprifigs, he says, is after rain (= Pliny XVII.256).

An alternative to caprification was simply to plant trees of the *erineos* next to the cultivated ones, early and late ripening types of the former by the corresponding variety of the latter, so that production of fig-wasps would synchronise with the appearance of the cultivated fruit (*CP* II.9.5; Aristotle *HA* 557b = Pliny XV.80).

It was believed that fig-wasps were initially produced in the fruits of the caprifig by spontaneous generation (Sextus Empiricus *Pyrrh.* I.41), specifically out of the seeds (*HP* loc. cit.) or from the decomposition of matter inside the receptacle (*CP* II.9.6,12 = Pliny XVII.255 and XI.118). Theophrastus (*HP* loc. cit.) cites as evidence for origin from the seeds that after fig-wasps have emerged there are no seeds left (= Pliny XVII.255). Their life history is described by Aristotle (*HA* 557b 25 ff.) as beginning with a small larva (σκωλήκιον): its skin later splits open, and the adult *psen* is released and makes its escape from the fruit. Theophrastus (*HP* loc. cit. = Pliny XVII.255) observes that 'most of them in coming out leave a leg or a wing behind'. It is surprising that this minor detail should

---

[109] McLean and Ivimey-Cook, op. cit., p. 1329; A. L. Peck, *Gen. An.*, pp.8–9.

have been noted, but it is true that the females do often lose wings or parts of antennae in struggling past the scales which bar the opening of the receptacle.[110]

Since Theophrastus did not realise that newly emerged fig-wasps entered other receptacles for the purpose of laying their eggs there, he had to provide some other explanation for their behaviour. He suggested therefore that having run out of food in the wild fig where they were born they set out in search of food material similar to that which produced them: he likened this to lice seeking after blood,[111] from the decomposition of which he believed they were generated (CP II.9.6 and 12). He seems to think that they eat their way into the fruit (HP loc. cit. = Pliny XV.2.80), though Aristotle speaks more correctly of their passing through openings already there (διὰ στομάτων; HA loc. cit.).

On the question of how exactly the entry of fig-wasps helps fruit to reach maturity, Theophrastus discusses in CP II.9 two rival theories. The first, which is the one he prefers, states that when the wasps penetrate to the centre of the fruit they consume excess fluid and allow the air to come in and ventilate the interior (=Pliny XV.80): it is this fluid along with τι πνεῦμα that cause the fruit to drop off the tree (ch. 6). Theophrastus (ch. 7) thinks that this idea of the wasp providing ventilation and removal of excess liquid contents would fit in with the fact that in certain situations their services are not required (either because no liquid is produced at all, or because it is removed in some other way).

The alternative theory (ch. 9) was that the effect of the wasp's entry was the exact opposite; that it caused the fruit to close up so that moisture could not get in from outside. The proponents of this view argued that, since more figs dropped when rain followed their appearance, it followed that the agent responsible was moisture from outside, and declared that in case of unavailability of caprifigs the desired effect could be obtained by sprinkling the fruit with dust thus making it close up. However, Theophrastus (ch. 10) argues that one could just as easily say that rain promotes formation of moisture within the fruit, and that the sprinkling with sand could be viewed as a drying agent.

3. Supposed alternatives to caprification

Since in either case the effect of the fig-wasps was judged to be a purely mechanical one, it was logical for classical farmers to believe that the same result could be obtained by means entirely different from caprification but which could be understood as having the same physical effect on the fruit. Sprinkling with dust or sand has already been mentioned, and Pliny (XV.81) states that when figs are grown next to a dusty road this will do the same job as the fig-wasps. Theophrastus states that the bag-like galls[112] found on the elm can be used in the same way as caprifigs, since they produce insects superficially similar to fig-wasps, as can the herb hulwort (polion), which he implies generates insects also. Similarly Palladius (IV.10.28) mentions the use of these elm galls, along with that of ram's horns, and scarification of the fig tree's trunk ut possit umor effluere.

---

[110] Proctor and Yeo, op. cit.
[111] Cf. 22.2 above.
[112] 47.4 below.

4. Enemies of the fig-wasp

Two supposed enemies of the *psen* are noted by Theophrastus. The first, called the κεντρίνης (*HP* loc. cit. = Pliny XVII.255 *genus culicum quos vocant centrinas*; cf. Hesychius s.v.) is said to be a type of fig-wasp itself just as the drone is a type of honey-bee; it is sluggish, kills the genuine *psenes* as they enter the fruit, and itself dies inside the fruit. There are in fact certain members of the family Torymidae, closely related to that of the fig-wasps, that live as parasites or inquilines in association with Agaonids, for example *Philotrypesis caricae*,[113] but it is more likely that Theophrastus is referring here to the wingless males which do die inside the fruit, and which might well have been thought to be predators upon the winged, and very different in appearance, females.

The other enemy is the κνίψ, concerning which Theophrastus states that when they are produced in fig trees they prey upon the fig-wasps; and that accordingly farmers fix up dead crabs for them to feed upon instead. *Knips* covers a range of small creatures,[114] so it is not clear what kind of insect is meant here. The name appears linked with *psen* in Aristophanes *Av*.590, where they seem both to be viewed incongruously as pests of figs. Hesychius refers (s.v. κεράμβηλον) to a type of beetle which was tied to fig trees in order to frighten off *knipes* by its stridulation.[115]

5. 'Fig-wasps' in the date palm and sycomore-fig

Herodotus (I.193) provides one of two examples of erroneous transference of the idea of caprification to other plants. He states that in Assyria the cultivators of date palms tie the fruit of the 'male' palm to the date-bearing tree, so that *psenes* from the former may enter the latter's undeveloped fruit. This account results from a confusion of caprification in figs with a process of artificial pollination, by shaking male flowers over the female ones, which was practised on date palms and is described by Theophrastus (*HP* II.8.4). Date palms are in fact wind pollinated, so that tying part of a male inflorescence onto a female fruiting tree is an effective process with this species.[116] Theophrastus would not have realised how the process operated, though in using the terms male and female he came quite close to the truth.[117]

A somewhat similar error can be seen in Physiologus (pp. 143 ff. Sb), where the Egyptian and Palestinian sycomore-fig (*Ficus sycomorus*) is described as having insects— here called σκνῖπες or κώνωπες—inside it which are released when incisions are made in the fruit. This is the idea behind the statement of Jerome (MPL 25.1077) to the effect that if the sycomore-fig is not incised it becomes corrupted by *culices*. The incisions here mentioned allude to the practice of scraping the fruit with an iron implement to cause it to ripen, described by Theophrastus, who thought the process had the same effect as

---

[113] Cf. Imms, op. cit., pp. 1221–2.

[114] 58 below.

[115] 30.1.3.iv above.

[116] Proctor and Yeo, op. cit., p. 351.

[117] Gossen, in PW Supp.VIII.237, not realising Herodotus' error, erroneously identifies his palm ψήν as a form of beetle.

caprification; viz. removing excess moisture and πνεῦμα (CP. I.17.9, II.8.4, II.9.8, HP IV.2.1. Cf. also LXX Am. VII.14, which is quoted by Physiologus here). The sycomore-fig does in fact have a group of species of fig-wasp which pollinate it, but these only live in the area where the tree is native; that is, outside the classical world, in tropical Africa and the extreme South of Arabia. Its presence in the Mediterranean region is a later development, and here, since it has no pollinators, it cannot produce viable seed and had to be propagated by cuttings.[118]

47. Gall wasps

1. Identification: general

The plant growths produced as a result of insect activity and known as galls belong more to the realm of classical botany than to that of zoology, and so will be dealt with rather summarily here.

Galls are produced by members of a number of insect families in order to provide shelter and nourishment for their offspring. These include certain aphids (Hemiptera: Aphidoidea), the gall midges of the family Cecidomyidae (Diptera), and the gall wasps of the Cynipidae (Hymenoptera),[119] the latter being those with which ancient authors are chiefly concerned.

Among the Cynipidae, the vast majority of species give rise to various types of gall on species of oak. The females lay their eggs on the living tissues of the foodplant, and it is the presence of the resultant larvae which somehow provides the stimulus in reaction to which a gall is produced by the plant tissues. This then provides a source of food for the growing larvae, which will pupate without emerging to the open air.[120]

Although this process was, not surprisingly, unrecognised in antiquity, it was often observed that minute insects were associated with various types of gall. These insects, as with other small winged creatures,[121] are commonly termed 'gnats' (κώνωψ, culex), or sometimes 'flies' (μυῖαι). The general view, as stated by Theophrastus (HP III.7.3 ff.), was that galls were natural products, like fruits or catkins, of the plants on which they occur, and that the associated insects arose from them subsequently by a process of spontaneous generation.

Galls were used in antiquity as sources of black dye,[122] in the tanning of leather,[123] and for numerous medicinal purposes.[124]

---

[118] J. Galil and D. Eisikowitch, 'Pollination Ecology of *Ficus sycomorus* in East Africa', *Ecology*, XLIX, 1968, pp. 259 ff.

[119] Cf. Imms, *Gen. Textbook*, pp. 717 ff., 995 ff., 1214 ff.

[120] Ibid., pp. 1214–15.

[121] 50 note 56 below.

[122] Cf. R. J. Forbes, *Studies in Ancient Technology* V, Leiden, 1957, pp. 7, 45 ff.; J. L. Cloudsley-Thompson, *Insects and History*, p. 205, dealing with the galls of the Middle Eastern *Cynips gallae-tinctoriae*, the *gallae Syriacae* of Pliny and others.

[123] Cf. R. J. Forbes, op. cit., III, Leiden, 1955, pp. 228–31.

[124] Cf. for example Pliny XXIV. 9–10.

2. Galls produced on oak

Theophrastus (*HP* III.7.4) states that the oak produces more things besides its own fruit than any other tree, which is in keeping with the fact that most of the Cynipidae are associated with this genus. In his account (*HP* III.7.4–5), which is followed by Pliny (XVI.26), he lists ten district varieties of oak gall: a small spherical pale one, a larger black resinous one, a mulberry shaped form, a form shaped like a penis, a small woolly ball, a hairy ball, a scarlet gall with white or black protuberances, a pumice-like form, a leaf-like oblong form, and a pale transparent ball found on the ribs of the leaves. The last variety is said sometimes to contain 'flies' (μύας):[125] Pliny speaks of *culices* being generated within them. The first two varieties are the most frequently mentioned elsewhere (for example Theophrastus *HP* III.5.2; Dioscorides *DMM.* I.107). They would include the well known marble and oak apple galls, produced respectively by the wasps *Andricus kollari* (Hart.) and *Biorrhiza pallida* (Oliv.).

3. Galls produced on rose

Pliny describes a sponge-like gall (*spongiolae quae in mediis spinis nascitur*) which is found on the wild rose (XXV.18) and contains a *vermiculus* (XXIX.94), the latter being erroneously described as developing into a form of blister beetle.[126] This is clearly to be identified as the robin's pincushion gall, a distinctive object covered with moss-like filaments and produced by the cynipid *Diplolepis rosae* (Linn.).[127] Also described from rose are a small ball (XXI.125) and a growth resembling a chestnut (XXIV.121).

4. Galls produced on elm

Certain bag-like objects (τὸ θυλακῶδες) known as κωρυκίδες and found upon the elm are described by Theophrastus (*HP* II.8.3, III.7.3, 14.1, IX.1.2) as producing gum and generating θηρίδια κωνωποειδῆ. Dioscorides (*DMM.* I.84) says that at first they are filled with moisture, and that as this dries it turns into the gnat-like insects described by Theophrastus. Similarly, Pliny (XIII.67) refers to *culices* generated from the gum of elms.

Because these galls were seen to produce insects which superficially resembled the fig-wasp, it was erroneously believed that they could be used for caprification.[128] They have been identified[129] as being produced by a species of aphid.

5. Galls produced on other plants

The terebinth is described by Theophrastus (*HP* III.15.4) and Pliny (XIII.54) as developing bag-like galls similar to those of the elm. These produce a sticky resin together with θηρίδια κωνωποειδῆ (Pliny *culices*), which have been identified[130] as aphids of the genus *Pemphigus*. Pliny also describes a leaf gall occurring on the beech (XVI.18), and a *pilula* found on cypress (XXIV.15).

---

125 Cf. 48 notes 1 and 4 below.
127 Imms, op. cit., pp. 1216–17.
129 Steier, *Aristoteles und Plinius*, p. 48, names them as *Tetraneura ulmi* (deG.).
130 Bodenheimer, *Geschichte*, p. 74; Steier, op. cit., p. 47.

126 33.1.2.v above.
128 46.3 above.

VIII

FLIES AND FLEAS

Insecta: Diptera (Flies)

48. Myia/Musca

1. Identification: general

Μυῖα[1] and *musca* are corresponding terms covering the larger two-winged flies or Diptera, as distinct from the smaller gnats, midges and related insects. More specifically they cover those species of the two families Muscidae and Calliphoridae which infest houses and/or have larvae which feed upon carrion or other decaying material.[2] A distinction is usually maintained between *myia/musca* and the various species of horse and cattle flies,[3] but the former names do sometimes seem to be extended to cover the latter as well.[4]

The family Muscidae includes most importantly the common housefly *Musca domestica* Linn., which infests houses and is attracted to human foodstuffs. It lays its eggs most especially upon fresh horse dung, but will do so upon any decaying organic material. There are also certain species of similar appearance which have biting habits and suck the blood of men and animals, the most notable of which is *Stomoxys calcitrans* (Linn.). This attacks chiefly horses, cattle and other domestic animals, being less attracted to man. The closely related family Fanniidae includes the lesser housefly, *Fannia canicularis* (Linn.)[5]

The Calliphoridae includes two main subfamilies. Of these, the Sarcophaginae includes the species known as flesh-flies which bear fully developed larvae and deposit them upon carrion or in open wounds and ulcers in man and animals. The genera involved here are *Sarcophaga*, for example *S. carnaria* (Linn.) and *S. haemorrhoidalis* (Fallén), and *Wohlfahrtia*, for example, *W. magnifica* (Schiner). The other division, the Calliphorinae, includes the metallic coloured bluebottles, *Calliphora* spp., and green-bottles, *Lucilia* spp. These lay their eggs upon meat—*Calliphora* especially infests human food—and carrion and also in the wounds of man and domestic animals. Some oviposit, for example, on the wool of sheep, and the larvae then bore directly into the flesh.[6]

[1] Attic μῦα, found in Theophrastus *HP* III.7.5 and Photius; Laconian μούα, in Hesychius.
[2] *Myia* is identified by for example Aubert and Wimmer, *Thierkunde*, I p. 168, Sundevall *Thierarten*, p. 223, Keller, *Ant. Tierwelt*, II p. 447, and Wellmann, in PW VI.2744–7, as covering particularly such species as *M. domestica*, *S. carnaria*, *Calliphora vomitoria* (Linn.), and *S. calcitrans*.
[3] 49 below.
[4] *Myia* is also used for gall-wasps (47 above) and bees (Fernandez, *Nombres*, pp. 61–2).
[5] Smith, *Insects*, pp. 251 ff.; Soulsby, *Helminths*, pp. 409 ff.
[6] Smith, op. cit., pp. 272 ff., 289 ff.; Soulsby, op. cit., pp. 429 ff., 440–1.

1.2. Identification: distinct forms

(i) The μεγάλαι μυῖαι referred to by Pseudo-Aristotle (*HA* 628b 34), which the hymenopterous *anthrenai*[7] hunt while hovering around dung.[8]

(ii) The *myiai* referred to by Pausanias (X.28.7) as being of a colour between blue and black, and as landing on meat, are undoubtedly bluebottles (*Calliphora*).[9]

(iii) The *muscae rufae* mentioned by Pliny (XXX.92) as a treatment for epilepsy.

(iv) Στρατιωτίς or κύων. Lucian includes in his *Muscae Encomium* (*ME* 12) a distinct variety of fly of which these are alternative names. They are described as large in size, rapid in flight, and notably long-lived, spending the whole winter without food, usually hiding in the roofs of houses. Fernandez[10] refers to Theophrastus' mention (fr. 174.1) of flies which appear in swarms in military camps, and suggests that Lucian's insect acquired its first name as a result of its appearance in such places, and its second from its persistence.[11] The reference to hibernation is a clue to their identity, since there are certain species of Muscidae and Calliphoridae[12] which overwinter in this fashion, often congregating in large groups in the roof spaces of houses.[13]

2. Life history

Sources for the life history are Aristotle (*HA* 552a 21 ff.) and Lucian (*ME* 4). There is a constant confusion between the old belief that the larvae were spontaneously generated in carrion etc. and the more accurate observation that these were the products of adult flies.

Mating is described by Aristotle (539b 7 ff., 542a) and Lucian (*ME* 6), who says that it takes place in flight, the female carrying the male about. In *HA* 539b 7 ff. and *GA* 721a 6 ff. (=723b 3 ff.) Aristotle, followed by Pliny (X.190) includes *myiai* among those creatures that are spontaneously generated from decaying solids or liquids, which, when they mate, produce not creatures like themselves but something imperfect (ἀτελές), from which nothing further is produced.[14] In the case of flies, *skolekes* are produced. It is curious that these are not equated with the *skolekes* in dung out of which flies are elsewhere said to develop. We seem to have two alternative accounts of the insect's life history here, which Aristotle has failed to harmonise. According to Augustine (MPL 42.363), flies are spontaneously generated in sewers.[15] Both Aristotle (552a 20 ff.) and Lucian state that flies derive from larvae or *skolekes*. The former does not explicitly state the origin of these

---

[7] 40.1.2 above.

[8] Wellmann, art. cit., identifies these as *Stomoxys*, and Keller, op. cit., II p. 447, as *Calliphora* or *Sarcophaga*, but the reference is too vague to permit an identification.

[9] Wellmann, art. cit., 2744.

[10] Op. cit., p. 167.

[11] Cf. the name κυνόμυια (52 below).

[12] E.g. *Musca autumnalis* DeG., *Dasyphora cyanella* (Meig.), *Pollenia rudis* (Fab.).

[13] Smith, op. cit., p. 489.

[14] He seems to mean that they are generated as adults, like the lice and fleas with which he groups them (cf. 22.2 above, 55 below).

[15] *Fannia* breeds in such locations.

larvae, and indeed juxtaposes them with various spontaneously generated creatures, but Lucian correctly states that they are the offspring of adult flies (*ME* 4).

Aristotle locates these larvae in dung, which fits *M. domestica*, while Lucian does so in human or animal corpses, which fits *Sarcophaga* spp. or one of the Calliphorinae. There is a clear reference to their being produced by adult flies from a far earlier period, namely in *Iliad* XIX.25–6,31, which speaks of *myiai* entering the wounds of corpses of men slain in battle and generating εὐλαί there. These corpse-feeding εὐλαί appear also in XXII.509 (where they are termed αἰόλαι, 'wriggling') and XXIV.414, and they are alluded to also by Herodotus (III.16), Plutarch (*Mor.* 165b, 337a), Diogenes L. (IX.79), *Anth.Pal.* VII.472 and ?Plato (*Ax.* 365c).

The term εὐλή[16] is restricted in usage to fly larvae, especially flesh-feeding, and corresponds to our English 'maggot'. It is also found in Herodotus IV.205, Hippocrates *Mul.* 1.75, *HA* 506a 30, and *Geoponica* XVII.27. More usually where larvae in carrion or corpses are referred to, the general terms *skolex* (for example Lucian *Asin.* 25; *AP* VII.480; LXX *Job* 7.5, *Sir.* 10.11, 19.3, *Is.* 14.11; Origen *c. Celsum* IV.57) and *vermis, vermiculus* (for example Lucretius III.719 ff.; Petronius 57; Apuleius *Met.* VI.32; Arnobius *adv.Nat.* VII.18; Augustine MPL 36.174) are employed. In such general references, words implying spontaneous generation are often found, in cases where any hint is given of the creatures' origin.[17]

Certain *skolekes* as large as the biggest *eulai* are described in the *HA* (506a 27 ff.) as being generated within the bodies of deer. They are found clustered in a group of about twenty in 'the hollow region under the tongue and near the vertebra to which the head is attached' (cf. also Pliny XI.135). These parasites would be the larvae of flies of the family Oestridae,[18] which infest the nasal and pharyngeal cavities of various mammals.[19]

As Aubert and Wimmer[20] observed, there is something seriously wrong with the text of *HA* 552a 21 ff. As it stands, the text gives us too many stages to the life cycle, with the larva switching about from mobility to immobility and back again. They suggest deleting the words from εἶτα to πάλιν in 27 ff., which seems the most obvious solution, disposing, as it does, of the superfluous stages. Assuming this correction, then, the cycle begins with a small motionless larva (σκωλήκιον) that later acquires motion.[21] The pupal state is ignored:[22] the larva is simply said to become immobile and the adult fly to emerge from it and take to flight.

According to Lucian's account (*ME* 4), the initial small larva puts forth little by little wings and limbs as it develops into an adult fly, which them gives rise to further larvae. He credits the adults with a very short life span.

---

[16] Cf. Fernandez, op. cit., p. 146.

[17] Cf. Pliny XI.114. For similar references to maggots in decaying food, cf. e.g. LXX *Ex.* 16.20,24; Philo *Vit. Mos.* II.260, *de Prov.* fr. 2.69, where ἕλμινς (Fernandez, op. cit., p. 144) is used.

[18] D'Arcy Thompson, *Hist. An.*, II.15 note.

[19] Cf. Imms, *Gen. Textbook*, pp. 1028–9.

[20] Op. cit., I p. 514.

[21] It is also said to become red, but since fly larvae are white this is unlikely to be an authentic part of the text.

[22] Since the larva pupates within its last larval skin.

3. Feeding habits

Aristotle (*HA* 532a 5 ff.) includes *myiai* among those insects that have a strong 'tongue' with which they can draw blood (cf. Lucian *ME* 3; Pliny XI.100); this 'tongue' is often loosely termed a sting (κέντρον), as in *HA* 490a 20 which associates *myiai* with gnats and horseflies (cf. also *PA* 616a20, 678b 15). They figure as insects sucking human blood in *Iliad* XVII.570–2, where they are said to keep returning however often they are driven away. Blood feeding is also mentioned in *Anth.Pal.* XI.191 and in the passages listed under 5 below.

In Lucian *ME* 6 *myiai* are said to be able to penetrate the skin of men, horses, oxen, and even elephants (cf. Pliny VIII.30). Phaedrus (III.6) refers to *muscae* stinging mules. In these latter two passages, *Stomoxys calcitrans* may be in view, as Wellman[23] suggests, or the terms *myia/musca* may be used here in a loose sense to cover species of Tabanidae also.

In *HA* 596b 13 *myiai* are classed among those insects feeding upon liquids only, but as being otherwise omnivorous. Their attraction to sweet substances such as honey is referred to by Apollonius Rhodius (IV.1453–4), who describes them clustering eagerly round a honey drop, Lucian (*ME* 8), *Aesopica* (80 Perry), and Pliny (XXI.79): cf. also Varro *RR* III.16.6 and Dio Chrysostom XXXII.49. Cicero (*Br.* 217) describes flies being attracted in numbers to the smell of medicinal ointments. *M. domestica* is primarily in view here.

The attraction of flies to milk is alluded to twice in the *Iliad* (II 469–71, XVI. 641–2), where they are described as buzzing round pails on a farm. This observation recurs in Plutarch (*Mor.* 750c) and Lucian (*ME* 3). The habit has been observed especially in the case of *M. domestica*.[24] Pausanias' reference to bluebottles feeding upon meat has already been noted (1.2.ii). The flies that caused problems by alighting on sacrificial victims (Lucian *Sacrif.* 9; Phaedrus IV.25; 8 below) would presumably have been primarily of this genus.

According to Lucian, *myiai* will feed upon all human foods (cf. Varro *RR* III 16.6) except oil, which is fatal to them (*ME* 4; cf. Alexander Aphrod. *Pr.* I.64). Lucian's reference (*ME* 7) to their holding food in their two front legs to feed upon it results from incorrect observation of cleaning behaviour. Plutarch (*Mor.* 473e) observes that flies are able to cling to any surface that has any degree of roughness, but that they slip off glass.

4. Popular beliefs and attitudes

There was a curious but often repeated belief that flies drowned in water could be restored to life by sprinkling with or covering in ashes (Pliny XI.120; Aelian II.29; Lucian *ME* 7; Varro *RR* III.16.38; Isidorus *Or.* XII.8.11).

Because of their constant and unremitting blood-sucking attacks on man and animals, and their infestation of homes and habit of alighting on human food, *myiai* were noted for boldness and persistence (*Iliad* XVII.570–2; Aelian II.29; Lucian *ME* 5) and for

---

[23] Art. cit., 2744.

[24] Smith, op. cit., p. 263.

intrusiveness and impudence (Oppian *Hal.* II.445 ff.; Xenophon *Mem.* III.11.5; Plautus *Poen.* 690; Pliny XXIX.106; Phaedrus IV.25; Arnobius *adv.Nat.* II.59; Symphosius *Aen.* 23B; John Chrysostom MPG 56.87; Gregory Magnus *Mor.* XVIII.43). The idea of flies as 'uninvited guests' in the home is found in Antiphanes (fr. 195) and Plutarch (*Mor.* 728a). Flies are also noted for idleness (Phaedrus IV.25) and for swiftness (Simonides fr. 32).

Aratus (*Phaen.* 975), Theophrastus (*De Sign.* 23) and the *Geoponica* (I.3.9) note that it is a sign of rain for *myiai* to be more than usually ready to bite.[25] Cassiodorus (*Var.* III.53), writing on water divining, says it is a good sign of the presence of water if a dense cloud (*spissitudo*) of very minute flies is observed flying to and fro above the ground. This is a good description of the communal 'dances' engaged in by certain small Diptera, for example Chironomidae,[26] for courtship purposes.

5. Relations with man

Fly larvae infesting human wounds are referred to in some medical contexts. They are described as being produced by dampness and suppuration, and various herbal and other preparations are listed as treatments (Pliny XXII.146, XXVI.142, XXX.114, XXXII.126; Dioscorides *Eup.* I.187; Galen XIII.733; Paulus Aegineta IV.42). Theophrastus (*Char.* 25.5) refers to *myiai* attracted to the injuries of battle casualties. The species involved here would be the various Calliphoridae noted in 1 above. Gruesome stories about living persons being infested and devoured alive by insects are frequent in ancient authors, since this was considered a fitting death for notorious evil-doers. Lice are most often mentioned in this regard,[27] but fly maggots (σκώληξ/*vermis*) figure also (Herodotus IV.205; Lucian *Alex.* 59; LXX 2 *Macc.* 9.9; Josephus *Ant.* XVII.169; NT *Acts* XII.23; Tertullian *ad Scap.* III.5; Lactantius *Mort.Pers.* MPL 7.247–8).

Myiai as pests sucking human blood figure in the *Iliad* (3 above), in Phaedrus (V.3), Callimachus (fr. 191), Chrysippus (*ap.* Plutarch *Mor.* 1039e, 1061a), Boethius (*Cons.Phil.* II.6), and in Oppian (*Hal.* II.445 ff.), where they are portrayed as relentlessly attacking the reapers at harvest. It is unlikely that these are references exclusively to *S. calcitrans*, as Wellmann suggests,[28] in view of the fact that man is not this species' primary victim. More probably the word *myia* is being used in an extended sense to cover tabanids, etc.[29] However, in the Aristotelian passages (3 above) the usage is probably to be taken as more precise; that is, Aristotle's blood-sucking *myia*, which he distinguishes from *myops/oistros*, is probably the result of his observing *S. calcitrans* and thinking it identical to *M. domestica* whose life history he later describes. General references to flies as troublesome pests are found in Aristophanes (*Vesp.* 597), Plautus (*Merc.* 361 ff., *Truc.* 65), and Seneca (*de Ira* II.25.3).

[25] The scholiast on Aratus suggests that temperature affects the skins of their victims and renders the blood more or less accessible. *S. calcitrans* is more persistent in rainy weather (Smith, op. cit., p. 266).

[26] Cf. Imms, op. cit., p. 990.

[27] 22.4.2. above.

[28] Art. cit., 2744.

[29] It may be that in normal usage it was a fly's host which determined its name, so that μύωψ/οἶστρος would be used only for those attacking animals.

Myiai as pests infesting human homes (primarily *M. domestica*) are mentioned by Plutarch (*Mor.* 94b, 728a), Antiphanes (fr. 195), Athenaeus (257b–c) and Lucian (*ME* 4), who speaks of them landing on the meal table. Fly-whisks (μυιοσόβη/*muscarium*) used to drive away *myiai* at mealtimes are alluded to by Menander (fr. 503), Martial (XIV.67), Anaxippus (fr. 7), Aelian (XV.14), Petronius (56.9) and Augustine (MPL 32.1039). Pliny (XX.184, XXIV.53, XXV.61) and the *Geoponica* (XIII.12) describe various decoctions of herbs in water, wine or milk to be sprinkled about the house to deter flies, as well as herbal or chemical fumigants, and the application of cassia to one's own person.

Cato (*De Ag.* 162.3) writes about how to store meat in order to prevent it being infested by *tiniae* or *vermes*, these being presumably nothing more than two names for the same thing; namely larvae of *Calliphora* spp. or other flies.[30] Larvae in meat are also referred to as μούια (Hesychius s.v.),[31] *teredo* (Pliny XXVIII.264),[32] and *tarmus* or *tarmes* (Isidorus *Or.* XII.5.12,18; Festus 495.1 Li; *C.Gl.L.* V.612.30, 637.1).[33]

6. Relations with domestic animals

Columella (*RR* VI.16.2–3), Vegetius (*Mul.* II.66, IV.20) and the *Geoponica* (XVII.27) mention certain *vermes* which infest open wounds of oxen if these are neglected, and recommend bandaging with appropriate herbal and other medicaments, after preliminary applications or pouring on of cold water to kill the creatures. These larvae are described as being generated by *muscae*. Both Varro (*RR* II.9.14), Columella (*RR* VI.13.1) and the *Geoponica* (XIX.2.10) prescribe applications of crushed almonds or pitch to prevent or treat attacks by certain *muscae* which cause ulcers between the toes or on the ears of dogs. It is possible that what is being referred to here is ulceration initially caused by ticks or mites being later infested by flies.

7. Medicinal and other uses

There is a single reference, in Martial V.18, to a fly being used as a fisherman's bait. Their use in traps to catch birds is also referred to (Augustine MPL 35.1386).

A number of supposed medical applications of flies, their 'blood', or ash of burnt flies are recorded by Pliny and Marcellus Empiricus for various skin conditions (XXIX. 106 = M.E.VI.21, XXX.21, 108); for eye ailments (XXVIII.29 = M.E.VII.52; cf. Theodorus Priscianus *Eup.* I.36,40); for eyebrows or lashes (XXIX.115, XXX.134); and for fevers (XXX.99). Fly larvae in general are prescribed by Hippocrates (*Mul.* I.75) to aid conception, and those from dung for disorders of the womb (*Superf.* 28). Pliny (XXIX.120) and Marcellus (VIII.102) mention the use of those generated in dead vipers to treat eye ailments. Flies were also occasionally used in veterinary medicine (Vegetius II.79.23; Chiron 998).

---

[30] *Tinea* (cf. 26.1 above) may also be used in this sense in *C.Gl.L.* II.198.24 (reading πέρνης σκώληξ).
[31] Laconian form of *myia*, defined here as '*skolekes* generated in meat'.
[32] Cf. 30.1.2.iv above.
[33] Cf. 30.1.2.v above.

8. Associations with religion

It is recorded by Pausanias (V.14.1), Pliny (XXIX.106, X.75) and Aelian (V.17) that at the time of the Olympic festival all the flies depart to the other side of the river Alpheus, according to the latter of their own free will out of respect for the god, according to the other as a result of sacrifices made to Zeus Apomyios. Pausanias states that the sacrifice was instituted by Herakles (Cf. Clement Alex. *Prot.* II. p. 33, Antiphanes fr. 229, *Etym.M.* 131.24). Similar accounts are given concerning the sanctuary of Apollo at Leukas (Aelian IX.8, where the author notes that the flies are inferior to those of Olympia because they require the bribe of a sacrifice; Clement II. p. 34), the festival of Athena at Aliphera in Arcadia, where sacrifices were offered to Myagros (Pausanias VIII.26.7), the sanctuary of Aphrodite at Paphos (Apollonius *Hist.Mir.* 8), and the shrine of Hercules in the Forum Boarium at Rome (Solinus I.11; Pliny X.79; Clement II. p. 33).

49. *Myops, Oistros/Asilus, Tabanus*

1. Identification: general

These corresponding pairs of Greek and Latin names cover the various species of horsefly or gadfly, family Tabanidae.[34] Most authors employ them as if they were synonyms, though some Greek sources attempt to distinguish μύωψ and οἶστρος by appearance or life history. There is some suggestion that *myops* was a more popular name than *oistros*, and that *asilus* was an older term than *tabanus*.

The most important genera of the Tabanidae are *Tabanus*, which includes the largest and most troublesome species, *Haematopota* and *Chrysops*. The females are notable blood-sucking pests, attacking primarily cattle and horses and also man, being active in warm sunny weather. Their eggs are laid on the leaves or stems of plants growing in the vicinity of water, the larvae dropping down on hatching. The larvae are found in various moist situations, such as mud at the margins of ponds and streams (so *Chrysops* spp.), damp soil or under stones in similar places (so *Tabanus* and *Haematopota* spp.), rotting wood, etc.[35]

Other insects that might perhaps have been included under the names in question are *Stomoxys calcitrans*[36] and, less likely, certain members of the family Oestridae, for example *Hypoderma* spp., which lay eggs upon the body hairs of cattle, the resultant larvae living as parasites beneath the skin.[37]

---

[34] Sundevall, *Thierarten*, pp. 223–4, and Aubert and Wimmer, op. cit. I p. 168, identify Aristotle's οἶστρος as *Tabanus* spp. and his μύωψ as its smaller relatives, *Chrysops* and *Haematopota*, but consider that in crediting the former with an aquatic larva he has confused true horseflies with the freshwater soldier-flies (Stratiomyidae). Gossen, in PW Supp.VIII.16–19 (also *Tiernamen*, no. 69) proposes a similar distinction between *oistros/tabanus* and *myops/asilus* as applying to classical literature in general, though outside Aristotle the terms appear as synonymous. Fernandez, *Nombres*, pp. 81–2, also accepts a real distinction between the two Greek names. (Οἶστρος is also used for drone bees in for example Pliny XI.47, Columella IX.14.4.)

[35] Smith, *Insects*, pp. 195 ff.; Soulsby, *Helminths*, pp. 406–9.

[36] 48.1,5 above.

[37] Imms, *Gen. Textbook*, pp. 1028–9.

(i) *Oistros* distinguished from *myops*. The authorities for a distinction between the two are Aristotle (*HA* 551b 21 ff., 552a 28), who separates them on the basis of their life history (2 below), Aelian (IV.51, VI.37), where appearance is the criterion, and a number of scholia (*in Od.* XXII.299, Theocritus VI.28a, and Apollonius Rhodius I.1265; also Eustathius *in Il.* 1928.16 ff.) which incorporate both sets of data. The source underlying Aelian and the scholia has been determined by Wellmann[38] to be Sostratus' work περὶ ζῴων. The difference is said to be that the *oistros* resembles a very large *myia*, robust and compact, with a strong sting, while *myops* resembles the *kynomyia*[39] and has a louder buzz and smaller sting than its relative. This supposed distinction, along with that of life history, is accepted at face value by Wellmann as proof that the names did originally refer to separate species, although being later confused. However, its extreme vagueness of expression, plus the mention of *kynomyia*, which is not a genuine insect species, do not give one any great confidence in Sostratus' accuracy. It is more probable that he was in fact inventing a distinction which did not exist in popular usage.

(ii) *Oistros* equated with *myops*. The earliest authority we have for regarding the two names as synonymous is Aeschylus. In the *Supplices* (307–8) he explicitly states that *myops* is another name for the *oistros*, the latter name being also used in 541 and—in compounds—17 and 573. In the *Prometheus* the gadfly is called *myops* once (575) and *oistros* twice (567,879) as a noun and four times in compounds (580,589,681,876). Nonnus (*Dion.* XI.191–2) likewise uses both terms as if identical in meaning. The equation is supported by a rival scholion (Schol. *Od.* XXII.299, Eustathius loc. cit.) to that deriving from Sostratus, which also describes the insect as being brassy or copper in colour. Both Callimachus (*Hecale* fr. 301 Pf.) and Apollonius (III.276–7) comment that *myops* was the name given by agricultural folk to the *oistros*, which implies that the two may be regarded as popular and literary names for the same insect.

(iii) *Myops/oistros* equated with *tabanus/asilus*. No surviving Latin author makes any attempt to transfer the Greek grammarians' distinction between *myops* and *oistros* to the pair of names existing in their own language. Instead, we are expressly told by Seneca (*Ep.* 58.2) that *asilus* was an older name which by his day had gone out of use in favour of *tabanus*. Pliny (XI.100) simply gives *asilus* as a synonym of *tabanus*.

Virgil (*G.* III.146 ff.) uses the older name and identifies the insect as that known to the Greeks as *oistros*. Servius, commenting on this passage, equates all four names in both languages on the authority of the grammarian Nigidius: *asilus est musca varia, tabanus, bubus maxime nocens. Hic apud Graecos prius* μύωψ *vocabatur, postea magnitudine incommodi oestrum apellarunt* (Cf. also Isidorus, *Or.* XII.8.15). A number of glosses likewise identify, in various combinations, the four terms (*C.Gl.L.* II.24.11, 194.28, 259.10; III.258.27; IV.21.52,264.25; V.169.17). The evidence

---

[38] M. Wellmann, 'Sostratos', *Hermes*, XXVI, 1891, pp. 344–8.
[39] 52 below.

therefore supports complete synonymy in popular usage. A third Greek synonym βουτύπος is found in Oppian *Hal.* II.529 and *C.Gl.L.* II.259.37.[40]

2. Life history

As has been noted above, Aristotle distinguished *myops* and *oistros* solely by their life cycle, otherwise classing them together. His data on *myops* consist only of the statement that they are generated out of timber (*HA* 552a28; = Schol. *Od.* and Apollonius loc. cit.). Sundevall[41] suggests that this idea is a deduction from the fact of the adults' common occurrence in woodland.[42] However, since some tabanid larvae may occur in rotting logs, there is a slight possibility that the statement is based on actual observation.[43]

Pliny (XI.113) reproduces Aristotle's statement, using the term *tabanus*. He ignores the Aristotelian distinction of species and fails to reproduce the passage on the development of *oistros*. He also refers in a medical context (6 below) to *vermiculi* from which horseflies develop.

Aristotle's details concerning the *oistros* are a little more extensive. He says (*HA* 551b 21 ff.; = Schol. *Od.* and Apollonius loc. cit.) that they develop out of the small flat creatures (πλατέων ζωδαρίων) that run about (ἐπιθεόντων)[44] upon the surface of rivers, which is why they are most numerous near water. It has been suggested by Sundevall and D'Arcy Thompson[45] that Aristotle has here confused true horseflies with one or other of the larger soldier-flies (Stratiomyidae), which frequent water in which their larvae breed, but that the supposed larvae he describes are in fact Whirligig-beetles (Gyrinidae), which skim about on the water surface. The latter half of this theory is almost certainly correct, since the gyrinids[46] fit the description both in habits and appearance, being ovoid and more or less flattened in shape.[47] However, it is unnecessary to bring the stratiomyids into the picture, since the Tabanidae are also especially abundant near their breeding places.[48]

Since in general usage *myops* and *oistros* were, as has been indicated above, equated, it is reasonable to assume that Aristotle, lacking a proper system of zoological nomenclature, is here, as in other places, using popular names in new specialised meanings for his own

---

[40] There is a curious scholion (*in Od.* loc. cit., Eustathius loc. cit.) from the lexicon of Pamphilus (cf. Gossen, in PW VIII.18) in which *oistros* is described as a creature resembling a grasshopper (ἀκρίς), speckled in colour, fast-moving, attacking cattle.

The term ὑσπληξ, found in Hesychius, *EM*, and Photius. is accepted by Gossen, *Zool. Glossen*, p. 120, as a synonym of μύωψ in its entomological sense, but is taken by LSJ as equivalent only to its secondary meaning of 'goad'. On the term χαλκῆ μυῖα, cf. 32 note 50 above.

[41] Op. cit., pp. 223–4.

[42] The genus *Laphria* to which he refers belongs to the Asilidae.

[43] Gossen, in PW VIII.18, proposes altering ξύλων to ξηρῶν (dry soil), but this would make for a less accurate statement.

[44] Scholia: ἐπιπλεόντων.

[45] Sundevall, op. cit., pp. 223–4; D'Arcy Thompson, *Hist. An.*, V.19 note.

[46] Cf. Imms, op. cit., p. 846.

[47] The only other insects living on the water surface are the pond-skaters etc. (18 above), whose appearance is different.

[48] Aubert and Wimmer's theory, op. cit., I p. 168, that the whole life history is based on that of *Stratiomys* corresponds less well with the account in Aristotle.

purposes. But since his life history data are so confused it is not possible to determine which species he intended to be understood by the two terms he uses. All that can be said is that he believed there were two kinds of horsefly distinguishable by their origin.

3. Feeding habits

Myops and oistros are classed together with gnats[49] by Aristotle as being among the class of insect with a sting-like 'tongue' (HA 532a10, 490a21, PA 661a24), strong enough to bore through the hides of quadrupeds (HA 528b 31). In HA 596b 14, they are noted as exclusive blood-feeders as opposed to the omnivorous myiai.

In the Odyssey (XXII.300) the oistros is described as αἰολός, a reference to its swiftness of flight when hunting its victims (cf. Eustathius loc.cit.). Lucian (Am.2) refers to its insatiable appetite and readiness to attack.

According to Philumenus (Ven. 25.1), the snake δρυίνας has rough scales harbouring bronze winged myiai which eventually destroy it. In Nicander (Ther. 417) these appear as myopes attacking the snake.

4. Popular beliefs

There appears in the Historia Animalium (553a 16) a reference to a belief that the myops died in the autumn as a result of its eyes becoming dropsical (ἐξυδρωπιώντων) (cf. Pliny XI.120 where it is said to die caecitate). The passage containing this sentence is rejected as spurious by Aubert and Wimmer.[50]

The name myops itself appears to mean 'short-sighted', and Sundevall suggests that it was applied to the insect as a result of a folk belief similar to that which has resulted in some modern European folk-names.[51] Fernandez,[52] however, denies that this is in fact the origin of the name, preferring to see it as a derivative of myia.

5. Relations with domestic animals

The myops or oistros was notorious for attacking cattle, which it is usually described somewhat imprecisely as 'stinging' (cf. (a) Callimachus fr. 301 Pf; Dio Chrysostom VII.14; Plutarch Mor. 458c; AP IX.739; Aristaenetus II.18; Photius and Suidas, all using μύωψ; (b) Varro RR II.5.14, using tabanus). It is described by Aelian (VI.37; cf. Eustathius in Il. 1648.42 ff., and Schol. Od. loc. cit.) as the worst enemy of cattle. There are frequent descriptions in literature of its effect upon its victims, how it would drive them to frenzy and cause them to rush out of control across the open countryside (cf. (a) using both names, Apollonius Rh. III.276–7; Nonnus XI.191–2; (b) using οἶστρος, Odyssey XXII.300–1; Nonnus XLII.185–93; Oppian. Hal. 521 ff.; Quintus Smyrnaeus XI.207–14; (c) using μύωψ, Colluthus 41–3; Tryphiodorus 359–64; (d) using asilus, Virgil G. III.146 ff.; Seneca Ep. 58.2; Valerius Flaccus Arg. III.581–3). The famous gadfly of

---

[49] 50 below.
[50] Op. cit., I pp. 516–17.
[51] Op. cit., p. 224.
[52] Op. cit., pp. 82–3. He is supported by Davies and Kathirithamby, Greek Insects, p. 160.

mythology, sent by Hera to torment Io in the form of a calf, and to pursue her across the world, has a prominent place in the *Supplices* and *Prometheus* of Aeschylus (1.ii above). Aeschylus terms it both *myops* and *oistros*, while later writers (Sophocles *El.* 5; Apollodorus *Bibl.* II.1.3) apply the latter name only. The popularity of the literary comparison between a maddened human being and an οἰστροπλήξ ox led to the development of various metaphorical usages divorced to a greater or lesser extent from thought of the actual insect. *Oistros* is normally found in such cases, but examples exist of the use of *myops* also, usually where the insect is still to some extent in view (e.g. Clement Alex. *Quis Dives* p. 949; Josephus *Ant.* VII.16.9; Achilles Tatius VII.3). Horseflies are said to cause especial distress by attacking the ears of cattle (Lucian *Cal.* 14; Eustathius *in Il.* 1928.16 ff.; Schol. *Od.* loc. cit.).

There are no mentions of the insect attacking man[53] and very few of its troubling animals other than cattle. There are, however, references to it affecting horses (Plato *Ap.* 30e; Xenophon *Equ.* IV.5, *Equ.Mag.* I.16).

As regards countermeasures, pasturing cattle late in the evening is mentioned by Virgil (*G.* loc.cit.) and Eustathius (*in Il.* 1648.42 ff.)[54] and keeping the animals penned in summer by Varro (*RR* loc. cit.). Various herbal preparations to be applied to cattle to deter attack are prescribed in the *Hippiatrica* (*Hipp.Par.* 919) and the *Geoponica* (XIII.12, XVII.7,11). In the first and last of these passages the *Geoponica* uses only the general word *myia*.

6. Medicinal uses

Pliny (XXX.101) gives as an amulet for fever *vermiculos ex quibus tabani fiunt, antequam pennas germinent*. Whether these *vermiculi* were correctly identified would seem rather doubtful, in view of the secretive development of genuine tabanid larvae.

7. Exotic species

Pliny (XXXII.10) refers, on the authority of Juba, to certain *asili* which attack camels in Arabia, and states that the animals were rubbed with the fat from fish to deter the insects.

50. Konops, Empis/Culex

1. Identification

The above names cover numerous species of mosquitoes, gnats and midges, belonging mainly to the superfamily Culicoidea. Their primary reference is to those kinds which suck the blood of man, but classical observers would have considered as identical the many non-biting species of similar appearance.[55] The two Greek names appear to have

---

[53] Cf. 48.5 above.

[54] Cf. Schol. *Odyssey* X.85, discussed by Davies and Kathirithamby, op. cit., p. 162.

[55] Sundevall, *Thierarten*, pp. 225–6, observes correctly that Aristotle's *konops* results from confusion between drosophilid flies (51 below) and a blood-sucking species; but suggests that the latter was *Stomoxys calcitrans* (Keller's view, op. cit., II pp. 451–2, is similar), whereas this insect is comprehended by Aristotle and others under *myia* (48.1,5 above). Sundevall, ibid., Aubert and Wimmer, op. cit., I pp. 163, 167, and Steier, in PW XVI.450–4, identify Aristotle's *empis* as *Culex pipiens*, but this seldom bites man. Fernandez, *Nombres*, pp. 26, 75, considers the two names more or less equivalent in ordinary usage.

been synonymous in general usage, though Aristotle gives each a distinct technical sense and some lexicographical sources attempt to distinguish them.[56] The most significant of the families involved here is the Culicidae, which includes the mosquitoes of the genera *Anopheles*, *Aedes*, and *Culiseta*, and the common gnat *Culex pipiens* Linn., though this latter seldom bites man.

These are the insects primarily in view when the above names were employed. As well as causing major discomfort by their biting habits, the members of the genus *Anopheles* also transmit malarial organisms in the process, though the ancient world was not aware of this fact. Their eggs are deposited on or near the surface of water, particularly stagnant water. The larvae live suspended from the surface film, head downward in the case of *Culex*, and feed upon small food particles in the water. The pupae float in a similar position.[57]

Other biting species include the blackflies (Simuliidae), the biting midges (Ceratopogonidae) and the sandflies (Phlebotomidae).[58]

Of the non-biting species, the most conspicuous, since they may swarm in immense numbers in the evening, are the midges of the Chironomidae. Their larvae are aquatic and include the well-known red pigmented 'bloodworms' which inhabit the bottom of ponds and small stagnant bodies of water, living in fixed tubes of mud, in which they pupate.[59]

(i) *Empis* distinguished from *konops*. In Aristotle, κώνωψ and ἐμπίς are of similar appearance and habits, being both blood-sucking flies (*HA* 532a 14, 490a 21), but are distinguishable by their life history, the former developing in sour wine (552b 5) and the latter in muddy bodies of water (551b 26 ff.). For him therefore *konops* is primarily a drosophilid fly,[60] which he either mistakenly assumed to have the same habits as the *empis* or actually equated with some distinct kind of biting dipteron with which he was familiar, and *empis* a culicid gnat or mosquito. Aristotle is here most probably not reflecting popular usage but rather giving loosely-employed common names a specific technical reference for his own purposes. Another author distinguishing the two is Artemidorus (*On.* III.8). The lexicographical tradition is divided on the question. Hesychius (s.v. *empis*) says that it is either another name for *konops*, or else it is a creature similar but larger, produced by water. The scholiast on Aristophanes *Av.* 244–5 agrees with the latter option, but adds that the *empis* is banded with white (cf. Schol. Aristophanes *Lys.* 1032; Suidas). This would fit the appearance of *Culiseta* or *Anopheles*. Schol. *Nub.* 157 says that according to some *empis* is a type of *konops*.

---

[56] *Konops* and *culex* are also used occasionally for other small insects which bear some superficial resemblance to gnats or mosquitoes, notably fig-wasps (46 above) and gall-wasps (47 above). Martial (XI.18) uses *culex* for a garden pest feeding on willow foliage, and Pliny (XVII.231) for pests troublesome to acorns, apples and other fruits, generated from juice beneath the bark of these trees. The gloss (*C.Gl.L.* IV.46.46, 501.57, V.187.7) *tinea de allece, quod in navibus nascitur* is unintelligible.

[57] Smith, *Insects*, pp. 37 ff., 490.

[58] Ibid., pp. 109 ff., 155 ff., 181 ff.

[59] Ibid., pp. 189 ff.

[60] 51 below.

(ii) *Empis* equated with *konops*. Outside the sources mentioned above, there is no evident difference in usage; the two names appear in similar contexts with similar attributes. That they are in fact synonymous is stated definitely by Aristophanes (fr. 707b), Hesychius (loc. cit.), and Schol. Aristophanes *Nub.* 157, who notes that '*empides* are what we call *konopes*'. *C.Gl.L.* III.258.33 equates both Greek names with the Latin *culex*.

1.2. Other names

The name βομβύλιος, normally used for a hymenopteron,[61] is given as a synonym of *konops* and *empis* in *C.Gl.L.* III.258.33. Βουτύπος, which appears elsewhere[62] as a name for the horsefly or gadfly, is defined by Hesychius as 'a kind of animal called *empis*'. Hesychius also lists σάβηττος[63] and σοίκιδες as synonyms of *konops*.[64] In late Latin, *zinzala*[65] is found as a synonym of *culex* (Cassidorus *in Ps.* 104 MPL 70.749; *C.Gl.L.* V.449.5, 526.1, 566.30).

2. Life history

The only detailed source for the life cycle of culicoids is Aristotle, who gives details under the name of *empis*. The life history data given to *konops* refer to drosophilid flies.[66] Only one of these two accounts, the second, is reproduced by Pliny, perhaps because he did not understand Aristotle's distinction between two species.

According to Aristotle (*GA* 721a 9 ff.) the *empis* belongs to that class of insects which are spontaneously generated and do not mate or produce any form of offspring. They arise (*HA* 551b 26 ff.) out of larvae which are generated in the mud at the bottom of wells or any other body of water containing earthy sediment. As the mud putrefies, it turns first white, then black, and finally to blood red, at which point there grow up what appear like small pieces of red seaweed (ὥσπερ τὰ φυκία μικρὰ σφόδρα καὶ ἐρυθρά).[67] These seaweed like objects move about in a fixed position for a time, before breaking off and becoming free-floating (φέρεται κατὰ τὸ ὕδωρ). At this stage they are known as ἀσκαρίδες,[68] which after a few days adopt an upright position in (or on) the water, becoming hard and immobile. Their outer covering then splits and the adult *empides*

---

[61] 42 above.   [62] 49 above.

[63] Fernandez, op. cit., p. 230–1, suggests it is related to the verb σαβάζω.

[64] Lycophron's δάπτης, mentioned by Fernandez, ibid., p. 109, is not a genuine synonym, but an oblique reference to mosquitoes. The locust names πάρνοψ and κορνῶπις are defined, probably erroneously (Latte rejects the mss reading in Hesychius s.v. κορνώπιδες), in some lexicographical sources as equivalents of κώνωψ (cf. 11.1.2.ii above).

[65] A word of onomatopoeic origin, Ernout and Meillet, *Dict. Ét.*, p. 759.

[66] Cf. 51.2 below.

[67] Aubert and Wimmer, op. cit., I p. 512, consider this reading impossible and propose ὥσπερ ὀκωλήκια. Peck, *Hist. An.*, II p. 179, suggests ὥσπερ σταφύλια. But a comparison with seaweeds would not be inappropriate in this context, red varieties being known in antiquity (cf. Theophrastus *HP* IV.6.5).

[68] This word is elsewhere used for internally parasitic worms. It is related to σκαρίζω, 'jump, palpitate': the second meaning would be appropriate to the motion of fixed chironomid larvae. Cf. Fernandez, op. cit., pp. 147–8.

emerge, sitting upon it until the sun or wind stirs them to take flight. Although not everything in this account is entirely clear, it is evident, as observed by Sundevall, Keller and D'Arcy Thompson,[69] that the first half of the account, concerning the red larvae developing in mud, is based on observation of the larvae of *Chironomus* spp. These blood-red larvae in their mud tubes fixed to the bottom have exactly the right appearance and habitat. It is, however, also clear that the latter part of the cycle given here does not correspond with the former, in that *Chironomus* larvae do not detach themselves from the substrate at any stage. Sundevall, Aubert and Wimmer and Keller[70] apply these latter remarks to the larva and pupa of *Culex* spp., making the phrase ἵστανται ὀρθαὶ ἐπὶ τοῦ ὕδατος refer to their posture as suspended from the surface film.

The reference in 601a 3 to the *empis* as an insect that sheds its skin may be simply alluding to the adult's emergence from the pupa. Sundevall,[71] however, thinks that an insect with an incomplete metamorphosis is in view here, and that the statement is possibly based on observation of small mayflies.[72] Further confusion is provided by the evidently corrupt text of 487b 3 ff., which as given by Bekker appears to make the *oistros* develop from aquatic *empides*. It is evident that the original wording must either have referred to the emergence of *empides* from *askarides*, or the *oistros* from its flat surface-living larvae. Karsch proposed to alter ἐμπίδων to ἀσκαρίδων and was followed by Dittmeyer, who brackets the incongruous following phrase γίνεται γὰρ ἐξ αὐτῶν ὁ οἶστρός. Peck[73] reconstructs the text on the basis of Michael Scot's early 13th century translation from the Arabic, which appears to preserve the sense of the Greek as it was before the corruption took place.[74] Following his restoration, the passage would translate as 'The *askarides* are an instance of this: at first they live in rivers, but later they change their form and out of them develops the *empis* which lives out of water'.

The development of the *empis* in water is also noted by Schol. Aristophanes *Av*.244, Hesychius and Suidas. Sextus Empiricus (*Pyrrh*.I.41) says that *konopes* are generated from water that is becoming putrid. Pausanias (VII.2.11) writes of *konopes* arising from a freshwater lake formed from a silted up inlet. Among Latin writers, the *culex* is described in similar fashion in *Culex* 183 as *umoris . . . alumnus* and by Columella (*RR* I.5.6), who advises against the siting of farmsteads near marshland because, for one thing, it generates creatures armed with troublesome stings.

3. Habitat and distribution

Aristotle (551b 29 ff.) describes the bodies of water in which *empides* are especially produced as those with an earthy or mixed sediment, giving as an example of their habitat

---

[69] Sundevall, op. cit., p. 226; Keller, op. cit., II p. 451; D'Arcy Thompson op. cit., V.19 note.

[70] Sundevall, op. cit., p. 226; Aubert and Wimmer, op. cit., I p. 163: Keller, op. cit., II p. 452. The larvae of *Anopheles* spp. lie along the surface rather than hanging suspended.

[71] Op. cit., p. 226, supported by D'Arcy Thompson, op. cit., VIII.17 note.

[72] Cf. 15 above.

[73] Op. cit., I pp. 10–11.

[74] *Sicut quod nominatur grece ambidas, nam ipsum manet in fluminibus prius, deinde mutatur forma eius et fiet ex eo animal quod dicitur astaniz et vivit extra.*

the water-meadows situated in the area between Athens and Megara.[75] This, he explains, is because in such situations the process of putrefaction necessary to generate the larvae operates more swiftly. He also states that they appear in greater numbers in autumn.

Marshy country is given as the habitat of the *empis* by Aristophanes (*Av.* 244), and as that of the *culex* by Columella (*RR* loc. cit.) and Ambrose (*Hex.* MPL 14.220). In the *Lysistrata* (1032) there appears a reference to ἐμπίς Τρικορυσία; for, as the Schol. ad. loc. explains, the Attic deme of Trikorythos was notorious for being infested by these insects, since the land there was damp and marshy (cf. also Suidas s.v.). Pliny (XIX.180) refers to *culices* infesting damp gardens.

4. Feeding habits

Both *konops* (*HA* 532a 14) and *empis* (490a 21) are classed by Aristotle among those insects, like *myops* and *oistros*, which suck blood by means of a 'sting-like' tongue. The first of these references is reproduced by Pliny (XI.100) as applying to the *culex*: he omits reference to any of the Aristotelian statements about *empis*. Aelian (XIV.22) notes that *konopes* are active both by day and night. The insect's blood-sucking habits are also alluded to in *AP.* V.151 and described by Isidorus (*Or.* XII.8.13). They are described as landing unnoticed on the skin due to their extreme lightness (Lucretius III.190).

As well as man, mosquitoes are also occasionally referred to as attacking animals, both domestic (Babrius *Fab.* 84) and wild (*Aesopica* 259; Achilles Tatius II.21–2).

5. Relations with man

Gnats and mosquitoes were considered in antiquity as among the worst of insect pests. Aelian (XIV.22) describes the *konops* as hostile (ἐχθρός) to man by day and night with its biting and humming. These two associated kinds of nocturnal disturbance are spoken of by Meleager (*AP.* V.151–2, ὀξυβόαι κώνωπες, ἀναιδέες, αἵματος ἀνδρῶν σίφωνες) as disturbing sleep, and the humming figures also in Aeschylus *Ag.* 891–3 (. . . λεπταῖς ὑπαὶ κώνωπος ἐξεγειρόμην ῥιπαῖσι θωύσσοντος) and Aristophanes *Plutus* 537–9. (Cf. also Horace *Sat.* I.5.14–15 *mali culices . . . avertunt somnos.* Other references to the insects' humming are found in Aristophanes *Nub.* 156–68; Achilles Tatius II.22; Martial III.90, VIII.33; Tertullian *adv.Marc.* IV.14, *de An.* 10; Paulinus Nol. *Ep.* V.13; and Gregory Magnus *Mor.* I.21. Other references to the *culex* as a blood-sucking insect are found in *Culex* 182 ff. and Hadrian *Carm.* 1.) All the Greek references which specifically mention blood-sucking or biting use the term *konops*, though there are some unspecific mentions of *empis* as a pest of man (Aristophanes *Av.* 244–5, *Lys.* 1031–2 and Scholia ad. loc.; Artemidorus *On.* III.8, mentioned together with *konops*; Lucian *Gallus* 31; Porphyry *de Abst.* III.20) or as an insect of extreme smallness (Lucian *Charon* 8, *Icarom.* 12). According to Pausanias (VII.2.11) the people of Myus in Achaia were forced to evacuate their town when a marine inlet silted up to form a lake in which vast swarms of *konopes* were produced.

---

[75] Adopting Peck's, op. cit., II pp. 180–1, conjecture ἐν ταῖς ὀργάσι for the meaningless ἐν τοῖς ἔργοις of the mss.

As has been mentioned, species of the mosquito genus *Anopheles* are responsible for transmitting malaria in the Mediterranean region, the organisms being conveyed to man when the insect bites. However, although what can be identified as malaria is described as a well known disease by classical medical sources,[76] no one in antiquity was aware of the connection between it and the mosquitoes they knew only as troublesome biting pests. They were, however, aware of the unhealthy nature of marshy country—unhealthy in reality because of their being the preferred breeding ground of *Anopheles* spp.—(cf. for example Hippocrates *Aer.* VII) and various speculations were brought forward as to why this should be so. The Latin agricultural and architectural writers all advise against building farms or settlements in the vicinity of marshes because of the disease prevalent in such localities. Vitruvius (*de Arch.* I.IV.1) took the view that the morning mist rising off the marshes became infected by the poisonous breath of noxious creatures (*spiritusque bestiarum palustrium venenatos*) living there, and noted (IV.11–12) that this was most likely to occur where the water was stagnant through having no outlet. Varro (*RR.* I.12) had a theory that in the marshes were produced organisms too small to be seen with the naked eye, that could be carried along in the air and enter the human body to cause disease (*crescunt animalia quaedum minuta, quae non possunt oculi consequi, et per aura intus in corpus per os ac nares perveniunt*). Columella (*RR* I.5.6) notes that in such areas are produced both mosquitoes which attack man in large swarms and the mysterious disease-producing organisms mentioned by Varro (*nantium serpentiumque pestes hiberna destitutas uligine caeno et fermentata colluvie venenatas emittit, ex quibus saepe contrahuntur caeci morbi*), but does not conceive of any link between the two. Cf. also Palladius (*RR* I.7.4): . . . *propter pestilentiam et animalia inimica quae generat.*

In order to kill or deter gnats that enter houses, various herbal fumigants were recommended (Pliny XX.184, XXII.157, XXVII.52, XXIII.114; Dioscorides *DMM.* I.174, *Eup.* 137; Galen III.121, XIV.537 K; *Geoponica* XIII.11.1,3,5,7,10). We also read of various preparations to be sprinkled around the house (*Geoponica* XIII.11.3,10) or smeared on walls (*Geoponica* XIII.11.6). Scattering horse hair about the building is also suggested (*Geoponica* XIII.11.1). There were herbal preparations recommended to be rubbed on the body, which were said to keep the person free from attack (Pliny XXVII.52; Dioscorides loc. cit.; *Geoponica* XIII.11.2,5) and the *Geoponica* (XIII.11.4,9) gives certain plants to be placed in one's bed as a deterrent at night. For gnats infesting gardens (Pliny XIX.180) the burning of *galbanum* resin is recommended.

Mosquito nets (Gk. κωνώπιον, latinised into *conopeum*—pure Latin *culicare* given in Schol. Juvenal. VI.80) were introduced into the Roman world from the East in the time of Augustus, being originally thought of an excessive luxury (Horace *Ep.* IX.16. Cf. Juvenal VI.80). There are three late epigrams concerning these articles in the Greek Anthology (IX.764–6).

---

[76] Cf. Kind, art. *Malaria*, in PW XIV.837–42; J. L. Cloudsley-Thompson, *Insects and History*, pp. 85 ff.

6. Uses to man

The *konops* is described by Aelian (XIV.22) as being employed as an angler's freshwater bait.

7. Exotic species

(i) *In Africa.* It was reported that in the territory of the tribe known as the Rhizophagoi in NE Africa vast swarms of mosquitoes (*konopes*) of greater size and robustness than those of the Mediterranean used to arise in mid summer (Aelian XVII.40; Agatharchides 50; Strabo XVI.4.9; Diodorus III.23).[77] The former writer implies that they do great damage, whereas the others, especially Diodorus, credit them with being inadvertently beneficial to the people of the region by attacking and driving away lions (cf. *Aesopica* 255) and other wild animals.

(ii) *In Egypt.* Herodotus (II.95) has an account of the marsh country of Egypt and the methods employed by the people to combat the attentions of the mosquitoes (*konopes*) infesting the region. South of the marshes, he writes, sleeping on raised structures is the stratagem employed, because the wind prevents the insects from flying at a height, while actually in the marshes the inhabitants use mosquito nets that in the daytime are employed for fishing.

　　Also located in Egypt is the σκνίψ (Latinised forms *sciniphes*, in for example Isidorus, Augustine *Civ.* XVI.43, and *ciniphes*, *-ifes*, in for example Vulg. *Ps.* 104, Cassiodorus)[78] of the Septuagint (*Ex.* VIII.16 ff., *Ps.* 104.31, *Wis.* 19.10; cf. Ezekiel *Exag.* 135) where it appears as some kind of insect pest of men and animals, translating the Hebrew *kinnim*, whose meaning is itself unclear. *Sknips* or *knips* appears in Hesychius as a synonym of *konops*, and it is presumably in this sense that the word is used here. Later commentators are divided in their interpretation of the passages in question. The more general opinion was that the *sknips* was a particularly noxious form of mosquito. According to Philo (*Vit.Mos.* I.107–8), it is a small winged creature which not only causes irritation to the surface of the body but also enters the ears and nostrils and flies into the eyes.[79] Others adopting this interpretation are more vague in their descriptions. Isidorus (*Or.* XII.8.14, *Qu.in Ex.* MPL 83.292; cf. Gregory of Nyssa *Vit.Moys.* MPG 44.350) describes the insect as a minute fly so small as to be hardly visible, which is therefore able to remain unnoticed until it has actually bitten, while Cassiodorus (*in Ps.* MPL 70.749) calls it a type of *culex*. Josephus (*Ant.* II.300), however, writes of the plague in question as involving lice (φθεῖρες) generated within the bodies of their victims.

(iii) *In Mesopotamia.* Ammianus Marcellinus (XVIII.7.5) gives a similar account from this region to Diodorus' report from Africa, concerning *culices* that attack the eyes of

---

[77] Gossen, *Tiernamen*, no. 68, identifies these as a species of *Anopheles*.
[78] For other uses of σκνίψ, cf. 58 below.
[79] Philo's description would fit the habits of Simuliidae.

lions so that they are driven into rivers and drown. Later on (**XXIV**.8.3) he describes the country as being full of swarms of flies and gnats (*muscarum et culicum multitudine referta*) so large as to obscure the light of day.[80]

51. *Oinokonops/Bibio*

1. Identification

Though the vinegar-flies of the family Drosophilidae were well known as pests in classical times, they were not clearly distinguished from gnats and mosquitoes and often seem to have been thought of as identical with these. The specific terms given above are rare, surviving only in the lexicographical tradition: more usually they are not distinguished by name from *konops/culex*. The species of *Drosophila* concerned here are those which are attracted, often in large numbers, to all kinds of sweet fermenting materials, such as wine, vinegar and decaying fruit, in which their eggs are deposited and upon which their larvae feed.[81]

It has been generally recognised by Sundevall[82] and later writers that what Aristotle writes concerning the life history of the *konops*, which is the only criterion by which he separates it from his *empis*, is based upon observation of the development of *Drosophila* spp. in old wine or vinegar. However, Aristotle also makes his *konops* a biting fly, not being able to free himself entirely from the popular confusion between drosophilids and the various gnats, mosquitoes and related pests.[83]

1.2. Nomenclature

As has been said, Aristotle gives the vinegar-fly the name of *konops*, which in its broader and more popular significance was a synonym of *empis*.[84] This usage is followed by Artemidorus Daldianus and by Plutarch. Pliny, reproducing Aristotle's data, translates *konops* by the corresponding Latin term *culex*. Sextus Empiricus calls the insect σκνίψ.[85] Aelian is alone among surviving sources in applying to the insect the name ἐφήμερον, which elsewhere is used to describe the mayfly.[86] It is suggested by *C.Gl.L.* II.296.30 (ἐμπίς: *musca vinaria*) that the word *empis* was also sometimes used to denote this insect. Οἰνοκώνωψ, a more precise name, is preserved only in *C.Gl.L.* III.258.34, where it is defined by the Latin name *bibo*. This latter name, which would appear to be derived from *bibere*[87] in allusion to the insect's attraction to liquids, is found also in Isidorus' list of entomological terms (*Or.* XII.8.16). He gives it in the form *bibio*, with a synonym *mustio* not found elsewhere (*Bibiones sunt qui in vino nascuntur, quos vulgo mustiones a musto*

---

[80] Herodian (III.9.5) has a story about the besieged citizens of Hatra in Mesopotamia throwing over the walls clay vessels full of small winged and venomous creatures, which would crawl unnoticed into the eyes and exposed parts of the skin of the besiegers and sting them. This account seems to be fictional.

[81] Cf. Smith, *Insects*, p. 491.

[82] Sundevall, *Thierarten*, p. 225: Aubert and Wimmer, *Thierkunde*, I p. 167; Keller, *Ant. Tierwelt*, II p. 451; Steier, in *PW* XVI.452–3.

[83] Cf. 50.1.i above.   [84] Cf. 50.1 above.   [85] 58 below.

[86] 15 above.   [87] Ernout and Meillet, *Dict. Ét.*, p. 70.

appellant), and provides an example of its use in the shape of a cryptic quotation from Afranius (fr. 407). Isidorus' second name appears also in *C.Gl.L.* V.187.6 as *muscio*, which is probably the correct form and related to *musca*:[88] here it is given as a synonym of *culex*.

2. Life history

The vinegar-fly was classed by Aristotle (*GA* 721a 9 ff.) among those insects that neither mate nor produce offspring and are themselves spontaneously generated from decaying matter. He explains in *HA* 552b 5 that they develop out of small larvae (σκωλήκια) that arise out of the sediment of sour wine or vinegar. Pliny's version of this statement appears in IX.160 and IX.118 (*Rursus alia genera culicum acescens natura gignit*). Sextus Empiricus (*Pyrrh.* I.41) includes vinegar-flies in his list of spontaneously generated creatures as being produced by wine that is turning sour.

Aelian (II.4) visualises the flies as being produced in closed wine vessels and as flying out when the vessel is opened. He states that they die very shortly on release and are hence termed ἐφήμερα.

3. Feeding habits

Aristotle, discussing the attraction of insects to their respective foods, says (*HA* 535a 3 = Pliny X.195) that the *konops* will not settle upon sweet substances but only upon those that are sour, an evident allusion to vinegar-flies gathering around such things as fermenting wine. Plutarch (*Mor.*663d) characterises inferior quality wine as being surrounded by a swarm of humming vinegar-flies (κωνώπων χορῷ περιᾳδόμενον). Elsewhere (*Mor.* 1073a) he says that they delight in sour wine or vinegar and the scum that develops on the surface of wine left to stand, but that fine palatable wine they avoid. Artemidorus has the same idea in mind when (*Onir.* III.8) he says that for winesellers to dream about *konopes* or *empides* signifies the degeneration of their stock into vinegar, since they delight in the latter. It is clear from his preceding sentence that he views vinegar-flies and biting-gnats and mosquitoes as of one kind.

4. Relations with man

Of the technical writers who deal with the manufacture and storage of wine, only Columella and the authors in the *Geoponica* deal with the insect pests liable to infest it. The former recommends the adding of salt to fig vinegar (*RR* XII.17.2) and to vinegar in which fruit is to be preserved (*RR* XII.10.3) to prevent the development of pests (*ne vermiculus aliudve animal innasci*): these *vermiculi* would be the larvae of *Drosophila*. In the *Geoponica* (VI.12.3–4), the reader is advised not to deposit the καθάρματα from new wine anywhere near the storage area, since it will be liable to breed vinegar-flies (τρεπομένων αὐτῶν κώνωπες ἀναφύονται), and (VI.13.3–4; cf. VI.7.3) to clean out the wine-press thoroughly after use for the same reason. In VII.15.1 it is suggested, in order to test whether wine is turning sour, that it be transferred to a new container, and the dregs in the old observed to see whether they produce flies or not (ζωογονοῦνται κώνωπες).

[88] Th.L.L. s.v.

These *konopes* are not distinguished by name from the domestic mosquitoes of XIII.11, and in one section there (11.8), where hanging up a vinegar-soaked sponge is said to attract *konopes* to it, the two kinds of insect are clearly regarded as identical.

52. Kynomyia

Κυνόμυια (Homeric κυνάμυια: Latin *cynomyia*, *musca canina*) is strictly speaking not a genuine insect name, but a purely literary creation which had no place in popular usage. In its original form as it appears in the *Iliad* (XXI.394,421: cf. Athenaeus 126a, 157a) it is, as Fernandez[89] points out, simply a term of abuse compounded from the names of two animals which in popular thought were taken as typifying shamelessness.[90] This is implied by the explanations given by Hesychius s.v. (ἀναιδής . . . καὶ θρασεῖα, ὁ μὲν γὰρ κύων ἀναιδής, ἡ δὲ μυῖα θρασεῖα) and Eustathius (*Comm. in Il.* 1243.20 ff.).

However, although the Homeric term was not intended to designate a particular species of fly, later commentators were clearly under the erroneous impression that it was. Sostratus, in his περὶ ζῴων (*ap.* Aelian IV.51, VI.37 and various scholia[91]), thought of it as a real insect and introduces it for purposes of comparison when distinguishing between the horsefly names *myops* and *oistros*.[92] Eustathius (loc. cit.) states that according to some the name refers to 'the fly which attacks dogs' or else the dog tick *kroton*.[93] *Kynomyia* figures in lists of insect pests in Lucian *Gall.* 31 and *AP* XI.265.

In later times the word came into prominence through being selected by the translators of the Septuagint (*Ex.* 8.21, *Ps.* 77.45, 104.31; cf. Ezekiel *Exag.* 138; Augustine *Civ.* XVI.43) to render the Hebrew *'arob*, which means simply 'a swarm of insects'.[94] Commentators on the LXX generally understand the word as referring to a particular kind of fly, especially noxious and persistent. Philo (*Vit.Mos.* I.130–1) gives the most vivid description, stating that it is a blood-sucking insect which possesses in a preeminent degree all those characteristics of fearlessness and persistence that were associated with the common *myia*. Unlike the *sknips*[95] with its silent approach, this creature rushes in from a distance with a whirr of its wings and hurls itself upon its victims like a javelin. Corresponding, though briefer, definitions are given by, for example, Gregory of Nyssa (*Vit Moys.* MPG 44.350), Cassiodorus (*in Ps.* MPL 70.749), and Isidorus (*Or.* XII.8.12; cf. *C.Gl.L.* III.319.60, V.178.19, IV.30.28 *musca venenosa*). Josephus (*Ant.* II.303) is aware that the Hebrew means something like 'swarm', but interprets it as referring to a mass invasion of 'wild beasts of every species and kind, the like of which no man had ever encountered before'. This is the 'Hebrew' interpretation contrasted with the 'Greek' by Origen (MPG 12.1542,1564).

---

[89] *Nombres*, p. 168.

[90] Cf. 48.1.2.iv, 4 above.

[91] Cited in 49.1.i above. Cf. *C.Gl.L.* II.259.37 where it is given as a synonym of *tabanus*.

[92] Gossen, *Tiernamen* no. 73, *Zool. Glossen*, p. 64, misses the point in identifying it as *Stomoxys calcitrans*.

[93] 8 above.

[94] Thus, contrary to the opinion of Keller, op. cit., II p. 425, no particular species is in view here.

[95] 50.7.ii above.

The correct explanation, however, is given by Jerome, who explains (*Ep.* 106.83) that the word means *omne genus muscarum*. (This is the rendering in Vulg. *Ex.* 8) (cf. Eucherius *Instr.* II.11 *coenomyia non musca canina, ut quidem putant, sed musca omnimoda*).

Insecta: Diptera or Odonata (Dragonflies)

53. *Hippouros*

An insect mentioned only by Aelian (XV.1) as living around the river Astraios in Macedonia. It is described as a kind of fly (μυῖα) combining the characteristics of various insects: the boldness of the fly,[96] the sound of the honey bee, the colour of the wasp, and the size of the ἀνθηδών.[97] It settles on the surface of the water to feed, and when it does so is frequently snapped up by certain fish of which it is a favoured article of diet. The natives of the region, who catch these fish, cannot use the flies themselves as bait since they immediately lose their colour when handled and thus cease to be attractive, so they manufacture artificial ones from coloured (φοῖνιξ) wool and feathers.

Fernandez[98] suggests that this insect, with its riverside habitat and its name of 'horse tail', is to be identified as a dragonfly (Odonata), a group of insects of which there are no explicit mentions in classical literature[99] and which contains black and yellow species. If this identification is correct, their approach to the water surface would not be for the purpose of feeding but for that of oviposition.[100] With their strong darting flight, these could be described as possessing the 'boldness' referred to.

An alternative suggestion, proposed by Gossen,[101] is that the insect is one of the soldier-flies or Stratiomyidae,[102] which again includes species with the appropriate habitat and colouration; although if ἀνθηδών does mean 'hornet' here these are perhaps rather too small.

54. *Mulio*

The *mulio* is listed by Pliny (XI.61) alongside wasps and hornets[103] as one of the predators of domestic honey bees. It appears also in his medical section (XXX.147), where it is said to live for only a single day, but the statement about its supposed properties has clearly been lost from the text. In both of these passages the insect is described as being *e culicum genere*.[104] Outside Pliny, the word is found only in Polemius Silvius' list of insect names.

[96] Cf. 48.4 above.
[97] Probably the hornet. Cf. 40.1.2.i above.
[98] *Nombres*, p. 478.
[99] Gossen's attempts, in PW VIII.257, to discover some are unconvincing.
[100] Imms, *Gen. Textbook*, p. 503.   [101] *Tiernamen*, no. 69.
[102] Imms, op. cit., pp. 1002–3.   [103] Cf. 40.5 above.

[104] *Culex* elsewhere refers to quite small insects, its usual meaning being 'gnat' or 'mosquito' (50 above), so that its use for a predatory insect is odd. It is not inconceivable that dragonflies, as a result of their similarity in shape, could have been thought of as giant relatives of the mosquito.

Ernout and Pepin[105] suggest that it derives its name from a habit of attacking mules. Leitner[106] similarly identifies it as a blood-sucking pest, but sees the bee predator as a second species comprehended under the same name. Possibly we are dealing here with one of the predatory Diptera of the family Asilidae. Alternatively, it is not impossible that the *mulio* is to be identified as a dragonfly (Odonata). Dragonflies do indeed prey upon honey bees, and are also believed in modern folklore to be capable of stinging man and animals, as their old name 'horse-stinger' testifies.[107]

Insecta: Siphonaptera (Fleas)

55. *Psylla/Pulex*

1. Identification and nomenclature

Fleas, along with lice[108] and bedbugs,[109] were well known in antiquity as pests infesting human habitations.[110] They are a distinctive group of insects, constituting the order Siphonaptera. Not surprisingly, they were thought of by the ancients as a single species, whereas in fact the number of species is quite large. They are all parasites sucking the blood of various mammals and birds, some restricted to a single host, others infesting a range of species. Many will feed readily on the blood of animals other than those with which they are normally associated. The most well known species in classical times, as today, was *Pulex irritans* Linn., the human flea, though there are some references to fleas infesting domestic animals. It may be noted that species of the latter kind, for example the dog flea *Ctenocephalides canis* (Curtis), and the cat flea *C. felis* (Bouche), may also become transferred to man.

Unlike insects such as lice, female fleas do not attach their eggs to the hair or feathers of their host, so that even when they do not leave the host to lay them they simply fall off and, in the case of animals, lie scattered about on the floor of its nest or sleeping place. The eggs of the human flea are found wherever dirt is allowed to accumulate in buildings. The larvae that emerge are active and worm-like, feeding upon scraps of organic material that lie about around them. When fully grown, they construct a silken cocoon to which debris adheres, and in this they pupate.[111]

The Greek name also occurs in the form ψύλλος; Hesychius' ψύλλαξ defined as ψύλλα is rightly taken by Fernandez[112] as a mere synonym of the more common name.[113]

---

[105] *Pline L'Ancien Histoire Naturelle XI*, p. 141.

[106] *Zool. Terminologie*, pp. 171–2. His specific identifications are very unlikely.

[107] The belief in a single day's life-span, true only in the case of mayflies (15 above), might perhaps have become applied to dragonflies through association of the two groups of insects as a result of their sharing the same freshwater habitat.

[108] 22 above.

[109] 17 above.

[110] Cf. Keller, op. cit., p. 401; W. Richter, in PW Supp.XV.101–6.

[111] Imms, *Gen. Textbook*, pp. 941 ff.; Smith, *Insects*, pp. 325 ff.

[112] *Nombres*, pp. 21–2.

[113] Rather than as the name for a distinct form, as Gossen, *Zool. Glossen*, p. 127 suggests.

The word *psylla* also had an extended usage, being applied to certain creatures regarded as resembling fleas because of their jumping habits or similar appearance and size.[114]

2. Life history

Aristotle (*HA* 539b 12) includes fleas among those insects that are themselves spontaneously generated and which, upon mating, produce offspring that are imperfect and do not give rise to anything further. The product of the mating fleas is described here as an egg-like larva, and in *GA* 721a 6 ff. (=723b 3 ff.), where the same point is being made, simply as a *skolex*. These *skolekes* are identified with the *konides* or nits (in reality eggs) produced by lice and bedbugs.[115] It is evident therefore that Aristotle was not familiar with or did not recognise the true larval stage of fleas described above. And it is possible that neither had he actually observed its eggs or recognised them as such, but that he simply deduced their life history on the analogy of lice, whose nits are obviously far more conspicuous.

The spontaneous generation of adult fleas was said to take place out of decaying material of unspecified nature (*GA* loc. cit.), and it is stated (*HA* 556b 25 ff.) that they may be found taking shape (συνίστανται) wherever there is any dry excrement. This idea is evidently the source of Isidorus' remark (*Or.* XII.5.15) that fleas are *ex pulvere magis nutriantur*. They were also believed by some to be generated from the human body in the same way as lice (Augustine MPL 32.1372).[116]

3. Habits

Fleas are described by Aristotle (556b 21 ff.) as feeding upon the juices of living flesh, an allusion to their blood-sucking habits. The flea's ability to leap for a remarkable distance in relation to its size is remarked upon several times in classical sources (Aristophanes *Nub.* 145–7, *Thesm.* 1179–80; Xenophon *Symp.* VI.8; Galen IV.362 K; Pliny IX.154). Aristotle includes it in his discussion of the jumping abilities of insects in *PA* 683a 34 ff.

4. Relations with man

There are quite frequent references to fleas as pests (Plautus *Curc.* 500; *AP* XI.264), biting man (Aristophanes *Nub.* 145–7; *AP* XI.432; Martial XIV.83; *Aesopica* 231, 272–3 Perry), sucking his blood (*Aesopica* 273), and thus averting sleep at night by their activities (Aristophanes *Pl.* 537–9). Pliny (IX.154) speaks of them as especially infesting inns in summer. Celsus (VI.7.9) deals with what to do if a flea should enter the ear.

The majority of the countermeasures prescribed for dealing with infestations of fleas in the home involve the sprinkling around of various preparations of herbs pounded up or boiled up in water (Pliny XX.172, XXII.27, 49, XXIV.53; Galen XIV.537 K; Dioscorides *DMM.* III.121, IV.15; Ps. Apuleius *Herb.* 103.4; *Geoponica* XIII.15.2–3, 5–6),

---

[114] 7b.1.2, 1.3.viii above; 63 below.
[115] Cf. 22.1.2, 2 above.
[116] Cf. 22.2, 4.2 above.

which were credited with having insecticidal or deterrent properties. We also read of a herbal fumigant (Pliny XX.155), the use of unslaked lime (*Geoponica* XIII.15.7; Ps. Theodorus Prisc. *Addit.* I.12) or sea-fleas[117] boiled up with the herb *psyllion* in seawater (Cyranides p. 123), or with goat's blood (Ps. Theodorus loc. cit.). Throwing down a glow-worm[118] is said to drive fleas away (Cyranides p. 91). Fleas are said to especially infest woolly carpets or mats (*Geoponica* XIII.15.11), but a bear skin is said to deter them (Cyranides p. 52). The *Geoponica* also includes a number of semi-magical folk remedies. These range from the making of a hole filled with chopped up *rhododaphne*, or a hole under one's bed in which goat's blood was to be poured (XIII.15.1.10; Ps. Theodorus loc. cit.) or a vessel rubbed with goat grease buried up to its lip (ibid.), as a kind of trap to gather the insects together, to the use of a verbal imprecation (15.9). It is reported (15.7; the method in 8 is similar, but part of the account is missing) that if a vessel is positioned in the centre of the house, a circle drawn around it with an iron implement, and the rest of the floor around the circle sprinkled with a herbal preparation, the fleas will all foregather in the vessel. Pliny (XXX.85) states that if earth from the print of one's foot where one is standing when the first cuckoo is heard is dug out and sprinkled, it will keep a place free from fleas.

5. Relations with domestic animals

Both Columella (VIII.5.3) and Varro (*RR* III.9.8) state that the straw provided for domestic hens to nest in needs to be regularly changed to prevent infestation by fleas and similar insects. Columella says that fleas are carried in by the bird, whereas Varro implies that they are generated in the straw itself. There is an erroneous reference in *Aesopica* 273 to fleas biting oxen.

Fles infesting dogs, *C. canis* being the species mainly in view,[119] are mentioned by Columella (VII.13.2) and the *Geoponica* (XIX.3.2). Application of brine, *amurca* or of herbal preparations are prescribed as a remedy. Varro (*RR* II.9.14) prescribes crushed almonds in water as a treatment to be applied to dogs' ears and between their toes, *quod muscae et ricini et pulices soleant . . . ea exulcerare.*

---

[117] The crustaceans known as sand-hoppers (Amphipoda).
[118] 35 above.
[119] So identified by R. Richter, art. cit., 104.

IX

UNIDENTIFIABLE AND FABULOUS INSECTS AND INVERTEBRATES

56. *Orsodakne*

The creature known as ὀρσοδάκνη, literally 'bud eater',[1] is referred to only twice in surviving sources. According to Aristotle (*HA* 552a 29–30) it results from the metamorphosis of small larvae which are spontaneously generated in (or on) the stalks of the cabbage. It is not, as one might otherwise suspect, to be equated with the *krambis* or cabbage white butterfly,[2] since the latter is dealt with elsewhere in the *Historia*. Hesychius, presumably using Aristotle as his source, identifies the insect as ζώυφίον τι ἐν κράμβῃ γινόμενον. Aubert and Wimmer[3] and Gossen[4] suggest that we are dealing here with some species of weevil (Curculionidae), though they disagree as to which one; while Sundevall[5] and D'Arcy Thompson[6] identify it as a flea beetle of the genus *Haltica*.[7]

57. *Prasokouris*

The πρασοκουρίς (Latin form *prasocoris* in Palladius) or 'leek-cutter' is included as a pest of garden vegetables by a number of authors, but without sufficient information to permit a conclusive identification. Although, as its name suggests, its preferred foodplant is said to be the leek (Pliny XIX.177; Aelian XIX.39) we are told that it also feeds on lettuces and various other plants (Theophrastus *HP* VII.5.4): other writers simply state that it attacks all garden vegetables (Hesychius; Palladius *RR* I.35.13). As regards its physical appearance it is described as green in colour (Hesychius and Photius). Like other horticultural pests it was believed to be spontaneously generated from its foodplant (Theophrastus loc. cit.; Aelian loc. cit.; Aristotle *HA* 551b 20): according to Aristotle, the adult insects are not produced directly but develop from spontaneously generated larvae;[8] he adds that the adults are winged. The botanical (Theophrastus loc. cit.; Pliny loc. cit.) and agricultural (Palladius loc. cit.; *Geoponica* XII.9) writers all give details as to how infestation by this pest may be dealt with, and two essentially similar methods may be

---

[1] Fernandez, *Nombres*, p. 140
[2] 23 above.
[3] *Thierkunde*, I p. 169.
[4] In PW X.1488.
[5] *Thierarten*, p. 197.
[6] *Hist. An.*, V.19 note.
[7] Cf. 63 below.
[8] He gives a name to these larvae, but the text is corrupt at this point. Peck, *Hist. An.*, II p. 179, suggests σινίδες.

distinguished. In the earliest of the above writers, Theophrastus, the text at the crucial point is very corrupt, rendering his account partially unintelligible: a number of emendations have been suggested, none of them entirely satisfactory,[9] but in any case the basic intention is clear from the later sources. Palladius and the *Geoponica* recommend that, since the creature is fond of dung (φιλόκοπρος), a sheep's intestine filled with this attractive substance should be placed on the ground near the infested plants and lightly covered with soil. The *prasokourides*, attracted by the dung, should then emerge from hiding and collect inside the intestine, where they may be conveniently destroyed. Theophrastus, on the other hand, who is followed by Pliny, evidently held that it was sufficient merely to place a heap of dung in an appropriate spot, whereupon the pests would leave their foodplant and burrow into it (if the conjecture accepted by Hort be taken into account, he also states that a pile of green fodder (κράστις) will have the same effect).

Aside from the above sources, which provide essentially similar details, the only independent information on the *prasokouris* is found in a passage from the comic dramatist Strattis (fr.66 *ap.* Athenaeus 69a). He portrays the creatures as wandering about in gardens on their fifty legs (πρασοκουρίδες, αἳ καταφύλλους ἀνὰ κήπους πεντήκοντα ποδῶν ἴχνεσι βαίνετ᾽)and 'dancing (χορoὺς ἑλίσσουσαι) upon the foliage' of lettuce, basil and celery. Fernandez[10] suggests that the σατυρίδια which figure rather cryptically in the centre of this same fragment (ἐφαπτόμεναι ποδοῖν υατυριδίων μακροκέρκων) are also some form of insect, but this seems unlikely.[11]

Various suggestions as to the identity of this creature have been put forward. Sundevall[12] concludes after a considerable discussion that it is not after all identifiable, a position followed by Aubert and Wimmer;[13] while Gossen[14] attempts to solve the problem by proposing that the word differs in meaning among the various sources. J. B. Gèze in a short article on the subject[15] cites some linguistic parallels from modern south European languages and dialects to support his contention that the insect in question is the mole cricket *Gryllotalpa*,[16] a subterranean pest feeding upon the roots of cultivated plants. Fernandez,[17] however, points out quite correctly that, however this may be, Strattis' description of a many-legged creature active upon the leaves of plants does not support this identification but rather suggests some form of caterpillar.

[9] Cf. A. Hort, notes ad loc.

[10] *Nombres*, pp. 218–20.

[11] J. M. Edmonds, *The Fragments of Attic Comedy*, Vol. 1, 1957, note ad loc., regards σατυρίδια as 'either a plant such as monk's rhubarb (cf. σατύριον) or satyr effigies used as scarecrows'.

[12] *Thierarten*, pp. 236–7.

[13] *Thierkunde*, I pp. 169–70.

[14] In PW X.1486–7 and 2A.573.

[15] 'La Courtilière existait-elle dans la Grèce Antique?', *CRAI*, 1931, pp. 47–9.

[16] Cf. 12 above.

[17] Op. cit., p. 142. LSJ Supp. also regard πρασοκουρίς as 'a kind of caterpillar'. E. K. Borthwick's comparison ('Notes on the Plutarch *De Musica* and the *Cheiron* of Pherecrates', *Hermes*, 96, 1968, pp. 71–2) of the Strattis fragment with Pherecrates fr. 145, with its undoubted reference to caterpillars (23.5 above), provides further support for this identification.

58. *Knips* or *Sknips*

Classical Greek contains a series of parallel monosyllabic insect names which in general terms seem to mean little more than 'a small insect pest' but which in actual usage tend to become attached to one particular type of creature. Thus *thrips*[18] customarily refers to what we know as the 'woodworm' and *kis*[19] to grain pests, and so forth,[20] although examples can always be cited to demonstrate that such words remain essentially interchangeable. Κνίψ or σκνίψ (the spelling differs in some cases even between different manuscripts of the same work, though the lexicographers tend to have separate entries under each) differs from the rest in apparently having no determinate meaning at all, no variety of small insect to which it particularly applies. Sundevall[21] is thus justified in his conclusion that the word can refer to more or less any pest species. In origin the name is related to σκνίπτω and contains the idea of 'to pinch or nip'.[22] The various contexts in which it occurs are as follows:

(i) *A wood-boring pest.*[23] In two passages of the *Historia Animalium* (593a 3, 614b 1) the *sknips* is a small insect living under the bark of trees and hunted by woodpeckers. Theophrastus (*HP* IV.14.10) also refers to *knipes* as insects generated in wood. Plutarch (*Mor.* 636d), discussing the subject of spontaneous generation, associates *sknipes* and *teredones* as creatures both produced in decaying wood. Both Hesychius (s.v. *knipes*) and Suidas (s.v. *sknips*) have 'small wood feeding creatures' among their several definitions, as does Zenobius (V.35).

(ii) *An insect attracted to honey.* Aristotle (*HA* 534b 19), dealing with the sense perception of insects, says that bees and *knipes* are capable of being attracted by the smell of honey from a great distance. He makes a similar comment in *de Sensu* 444b 12.

(iii) *A form of ant.* In the last mentioned passage, Aristotle describes the *knips* as being a type of small ant (τὸ τῶν μικρῶν μυρμήκων γένος οὓς καλοῦσί τινες κνίπας). The word is also associated with ants by Sextus Empiricus (I.57), who informs us that although ants and *sknipes* cause severe discomfort if swallowed by human beings sick bears are able to cure themselves by licking them up. *Sknips* is applied to one of the seven species of ant listed by the Cyranides (pp. 68, 294).[24]

(iv) *A pest of figs.* *Knips* appears in Theophrastus (*HP* II.8.3) as a predator of the beneficent fig-wasp.[25] Aristophanes (*Av.* 590) seems to be under the erroneous impression that both the *knips* and the *psen* or fig-wasp are equally pests of figs, while the Physiologus (48) uses *sknips* as his name for the latter.[26]

---

[18] 37 above.   [19] 36 above.

[20] Cf. also κορίς (17 above), φθείρ (22 above), ἴψ (25 above), σής (26 above).

[21] *Thierarten*, pp. 193–4. His conclusion is accepted by Fernandez, *Nombres*, pp. 112–14, though not by for example Gossen, who has nonetheless attempted to identify it in specific terms.

[22] Fernandez, op. cit., p. 114.

[23] Cf. 37.1.2.v above

[24] 44.1.2.ix above.   [25] 46.4 above.   [26] 46.5 above.

(v) *A pest of vines.* Galen (XII.186 K) uses the name *sknips* for the lepidopterous larva known as *ips.*[27]

(vi) *A green winged insect.*[28] Hesychius (s.v. *sknips*) includes the definition ζῷον χλωρόν τε καὶ τετράπτερον. This is perhaps to be identified as an aphid or greenfly.[29]

(vii) *An insect preyed upon by birds.* Aristotle (*HA* 593a 12) as well as the references to woodpeckers cited above characterises a certain insectivorous bird as κνιπολόγος. The verb parallel to this adjective is used by Aristophanes of Byzantium (*HA Epit.* II.440) in describing the feeding habits of the bat.[30]

(viii) *The vinegar-fly.* Sextus Empiricus (*Pyrrh.* I.41) uses *sknips* on a second occasion to refer to the vinegar-fly.[31]

(ix) *A biting fly.* Hesychius' third definition (s.v. *knips*) ζῷον πτηνόν ὅμοιον κώνωπι is paralleled by Suidas' (s.v. *sknips*) and Herodianus' (*Tech.* II.718.18) ζῷον κωνωπῶδες. Suidas and Herodianus (*Tech.* II.423) note the existence of an alternative spelling σκίψ. In this sense the word was used by the Septuagint translators to render one of the Egyptian plagues, and consequently figures in a number of authors dependent upon the LXX.[32]

(x) *The bed-bug.* The latinised *sciniphes* is found in Petronius (98.1) as an alternative to the usual *cimex.*[33]

(xi) *A jumping insect.* The Greek paroemiographers (Zenobius V.35; Diogenian V.91, VII.25; cf. Suidas s.v. *sknips*) include in their compilations a certain proverbial expression involving the *knips* which is said to have been used by the comedian Strattis (fr. 70). It is variously quoted as ὁ σκνὶψ ἐν χώρᾳ (Zenobius; Diogenian VII) and κνὶψ ἐκ χώρας (Diogenian V; Suidas), and it is not clear how it should be translated: Edmonds[34] gives two suggestions and LSJ another. Nevertheless the explanatory comments given by the proverb anthologists are clear enough, stating as they do that the expression is used 'for those who jump rapidly about from place to place' (ἐπὶ τῶν ταχέως μεταπηδώντων or μεταπιπτόντων or μετακινούντων) since the creature involved is 'a little wood feeding creature leaping about from one spot to another'. It has been suggested,[35] not without reason, that the word should be translated here as 'flea',[36] but this would involve the rejection as misplaced of the description ξυλοφάγον.[37]

---

[27] 25.1.2.ii above.
[28] Gossen, in PW Supp.VIII.240, erroneously identifies this as a species of weevil.
[29] Cf. 64 below.
[30] LSJ render this word as 'catch fleas'.
[31] 51 above.
[32] 50.7.ii above.
[33] Cf. 17.1.2 above.
[34] J. M. Edmonds, *The Fragments of Attic Comedy*, note on Strattis fr. 70.
[35] Edmonds, ibid.; LSJ s.v.
[36] 55 above.
[37] The superficially similar name κνώψ, used by Nicander (fr. 74) for an insect pest damaging poppy capsules, seems (cf. Fernandez, op. cit., p. 51) to be an abbreviated form of κινώπετον.

59. *Skolex tes chionos*

In a probably inauthentic passage of the *Historia Animalium* (552b 7 ff.) dealing with unusual examples of spontaneous generation, we are informed that snow which has been lying on the ground for some time may take on a reddish colour and give rise to worms or larvae (*skolekes*) of the same shade and coated with hair. In Media large white worms are produced in the same way. Both varieties are sluggish in movement and soon perish if removed from their natural habitat. These creatures are referred to briefly in the pseudo-Aristotelian treatise *De Plantis* (825a 2), where they are called ἔλμινθες,[38] and by Antigonus of Carystus (*Hist.Mir.* 90). The details supplied by the *Historia Animalium* are reproduced in a somewhat muddled form by Pliny (XI.118), who, not realising that *Media* is a proper name, has rendered the phrase in which it occurs as *in media . . . altitudine*.

An independent account of the animals in question is provided by Strabo (XI.14.4), who cites as his source Apollonides and Theophanes. He locates them in a different though adjacent geographical region, namely the northernmost provinces of Armenia bordering the Caucasus mountains, and associates them with certain hollow masses of ice which form in the snow and contain pockets of liquid water. He gives nothing by way of a physical description, but records that the former of his sources named the creatures *skolekes* and the latter *thripes*.[39]

It has been generally agreed that some degree of actual observation lies behind these accounts, though it is a matter of debate as to how much in them is purely fabulous. As has been pointed out by Aubert and Wimmer and Peck,[40] part of the explanation for the story is undoubtedly the fact that long lying snow can acquire a reddish discolouration due to the proliferation of microscopic algae. For the *skolekes* themselves, however, no satisfactory explanation has been provided, and it is probably futile to seek to identify them, at least not in terms of the description given by the *Historia*. If we ignore the latter, the wingless scorpionfly *Boreus hyemalis* Linn., known as the 'snow-flea', and its relatives,[41] and certain springtails (Collembola), for example *Hypogastrura* and *Isotoma*,[42] suggested by D'Arcy Thompson[43] could perhaps be relevant here, since the term θρίψ considered in isolation might well be used to describe such small active insects as these.[44]

60. *Pyrigonos*

Associated with the classical belief in the existence of four basic physical elements was the

---

[38] Cf. 1.1 above.

[39] On σκώληξ and θρίψ, cf. 37.1 above.   [40] Aubert and Wimmer, op. cit. I p. 515.

[41] *Boreus* spp., some of them brightly coloured, are well known for running about over the surface of snow in mountain regions.

[42] On Collembola cf. 10 above. Springtails such as those cited are also conspicuous alpine insects: cf. Imms, *Gen. Textbook*, p. 470.

[43] *Hist. An.*, V.19 note.

[44] Sundevall's suggestion, *Thierarten*, pp. 238–9, followed by Steier, *Aristoteles und Plinius*, p. 50, Gossen, in PW Supp. VIII.237, and Leitner, *Zool. Terminologie*, p. 216, of soldier beetle larvae is very unlikely.

idea that each should have a group of living creatures corresponding to it. In one of the probably interpolated passages of the *Historia Animalium*, immediately following the description of the 'snow worms', we have an account of certain winged creatures, a little larger than large flies, which are spontaneously generated in copper smelting furnaces in Cyprus, where the fire is continuously alight for several successive days. These jump and crawl (πηδᾷ καὶ βαδίζει) about through the fire completely unharmed, and indeed soon perish if removed therefrom (552b 10 ff.). Aristotle himself was either unaware of or refused to believe in the reputed existence of these remarkable animals, since in his genuine works he clearly states that fire is an element which is not capable of generating life (*GA* 737a 1, *Met*. 382a 6 ff.).[45]

The details given in the *Historia* are followed by Antigonus (*Hist.Mir*. 90) and Aelian (II.2,31). The latter applies to the creatures the name πυριγόνος and embellishes his account by saying that they flourish (τεθήλεναι) in the fire, that they fly about in it, and that if they stray beyond its range the cooler air immediately destroys them. Pliny (XI.119) gives a similar description and records two Greek names *pyrotocon* and *pyrallis* otherwise unknown: his statement that the creature has four legs is clearly a detail misplaced from the description of the *ephemeron* which follows in the *Historia*. The *pyrigonos* figures anonymously in Sextus Empiricus' list (*Pyrrh*. I.41) of examples of spontaneous generation (τὰ ἐν ταῖς καμίνοις φαινόμενα ζῳύφια), not ascribed to any particular geographical location, and likewise in Seneca's *Naturales Quaestiones* (V.6) and, under the name πυρίβια, in Diogenes Laertius (IX.79). Strabo (XI.14.4), like the unknown interpolator of the *Historia*, compares the *pyrigonos* with the 'snow worm' but displays his independence of that source by describing it as a form of gnat (κώνωψ) produced 'from the flames and sparks in mines'. Philo was evidently somewhat fascinated by the creature, as he refers to it under the name of τὰ πυρίγονα three times in his surviving works (*de Gig*. 7, *de Plant*. 12, *de Aet.Mund*. 45), always in the context of the idea that each element has its own appropriate form of life: he bears witness to a source other than those previously mentioned by adding in the first two passages the comment that the 'fire-born' are found especially in Macedonia.[46]

The earliest commentators such as Sundevall[47] and Aubert and Wimmer[48] maintain the obvious conclusion that the story of the *pyrigonos* is as fabulous as the reported habits of the salamander, which Ps. Aristotle cites as a confirmatory example of an animal capable of existing unharmed in fire. According to Aelian, the habitat of the salamander is the same as that of the creatures we are considering. Although Gossen[49] has suggested that the origin of the story can be explained by moths being attracted to the light of the furnaces, the only serious attempt to provide a plausible explanation and to uphold the classical data as being the product of accurate observation is the ingenious if not entirely

---

[45] Cf. Peck, *Gen. An*., note ad loc.

[46] Talking of animals in fire, Augustine (*Civ*. XXI.2) refers to a certain kind of *vermis* which lives unharmed in the waters of hot springs and dies if removed from this habitat.

[47] *Thierarten*, p. 239.

[48] *Thierkunde*, I p. 515.

[49] *Tiernamen*, no. 66.

convincing theory put forward by Janssens.[50] He draws attention to the fact that certain species of insect are attracted by the smoke of burning timber and suggests that it is this habit which the Cypriot and other copper workers observed. They would have noted that the insects mysteriously appeared and disappeared according to whether the furnaces were in operation and would have concluded that their existence was dependent upon the fire.

61. *Myrmekoleon/Formicoleon*

Although most of the interest in the so-called 'ant-lion' in antiquity comes as a result of its appearance in the Septuagint, stories about creatures apparently combining the natures of ants and lions were earlier current among ancient geographers. Agatharchides (68–9) and Strabo (XVI.4.15) both write of 'lions called ants' in terms which are reminiscent not only of the LXX and patristic creature but also of the gold-digging ants of Herodotus and later sources.[51] These authors locate the creatures in Arabia and Ethiopia, and describe them as having gleaming fur which shines like gold, and as having their genitals reversed. As described here, they sound more like mammals than insects, and that is probably how they were originally envisaged, but Aelian (XVII.42) understands them unequivocally as ants, omitting any reference to 'lions' and characterising them only by their oddly placed genital organs: he situates them in Babylonia, where Agatharchides mentions that they also occur.[52]

The translators of the Septuagint, with their customary disregard for zoological accuracy, selected the term μυρμηκολέων—the compound is not attested in any earlier writing—somewhat arbitrarily to render the Hebrew word for 'lion' in *Job* IV.11, having already used λέων, λέαινα and the totally incongruous δράκων for three virtual synonyms in the preceding verse. And the result of their choice was to provide an excuse for much fantastic speculation and misplaced ingenuity to be exercised by later Biblical commentators who display, in addition to an unwarranted trust in the zoological and linguistic competence of the 'Seventy', considerable interest in the identity of scriptural animals and in drawing moral lessons from their habits, real or supposed.

The moralising compendium of popular natural history known as the Physiologus, which is largely based upon the fauna and flora of the LXX, devotes a chapter to the ant-lion (20, pp. 73–6 Sb.) in which it is deduced from the text that the creature is a composite monster with the forepart of a lion and the hindpart of an ant, a hybrid in fact of these entirely disparate species. The Biblical verse in question speaks of the animal perishing for lack of food (Μυρμηκολέων ὤλετο παρὰ τὸ μὴ ἔχειν βοράν), and the Physiologus goes on to explain that the creature meets with this unhappy fate owing to the fact that it

---

[50] E. Janssens, 'Le Pyrotocon de Pline L'Ancien', *Latomus*, IX, 1950, pp. 283–6. He suggests a Buprestid beetle of the genus *Melanophila*.

[51] 45 above.

[52] These ant-like mammals are perhaps the *myrmekes* whose young are said in Aelian VII.47 to be known as σκύμνοι.

cannot eat the natural food of either of its parents and so inevitably starves to death. These details are reproduced in the *Hexaemeron* ascribed to Eustathius (MPG 18.745), and in Olympiodorus of Alexandria (*in Job* 4.11 MPG 93.72). However, in the comments of Gregory Magnus (*Mor.* V.20,22) on the book of *Job* we find a more rational interpretation of the mysterious LXX word. Here the ant-lion is depicted as an insect which preys upon the industrious ant, cunningly concealing itself in the dust and pouncing on its victims as they carry grains of corn home to their nests (*parvum . . . animal, formicis adversum, quod se pulvere abscondit et formicas frumenta gestantes interficit interfectasque consumit*): this account is not only a gain in intelligibility but also allows the drawing of a moral lesson more suited to the surrounding context of the verse. The same definition appears in the works of Augustine (*Ann. in Job* 4.11 MPL 34.828), though not without evidence of some mental confusion, and is reproduced more or less verbatim by Isidorus (*Or.* XII.3.10) who uses the latinised name *formicoleon* (Gregory and Augustine simply transliterate the Greek, as does the Old Latin version of *Job*: the Vulgate disposes of the animal entirely and renders the Hebrew as *tigris*).

Apart from the above, the only other example of the use of the term *myrmekoleon* is found in the Cyranides (pp. 68, 261, 294), where it is the name applied to the seventh and last in a list of varieties of ant.[53] The insect is here described as larger than other ants and speckled (ποικίλος) in appearance, as being carnivorous (σαρκοφάγος), and as having a short life span (τάχιον ἀποθνήσκοντες). It has been suggested that this account is either dependent upon the Physiologus or that both derive from a common source, but it is far from evident that there is any relation between the two texts at all. The Cyranides could, however, be harmonised with the description given by Gregory. It may be mentioned here that a possible synonym, λεοντομύρμηξ, is noted without definition by the grammarian Herodian (*Tech.* I.46).   In considering the relationships between the above mentioned texts, it is reasonable to conclude with Fernandez[54] that in seeking to supply the need for an additional synonym of λέων the LXX translators had in mind the 'lions called ants' of the geographers; although there is no need to suppose that they actually invented the compound name for the purpose. There is a striking parallel example in the LXX text of *Leviticus* of the name of a fabulous creature being introduced to supply a shortage of names for the locust.[55] The Physiologus account may then be seen not as reproducing some pre-existent popular fable, but as simply an imaginary construction based on deduction from the text of *Job* IV. G. C. Druce credits the LXX translators with more care in their selection of zoological nomenclature than is warranted by their general procedure elsewhere, when he supposes that the story of the unfortunate starving hybrid predates their work and lies behind their choice of the word in question.[56]

The origin, however, of the new definition provided by Gregory is rather more debatable, since it is necessary to take into account the insect to which modern

---

[53] Cf. 44.1.2.ix above.

[54] *Nombres*, pp. 56–60.

[55] 11.1.2.xxiii above.

[56] G. C. Druce, 'An Account of the Myrmekoleon or Ant-lion', *The Antiquaries Journal*, III, 1923, pp. 347 ff.

entomology has applied the ancient name of ant-lion. The modern ant-lions are a family of lacewings (Neuroptera: Myrmeleontidae) whose larvae construct pits in the sand and conceal themselves at the bottom to seize with their powerful jaws any insect which may fall in. If we take the phrase about hiding in the dust to be a reference to these pits, Gregory's description could then be seen as based on actual observation of this insect, as is the opinion of Fernandez[57] and Gerhardt,[58] following Wellmann.[59] In this case we would have to suppose that our present day ant-lion was known in antiquity under that name, and that Gregory and others were under the impression that this was the creature to which the LXX was referring. On the other hand, it may be added that these patristic descriptions are sufficiently vague as not to rule out the possibility that they may simply result from a rather more intelligent deduction from the text than that of Physiologus, a deduction which purely by chance happens to coincide in some measure with a genuine non-fabulous insect.

In this regard, the Cyranides is of interest in presenting the *myrmekoleon* as simply a carnivorous insect rather than as a hybrid monster. A number of authors, from Wellmann onwards,[60] have discerned similarities between the details given here and the Physiologus chapter, and have contended that they derive from a common source, or that the former derives from the latter. But in view of the great difference between the two texts—in the one we are in the realm of natural history, while in the other we are in the realm of fable—it seems more reasonable to suppose that their descriptions have only the name of *myrmekoleon* in common. Gossen firmly[61] and Gerhardt[62] tentatively identify the creature depicted by the Cyranides with our modern neuropterous ant-lion larva. If this is so, the Cyranides reference should then be placed in relation to the later patristic descriptions rather than to the Physiologus, as further supporting evidence that classical natural history knew the *myrmekoleon* as a common predacious insect and not only as the name for a creature of fable.

62. *Serphos*

Σέρφος or σύρφος is an insect name which, like *knips*[63] appears to have no clearly determinate meaning. It appears several times in Attic comedy, and in the works of Plutarch and Aelian, but with no indication as to which, if any, particular kind of insect is in view. The ancient scholiasts and lexicographers do indeed attempt to provide the word

[57] Op. cit., p. 58.

[58] M. I. Gerhardt, 'The Ant-Lion', *Vivarium*, III, 1965, p. 13.

[59] M. Wellmann, 'Der Physiologus: eine Religionsgeschichtlich-naturwissenschaftliche Untersuchung', *Philologus*, Supplementband XII, Heft 1, 1930, pp. 37–8.

[60] M. Wellmann, ibid., pp. 37 ff.; M. I. Gerhardt, art. cit., pp. 8–9, with further references.

[61] In PW Supp.VIII.3.

[62] Art. cit., pp. 9–10. He discusses the possibility of details from adult as well as larval stages being combined in the text, but rejects this on the grounds that the ancients would have been unlikely to have associated larva and imago. The addition in the Latin version of Cyranides of *alatae* is probably a mistake, as a winged ant appears as one of the earlier listed varieties.

[63] 58 above.

with a firm definition, but in the process display wide disagreement among themselves, suggesting that they are seeking to define it with more precision than is warranted by its actual usage. It seems reasonable, therefore, to conclude with Venmans[64] that we are dealing here with a popular term that could be applied to more or less any small insect, perhaps especially those of a worm like nature. Fernandez,[65] like LSJ, takes more seriously the lexicographical tradition and defines the word as 'a winged ant or mosquito, creatures which could easily be confused in popular terminology'. In the opinion of Venmans,[66] the word is related etymologically to ἕρπω and thus has a literal meaning no more precise than 'creeping thing', while Fernandez[67] regards it as more probably connected with σύρω and having reference to the insect's small size. Either of these etymologies would support the conclusion that in general usage the meaning of serphos was wide and undefined.[68]

In two of its three appearances in the plays of Aristophanes (Av. 82, 569–70) the serphos is depicted as constituting part of the diet of birds, as also in fr. 1 of Nicopho, where it is grouped with σκώληκας, ἄκριδας, πάρνοπας. Aelian (IX.3) and Plutarch (Mor. 982d), in parallel passages, use the word in the context of a list of food items of young crocodiles. The third reference by Aristophanes (Vesp. 352 οὐκ ἔστιν ὀπῆς οὐδ' εἰ σέρφῳ διαδῦναι) characterises the serphos as notable for its small size, as does the popular proverb Ἔνεστι κἄν μύρμηκι κἄν σέρφῳ χολή (Suidas and scholia ad Vesp. 352 and Av. 82; AP X.49). Aelian (XIV.22) mentions the serphos in passing as a possible bait for fish, clearly distinguishing it from konops. In the philosophical writings of Philodemus (de Morte 34, de Deis I.25, cited by Venmans p. 68) the word is used in a metaphorical sense for contemptible human beings.

The ancient commentators on Aristophanes were evidently very uncertain as to the meaning of the word serphos, and seem therefore to have resorted largely to deductions and guesses from the text, citing also the passage from Nicopho and the proverb cited above. Some, we read (Schol. Vesp. 352, Av. 569), were even under the impression from the context that it referred to a kind of fruit or seed. It is variously described as meaning an ant (Crates ap. Schol. Vesp. 352 is said to have equated it with myrmex) or a creature like an ant (μυρμηκῶδες Schol. Av. 82 and Suidas), or like a gnat or mosquito (κωνωπῶδες Schol. Vesp. 352), or like a worm or larva (σκωληκῶδες Schol. Av. 82). According to Photius, serphoi are 'the winged ants which we call nymphai' (cf. Schol. Av. 569 Σέρφος μύρμηξ πτερωτός),[69] at least in the opinion of Didymus Chalcenterus; but he adds that Cassius Longinus explained them as being 'a small winged creature resembling a konops in size' and stated that the winged ants of Didymus were properly termed στέρφνοι. Hesychius gives a variant form σύρφος, which he defines as 'a small creature like the empis'.[70]

---

[64] L. A. Venmans, 'Serphos', Mnemosyne, LVIII, 1930, pp. 58–73.

[65] Nombres, p. 97.   [66] Art. cit., p. 72.   [67] Op. cit., pp. 98, 225.

[68] The name σέριφος, as found in a compound name for the praying mantis (14 above) is probably a variant of σέρφος; Fernandez, ibid., p. 192.

[69] Cf. 44.2.1 above.   [70] On empis and konops cf. 50 above.

63. Psylla/Pulex

We are concerned here not with *psylla* the domestic pest,[71] but with certain horticultural pests of the same name. These are first referred to in Theophrastus' list of the characteristic insects spontaneously generated among particular varieties of garden vegetable (*HP* VII.5.4) where they are especially associated with the radish. We are told that there is no treatment that will prevent these creatures being produced, but that they may be combatted by sowing vetch among the crop (we read the same in *CP* II.18.1, where the effect is cited as an example of the property possessed by some plants of protecting others, in this case by 'keeping them free of the pests that arise': but it is not explained how the effect operates). Later on in the same work (VIII.10.1) we read that the chickpea is especially liable to being eaten by caterpillars and *psyllai*. Pliny, in reproducing Theophrastus' list of garden pests (XIX.177), translates *psylla* by the corresponding term *pulex*. Under this name, the insect is mentioned twice by Columella (*RR* X.3.21, XI.3.60), who states that its depredations are promoted by the drying out of the plant and depicts it as consuming the young shoots. Palladius (*RR* I.35.5) notes the *pulex* as a pest of garden vegetables in general and recommends the sprinkling of infested plants with a preparation of henbane in vinegar as an insecticide. The *Geoponica* includes it as a spontaneously generated insect infesting radish, cabbage and similar vegetables (XII.7, XII.4.1) and figs (XII.19.9), and suggests similar methods of combatting them.

Since the creature is never actually described, the only real clue to its identity lies in its name. It is probable that it would have acquired the name of 'flea' through being a jumping insect, and Gossen,[72] Richter[73] and Fernandez[74] therefore identify it, reasonably enough, as a flea beetle of the genus *Phyllotreta* (Chrysomelidae: Halticinae), these being important pests of cruciferous plants.

64. Phtheir

According to a similar verbal usage to the preceding, other small plant pests were known by the name of 'lice',[75] either as a result of a superficial similarity in appearance, or of the word's original etymological meaning of 'destroyer',[76] which would make it an appropriate term for any pest.[77]

Phtheires appear in the *Geoponica* (XII.7) as pests of vegetables, and the same countermeasures are recommended as for the 'flea'. LSJ suggest that these insects may be red spider mites (*Tetranychus* and allied genera). They could also be aphids (Hemiptera: Aphidoidea), an important and well known group of pests to which there are surprisingly no unequivocal references in ancient literature.[78]

---

[71] 55 above.   [72] In PW Supp.VIII.239.   [73] In PW Supp.XV.104.

[74] *Nombres*, p. 64.   [75] Cf. 22 above.   [76] Fernandez, *Nombres*, p.118.

[77] Cf. 9, 22 note 103, 25.1.2.iii, 36.1 above. Pliny refers to *pediculi* generated from the herb basil (XX.121), and Galen to *phtheires* from dried figs (VI.572,793 K).

[78] Apart from the specialised gall aphids (47.4–5) above). The *knips* of Hesychius, and therefore perhaps of other authors (58 above), may be the aphid, as may the *culex* of Martial XI.18 and Pliny XVII.231 (50 note 56 above), in view of the gall aphids being termed 'gnats'.

65. *Vermis caeruleus*

Pliny reports on the authority of Statius Sebosus (IX.46; cf. Solinus LII.41; Isidorus *Or.* XII.6.10) the existence in the river Ganges in India of certain monstrous worms, blue in colour with a pair of gills sixty cubits in length, which prey upon elephants coming to drink, seizing them by the trunk and dragging them into the water.

66. *Skolex leukos*

Equally fabulous is the 'creature resembling a white worm' (θηρίον σκώληκι εἰκάσμενον λευκῷ) described by Philostratus (*Vit. Apoll.* III.1) as inhabiting the river Hyphasis in the same country. These, we are told, are caught and melted down (τήκοντες) to produce a type of inflammable oil (ἔλαιον), the flame given off by which can only be contained within glass. Accordingly, the capture of them is the prerogative of the king, the oil being employed to set fire to the battlements of enemy cities, it being impossible to extinguish the resulting conflagration by any normal means.

A similar report is given by Ctesias (*Ind.* 27) and Aelian (V.3) concerning a gigantic *skolex* found in the river Indus, which is described in highly coloured terms as requiring whole animals as bait and many folk to drag it to shore when caught. It is said to resemble the *skolekes* generated in timber.

67. *Galba*

We hear of this popular insect name only as a result of Suetonius' explanation of the origin of the name of the emperor Galba (*Galba* 3). He states that there are contradictory explanations of why his ancestor should have adopted this name, but that according to one it was because he was 'as slender as the creatures which are produced in oak trees and are called *galbae*'. These would probably be some form of caterpillar, as Gossen[79] suggests, but a more precise identification is hardly possible. Contrary to Suetonius' statement, the name seems to mean 'fat' or 'stout', with reference to the insect's shape.[80]

68. *Rauca*

In the works of Columella (*RR* XVII.3) and Pliny (XVII.130), the *rauca* is a pest of young olive seedlings. It is described as a kind of worm (*vermis*) produced (*nascuntur*) in the roots of oak trees, which transfers itself to the olive if one happens to be planted where an oak has been dug up.[81] In the only other surviving reference (Ulpian *Dig.* XIX.2.15.2), *rauca* is used in a more general sense as a name for pests of field crops (*si raucis aut herbis segetes corruptae sint*).

[79] In PW 2A.580.

[80] Ernout and Meillet, *Dict. Ét.*, s.v. Cf. the insect name *cossus*, also used as a personal name (30.1.2.vi above).

[81] Steier, *Aristoteles und Plinius*, p. 50, identifies it as a gall wasp, but such an insect would be too small to attract attention in this way. For other olive pests, cf. 25.4, 30.3 above.

69. *Biurus*

Pliny notes the fact that (XXX.146) he had discovered in the course of his researches references to animals whose identity he could not discover. One such was the *biurus* or 'two-tailed', said by Cicero to be a pest of vines in Campania (*M. Cicero tradit animalia biuros vocari qui vites in Campania erodant*).[82]

70. *Phryganion*

A similar case is provided by the *phryganion* (diminutive of φρύγανον, a dry stick or twig, or an undershrub), which Pliny (XXX.103) describes as a creature recommended by Chrysippus to be worn as a amulet against quartan fever. 'But', he continues, 'what the animal is Chrysippus has left no account, and I have met nobody who knew. Yet a statement made by so great an authority it was necessary to mention, in case somebody's research should meet with better success.' Ernout[83] suggests, on the basis of the word's literal meaning, that it is to be identified as the aquatic larva of the caddis flies (Trichoptera), most of which construct protective cases from pieces of twig and other items as do the terrestrial larvae of Psychid moths.[84]

71. *Names listed by Greek lexicographers*

(i) Ἀστάλη. According to Hesychius, a form of worm or larva with a tail (ἔνιοι σκώληκα οὐρὰν ἔχοντα).

(ii) Αἰγιόνομος. The entry in Hesychius Αἰγιόνομοι· ζῷα οὕτω καλούμενα is considered by Fernandez,[85] who takes the word to mean 'eaten by goats', probably to refer to some form of insect.

(iii) Δίκηλον. Hesychius has ζῴδιον among several definitions for this word (a variant of δείκηλον, 'image', 'phantom'). Gossen[86] sees a reference to an insect here, but this is purely speculative.

(iv) Ἔνθριον. Gossen[87] interprets the entry in Hesychius which appears in the mss as Ἔνθρια·ζῴδια as referring to a species of insect, seeing a connection with θρῖον 'leaf'. However, Latte considers the text to be corrupt at this point and suggests the reading Ἔνθριπα· ζῴδιοις βεβρωμένα, with a reference to the insect name *thrips*.

(v) Εἰλύιος. The name variously spelt ἐλείος, εἰλύιος, and ἴηος (Hesychius has separate entries under each of these) is related etymologically to ἕλμινς, εὐλή, the

---

[82] For other vine pests, cf. 25 above.
[83] A. Ernout, *Pline L'Ancien Histoire Naturelle XXX*, p. 97.
[84] Cf. 29 above.
[85] *Nombres*, p. 139.
[86] *Zool. Glossen*, p. 24.
[87] Ibid., p. 29.

verb ἰλυσπάομαι, and a number of other words expressing the idea of wriggling movement and associated with worms of various kinds.[88] It may therefore originally have been a fairly general term for worms or worm-like animals, such as, for example, εὐλή and σκώληξ.[89] There are, however, no surviving examples of its usage, and the definitions of the lexicographers give it a more specific meaning, although displaying considerable uncertainty as to what that meaning in fact was. Citing the example of two of his predecessors, Hesychius (s.v. ἐλειούς) states that 'Aristarchus says that it is generated among undershrubs (or twigs ἐν τοῖς φρυγάνοις) like lizards; but Callistratus says it is a worm-like (σκωληκοειδές) creature which fishermen use as bait and which occurs among oak trees (ἐν ταῖς δρυσίν)' (the definition s.v. ἰληοί is abbreviated from this). Elsewhere (s.v. εἰλύιος) the two conflicting definitions are combined to give us a θηρίον ἀπὸ φρυγάνων σκωληκοειδές ᾧ χρῶνται πρὸς δέλεαρ. There is also evidence of confusion (s.v. ἐλειούς and ἰληοί) with the rodent known as ἐλείος μῦς, confusion which provides the probable explanation for Eustathius' (Comm.in Il. 295.33) 'type of fly (εἶδος τι μυιῶν) which is known as ἐλείος'. The Latin illa (hilla Lindsay), preserved only in C.Gl.L. II.77.10, where it is defined as σκώληξ κλεινης [sic], has been equated with this word.[90]

(vi) Ἔμβολος. The name, according to Hesychius, of a pest of vegetables (εἶδος θηρίου ἐν λαχάνοις).[91] Fernandez[92] sees the word as containing an allusion to the creature's physical appearance.

(vii) Κασιοβόρος. In Hesychius, the name of a form of larva believed to be generated in the exotic cassia tree (ἐν κασίᾳ γινόμενος σκώληξ). Possibly this is the larva of some wood boring beetle,[93] as Gossen[94] suggests, followed by Fernandez.[95]

(viii) Κόλυμβος. The term 'diver' in a zoological context is normally applied to a species of bird. Hesychius is unique in stating that it can also refer to what he describes as ζῴφια ἐν κολυμβήθραις. It is possible that, as Gossen suggests,[96] these should be identified with the various species of water beetle (Dytiscidae etc.)

(ix) Κυνόπρηστις. Κυνόπρηστις, which Hesychius defines in a distinctly uninformative way as ζῷόν τι, is clearly parallel with βούπρηστις[97] though not necessarily synonymous, as Fernandez suggests.[98] It therefore suggests some venomous creature which was believed to be eaten by dogs and subsequently to cause them to swell up and die.[99]

---

[88] Fernandez, Nombres, p..146, LSJ suggest 'woodworm' as a definition (cf. 30 above).
[89] Cf. 23.1.2, 48.2 above.
[90] C.Gl.L. index and Th.L.L. sv.
[91] Cf. the cabbage white butterfly (23 above).   [92] Op. cit., p. 35.
[93] Cf. 30 above.   [94] Art. cit., p. 46.   [95] Op. cit., p. 140.
[96] Art. cit., p. 58. Cf. Fernandez, op. cit., p. 157. Cockroaches (13 above) and house crickets (12 above) are referred to as inhabiting bath houses, but the name suggests an aquatic creature.
[97] 34 above.
[98] Op. cit., p. 137. Gossen, art. cit., p. 64, suggests a louse.
[99] As it is defined by LSJ.

(x) Λάθαργος. Defined by Hesychius as a form of, or a synonym of, *skolex*.[100] Gossen[101] attempts to identify it as a species of caterpillar.

(xi) Λυρίτης. According to Hesychius, this is the name of 'a certain creature generated by oak trees (ζῷον τί ἐν ταῖς δρυσὶν ἐντίκτον). It has been suggested that the true reading may be δρυίτης.[102]

(xii) Μυρτίλωψ. Μυρτίλωψ, another of Hesychius' vaguely defined 'certain animals' is considered by Strömberg[103] and Fernandez[104] to refer to an insect. The former regards the word as a compound of μύρτος 'myrtle' and λέπω 'to peel or remove bark from', and the insect concerned therefore as a pest of this plant. This is likely enough, but to go on from this point to identify it with a particular species, in this case the stag beetle,[105] is to build too much on insufficient data.

(xiii) Ξίφος. Presumably named from its sword-shaped appearance, Hesychius defines this creature under the heading Τὸ κατὰ γᾶς ξίφος as 'a small animal resembling a centipede' (θηριδίον σκο.λοπένδρᾳ ὅμοιον).[106]

(xiv) Συβώτας. Gossen,[107] followed by Fernandez[108] takes this word, in view of its literal meaning of 'swineherd', to refer to some external parasite, a louse or tick[109] of domestic pigs. It is another of Hesychius' 'certain animals'.

(xv) Σφηκαλέων. The 'wasp-lion' is mentioned only in one of the magical papyri (*Pap. Mag.Leid.* VIII.5), where it is prescribed as the chief ingredient in a love potion. The creatures are described as being found in spiders' webs (σφηκαλέοντες τοὺς ἐν τῇ ἀράχνῃ), and it is stated that after capture they should be ground up and given to drink. In view of the obvious parallel with μυρμηκολέων,[110] the 'wasp-lion' may either have been a creature believed to prey upon wasps, or a variety of wasp considered to have some resemblance to a lion.[111] The reference to spiders' webs, however, is not readily explicable. Dieterich[112] takes it to be a form of wasp so named because of its fierceness, like the spiders known as λύκοι.[113]

(xvi) Ὑάλη. This term, defined as σκώληξ by Hesychius, is evidently a dialectical form of εὐλή, a worm or fly larva[114] as LSJ[115] point out. Hesychius also lists a verb ὑάλεται, defined as σκωληκιᾷ.

(xvii) Ὑλομήτρα. The ὑλομήτρα or ὕλης μήτηρ (Hesychius gives both forms) is defined as a type of *skolex* or alternatively as a synonym of *bombyx*.[116] The word is similar in form to τεττιγομήτρα[117] and ἐχινομήτρα (large sea urchin) and is

---

[100] Cf. 23.1.2 above.   [101] Art. cit., p. 67.
[102] Latte, ad loc. Gossen, art. cit., p. 71, sees this word as a wood boring larva (cf. 30 above).
[103] *Gr. Wortstudien*, p. 20.   [104] Op. cit., p. 140.
[105] Cf. 30.1 above.   [106] 3 above.   [107] In PW Supp.VIII.356.
[108] Op. cit., p. 177.   [109] Cf. 22.5 above.   [110] 61 above.
[111] M. Wellmann, art. cit. under 61 above, p. 37, takes the two names as parallel formations.
[112] A. Dieterich, *Abraxas*, Leipzig, 1891, p. 192.   [113] 7b.1.3.vii above.   [114] 48.2 above.
[115] Also Fernandez, op. cit., p. 146. The word has nothing to do with glass, as Gossen, *Zool. Glossen*, p. 118, assumes.
[116] Cf. 27.1.2.v, 42 above.   [117] 16.2 above.

evidently a popular name.[118] Fernandez[119] sees the first element in the compound as related to the preceding and to εὐλή etc. rather than as having any connection with ὕλη, as Strömberg[120] suggests.

(xviii) *Χιλαάγρα*. Defined by Hesychius merely as ζῳύφιον τι, this term is taken by Gossen[121] to refer to an insect.

72. Names listed by Polemius Silvius

The catalogue of names of insects and other creatures compiled by the early fifth century author Polemius Silvius[122] has been considered in detail in an article by A. Thomas.[123] The section with which we are concerned here is entitled *Nomina insectorum sive reptantium*, and consists of a bare enumeration of some sixty names arranged in no particular order. Those which are identifiable are, with three exceptions, all terrestrial insects or other arthropods (plus the earthworm). However, the list's main interest lies in its inclusion of a large proportion of words which are either not attested elsewhere at all, or else are preserved only in one or two other sources. The spellings given for these are not necessarily reliable, judging by the form in which some of the known words appear.[124]

Thomas[125] argues that there is no reason to suppose that the catalogue has been interpolated subsequent to Polemius' time (indeed some very common insect names are absent from it), but that it has been reliably compiled by Polemius from both written and contemporary oral sources. At least some of the names included are not native Latin but Gallic words,[126] Polemius being an inhabitant of Gaul.

Apart from those dealt with elsewhere,[127] the following unusual words are listed: *ablinda*,[128] *acina*,[129] *asio, cabarus, gristus, liscasda*,[130] *lucalus, minerva, musomnium*,[131] *petalis*,[132] *piralbus, popia, ruscus, sexpedo*,[133] and *stillo*.

---

[118] Fernandez, op. cit., pp. 190–1.

[119] Ibid.

[120] *Gr. Wortstudien*, p. 23.

[121] Art. cit., p. 125.

[122] *Monumenta Germanica Historica, Auctores Antiquissimi* IX, *Chronica Minora* 1.2., pp. 543–4.

[123] A. Thomas, 'Le Laterculus de Polemius Silvius et le Vocabulaire Zoologique Roman', *Romania*, XXXV, 1906, pp. 161–97.

[124] For example *oester* for *oestrus*, *pedusculus* for *pediculus*.

[125] Art. cit., p. 168.

[126] For example *delpa* (23 note 8 above).

[127] Viz. *bubo* (11.1.2.xxv above), *cervus* (30 note 215 above), *corgus* (36 note 125 above), *delpa, laparis* (35 note 106 above), *lanarius* (27 note 137 above). *Ficarius* is presumably *culex ficarius* (46 above).

[128] This is probably a reptile: cf. Thomas, art. cit., p. 168.

[129] Related by Ernout and Meillet, *Dict. Ét.*, p. 6, to *acinus*.

[130] Cf. Th.L.L. s.v.

[131] Cf. Latin *musmo, musimo*.

[132] Cf. Greek πετηλίς (11.1.2.xvi above).

[133] Thomas, art. cit., p. 167, suggests this refers to the ant.

INDEX

1. Greek

ἀγριομύρμηξ   178 n. 132, 200
ἀγρώστης   35, 47, 52
ἄδιγορ   79
αἰγιόνομος   255
αἰλουρόμορφος   159
ἀκανθίας   94
ἀκαρί   34, 60–1, 118
ἀκατίς   16 n. 41
ἀκορνός   64, 66
ἀκρίς   62–79, 82–3, 85–8, 91–2, 103, 105 n. 73,
   227 n. 40, 252
ἀλακάται   2, 189
ἀλίβας   66
ἀμφισδεσφάγανον   12
ἀνθηδών   188, 239
ἀνθρηδών   188, 195
ἀνθρήνη   187–96, 220
ἄρασιν   35
ἀράχνη   39–45, 51, 55
ἀράχνηξ, ἀράχνης   35
ἀράχνιον   35–6, 38, 45 n. 55
ἀραχνός   34
ἄρκυμα   66
ἀρουραία   86–7
ἀσείρακος, ἀσίρακος   67–8, 77
ἀσκαρίς   231–2
ἀστακός   68
ἀστάλη   255
ἀστέριον   47
ἀττάκης, ἄττακος, ἀττακύς   68
ἀττάλαβος, ἀττέλαβος, ἀττέλεβος   62–6, 69–70,
   79
ἀχέτας   92, 94–5, 100
ἀχραδίνης   183
ἄχωρ   114
ἄψοος   134

βάβακος   95
βαῖτυξ   5
βασκανία   86
βδέλλα   4–10
βέμβιξ, βεμβίς   195–6
βερκνίς   65
βλέτυες   5
βομβύκιον   141, 197–8
βομβύλιος   141–3, 195–8, 231

βομβυλίς   141–3
βόμβυξ   140–9, 195 n. 44, 197, 198 n. 59, 257
βόρμαξ   199
βόστρυχος   177
βουδάκη   174
βούπρηστις   148, 168, 171, 173–5, 185, 256 n. 97
βοῦς, ξυλοφάγος   153
βουτύπος   227, 231
βραύκη, βραῦκος   65
βραύλα   114
βρέκος   65
βρέττανα   66
βρόκος, βρούκα   65
βροῦκος, βροῦχος   63, 65–6
βρύτον   158 n. 15
βρύχος   65
βρωστήρ, βρωτήρ   137
βύρμαξ   199
βύρρος   158

γαφάγας   2
γῆς ἔντερον   1–4
γραῦς σέριφος, σερίφη   86, 87 n. 117, 252 n. 68

δάπτης   231 n. 64
δέλλις   189
δεμβλεῖς   5
δεμελέας   5
δερμηστής, δερμιστής   138
δήξ   152
δίκαιρον   160 n. 30
δίκηλον   255
δόρκα   113
δρίλαξ   5
δρῖλος   2, 5

ἑαρίς   169
εἰλύιος   255–6
ἐλείος   255–6
ἕλμινς   1, 221 n. 17, 247, 255
ἔμβολος   256
ἔμβρυλλαι   2
ἐμπίς   89 n. 125, 123 n. 25, 229–36, 252
ἔνθριον   255
ἔντερον γῆς   1
ἐξυδρίς   107

ἔρπηλα 11
ἐρπυλλίς 95
ἐρυσίβη 69, 178
εὐλή 151, 221, 255, 257–8
ἐφήμερον 88–9, 236–7, 248

ζειγαρά 95

ἡλιοκάνθαρος 159
ἡλιοκεντρίς 50–1
ἡμερόβιον 89
ἠπίαλος 130 n. 55
ἠπίολος 130
ἠχέτης 92

θήραφος 35
θρίψ 104 n. 69, 133, 137–8, 149–52, 155, 177 n. 120, 181–4, 245, 247, 255

ἰβιόμορφος 159–60
ἴληος 255–6
ἴξ 133
ἴουλος 2, 11, 13–19
ἱππεύς μύρμηξ 199–200
ἵππουρος 239
ἰχνεύμων 189
ἴψ 14 n. 33, 104 n. 69, 132–5, 137, 139, 177 n. 119, 246

καλαμαία 86–7
καλαμαῖον 94–5
καλαμίς 95
καλαμῖτις 67
καματερή 36
κάμπη 122, 124–8, 130, 133–6, 141, 143, 149, 170–1, 177
κάμπη, πιτυίνη 148
κανδήλα 176 n. 109
κανδηλοσβέστης, -έστρια 130
κανθαρίς 52–3, 135, 148, 157 n. 4, 158, 166 n. 47, 168–74, 176, 184
κανθαροειδής 52, 54
κάνθαρος 79, 111, 144, 153–4, 157–65, 166 n. 47, 168–9, 171 n. 84, 176 n. 114, 183
κάνθαρος, Αἰτναῖος 158, 160
κάνθαρος ἡλιακός 159, 163
κάνθαρος σεληνιακός 159, 163
κάνθων 158
κάρ 114
κάραβος 149–56, 158
καράμβιος 153, 158
κάρνος 114
κάρον 64, 66–7
κασιοβόρος 256
καταχήνη 87
κεντρίνης 215

κεραίς 137 n. 94
κεράμβηλον 154, 215
κεράμβυξ 153–4
κεράστης 149–56
κέρκα 67, 94
κέρκος 134
κερκώπη 93–5
κεφαλοκρούστης 53
κηροδύτης 60
κικνίον 114
κίκους 95
κίλλος 95
κίξιος 95
κίς 104 n. 69, 133, 177–80, 182–3, 200, 245
κλῆρος 43, 123, 186
κνίψ 104 n. 69, 133, 154, 183, 215, 235, 245–6, 251, 253 n. 78, see also σκνίψ
κνώψ 246 n. 37
κόβαρος 16
κοίελος 180
κόκκος 47, 108–11
κολεόπτερος 157
κόλυμβος 256
κόνις 105, 113–14, 241
κόρις 104–6, 114, 137, 177 n. 119
κόρνοψ 64, 66, 75
κορνώπιδες 64, 231 n. 64
κουβαρίς 16
κραμβίς 126, 243
κρανοκολάπτης 24, 47, 53–4, 131–2
κροτών 56–60, 238
κύαμος 15
κυάνεον 47
κυνακρίς 79
κυνάμυια see κυνόμυια
κυνόλφη 83
κυνόμυια 220 n. 11, 226, 238–9
κυνόπρηστις 256
κυνοραιστής 59
κυσολαμπίς 175–6
κύων 220
κῶβαξ 95
κώνωψ 64, 109, 123 n. 25, 178 n. 132, 215–7, 229–38, 246, 248, 252
κωριδάμνας 66

λαέρτης 200
λάθαργος 257
λακέτας 93
λάκκος 111
λάμπουρις, λαμπυρίς 175–7
λεοντομύρμηξ 250
λιγάνταρ 95
λύκος 34, 36, 45, 47, 51–2, 257
λυρίτης 257

μάντις 85–8
μαργαρίτης χερσαῖος 111–12
μάσταξ 25, 64, 65–6
μέλισσα, ἀγρία 64, 188, 196
μελουρίς 81, 83
μέμβραξ 93
μεμφίδες 123
μεταλλεύς 200
μηλάνθη 165
μηλολάνθη, μηλολόνθη 107, 157 n. 4, 164–8, 176 n. 114, 198 n. 60
μηλόνθη 165
μίδας 180
μολουρίς, μολυρίς 81, 83
μονήμερον 89
μούια 219 n. 2, 224
μῦα 217, 219 n. 2
μυῖα 51, 71, 166 n. 50, 198, 216–17, 219–26, 228, 229 n. 55, 238–9, 256
μυλαβρίς, μυλακρίς 78, 81–4
μυληθρίς 83–4
μύλοικος 82, 84
μυριόπους 15
μύρμαξ 199
μυρμηδών 199
μυρμήκειον 48
μυρμήκιον 200–1
μυρμηκοειδές 48
μυρμηκολέων 249–51, 257
μύρμηξ 198–209, 249, 252
μύρμηξ ἡρακλεωτικός 48
μύρμηξ χρυσωρύχος 209–11
μύρμος 199
μυρτίλωψ 257
μύωψ 166 n. 50, 188 n. 7, 223, 225–9, 233, 238

νεκύδαλος 142–3
νίρμος, νίρνος 114
νύμφη 192, 201, 252

ξίφος 257
ξυλοφάγον 151
ξυλοφόρον 148–9

οἰνοκώνωψ 236
οἶστρος 223, 225–9, 232–3, 238
ὀκορνός 64, 66
ὀλίγιας 67
ὀλκός 52
ὄνιννος 15
ὀνίσκος 13–19
ὄνος 13–19, 68
ὄνος ἰσόσπριος 15
ὀξύγη 69
ὀρειβάτης 52
ὅρμικας 199

ὄρπας 67
ὀρσοδάκνη 243
ὀφιοκτόνη 12
ὀφιομάχης, -ος 12 n. 22, 68–9, 250

πάρνοψ 63–6, 75, 78, 189, 231 n. 64, 252
πεμφρηδών 93, 195
πετηλίς 67, 258 n. 132
πετηνίς 105
πηνίον 123
πιθήκη 52
πιτυοκάμπη 148, 171, 194
πολύπους 15
πόρνοψ 64, 75
πρανώ 64
πρασοκουρίς 243–4
πυγολαμπίς 175, 176 n. 111
πυραύστης 130, 186 n. 181
πυρίβια 248
πυριγόνος 79, 247–8
πυριλαμπίς, πυρολαμπίς 175–6

ῥάγιον 47
ῥάξ 47
ῥόμοξ, ῥόμος 152
ῥώξ 47, 55

σάβηττος 231
σάθραξ 114
σακοδερμηστής 138
σαλαμίνθη 36
σαλαφίον 185
σάραξ 138
σειρήν 190, 195, 198
σέρφος 86, 201, 251–2
σηνίκη 16
σήρ 143
σήραμβος 158
σής 61, 104, 122, 131, 133, 136–40, 150, 177 n. 119, 178, 182–3
σητοδοκίδες 122
σήψ 12
σιβρίται 33
σιγαλφός 95
σίγιον 94–5
σίλφη 78, 80–5, 107, 159, 185
σινίδες 243 n. 8
σίφων 48
σκήν 123, 186 n. 183
σκίψ 246
σκληροκέφαλον 54
σκληρός 186
σκνίψ 105, 133, 177 n. 119, 178 n. 132, 183, 200, 215, 235–6, 238, 245–6, see also κνίψ
σκολόπενδρα 10–14, 257
σκορόβυλος 158

σκορπιομάχος 68
σκορπίος 21–34
σκορπίος κοινός 29
σκορπίος πτερωτός 24–5
σκορπιῶδες 34
σκρόφα 17 n. 43
σκυταλίδες 122
σκυταλωτούς 122
σκωλήκιον 26, 38, 54, 108, 131, 148, 153, 157 n. 3,
    161, 166, 186, 201, 221, 237
σκώληξ 1–2, 14, 70–1, 82, 96, 108, 110, 122–8,
    133–4, 136, 138–9, 141, 143, 150–6, 167, 170–1,
    176, 178, 180–2, 186, 190, 192, 220–1, 223–4, 241,
    247, 252, 254–7
σκωλοβάτης 178
σοίκιδες 231
σπονδύλη see σφονδύλη
σταφυλῖνος 173, 185
στέρφνος 252
στιτθόν 67
στρατιωτίς 220
συβώτας 257
σύρφος 86, 251–2
σφάξ 187
σφηκαλέων 257
σφήκειον 48
σφηκίον 189, 194
σφήξ 187–96
σφήξ, ἄγριος 188, 190–1
σφονδύλη 12, 184–5
σχαδών 192 n. 31

ταυροειδής, ταυρόμορφος 159, 160 n. 26
τέκτων 45
τενθρηδών 195–6
τενθρήνη 196
τερηδών 130, 150–1, 182–3, 186, 245
τετραγνάθον, -ος 45, 49–51, 55
τετραπτερυλλίς 67
τεττίγιον 94
τεττιγομήτρα 96, 102, 257
τεττιγόνιον 92, 94–5, 100
τέττιξ 71–2, 78 n. 77, 91–103
τεφράς 93, 95
τίλφη 81
τιτιγόνιον 94

τίφη 81–2, 106–7
τοξαλλίς 79
τριγόνιον 94–5
τριξαλλίς 79–80
τριξέλλας 79
τριοπίς 67
τριχόβρως 138
τριχοτρώκτης 138
τρώξ 79, 180
τρωξαλλίς 64, 72, 78–80

ὑάλη 257
ὕλης μήτηρ 257–8
ὑλοδρόμος 52
ὑλομήτρα 257–8
ὕπερον 123
ὕσπληξ 227 n. 40

φαλάγγιον 22, 26, 34–6, 38–9, 44–56, 173, 189, 208
φάλαγξ 35, 43, 45
φάλλαινα 43, 123, 129–32
φάλλη 129
φαρμακίς 67, 87
φερέοικος 189
φθείρ 15, 57, 60, 112–20, 133–4, 177 n. 119, 178,
    235, 253
φθείρ, ἄγριος 113–14

χαλκῆ μυῖα 166 n. 50
χαλκομυῖα 166 n. 50
χαμαισκώληξ 2
χελωνίας 170
χιλαάγρα 258
χρυσαλλίς 83 n. 105, 122, 125, 149, 166
χρυσοκανθαρίς 166
χρυσοκάνθαρος 80 n. 97, 165–6
χρυσολαμπίς 175–6
χρυσομηλολόνθιον 166

ψήν 212–6, 245
ψύλλα 45, 52, 240–2, 253
ψύλλαξ, ψύλλος 240
ψυχή 121–9, 130 n. 55, 142
ψώμηξ 167
ψώρα 130

2. Latin

ablinda 258
acina 258
anima, animula 127 n. 36
apis 188
araneola, -us 35
araneus 34–45, 48, 50, 142 n. 135, 173, 186
araneus muscarius 37
asellus 15–16
asilus 225–9
asio 258
attacus 68
attelabus, attelebus 63–4
avenaria 92

bambis 142 n. 135
bibio, bibo 236–7
biurus 255
blatta 80–5, 157 n. 5
bombites 142 n. 135
bombycini 142 n. 135
bombylis 141
bombylius 141
bombyx 140–9, 197
bruchus 63–5
bubestris 174
bubo 69, 258 n. 127
bubrostis 174
bufo 69
bullus 166 n. 52
bumbix 142 n. 135
buprestis 173–5

cabarus 258
campe 134
cantareda, cantharida 169
cantharis 168–73
centipeda 12, 16
centipedium 12
centipes 10–13
centrinae 215
cerastes 153
cervus 154 n. 215, 258 n. 127
chantari 79
chrysallis 122
cicada 91–103
cicindela 175–7
cimes 105 n. 72
cimex 13, 104–6, 246
cinifes, ciniphes 235
claros 186 n. 182
coccus 108–11
coenomyia 239

conculio 178
contifex 12
convolvolus 14 n. 33, 132–5
corgus 178 n. 125, 258 n. 127
cossis 152
cossus 149–56, 254 n. 80
crabro 48, 187–95
cufo 69
culex 183, 216–7, 229–39
culex ficarius 212–6, 258 n. 127
curculio 122 n. 13, 177–80, 200
curcurio 178
cusculium 108
cutio 16, 19
cynomyia 238

delpa 122 n. 8, 258 n. 126
dolva 122 n. 8

eruca 122, 134, 148, 151, see also *uruca*
erudo, erugo 5

falangium 45
ficarius 258 n. 127
formica 198–209
formica Herculanea 200
formica Indica 209–11
formicoleon 250
frumentaria 92
fullo 166, 168

galba 254
gesentera, -us 1
grillus 79, 81
gristus 258
gryllus 78–80
gurgulio 178

herudo, herugo 5
hirudo 4–10

illa 256
involvolus, involvus 134
irudo 5
iulus 14

lampyris 175
lanarius 143 n. 137, 258 n. 127
laparis 175 n. 106, 258 n. 127
lendis 113
lens 113, 116
lindines 113

liscasda 258
locusta 62–78, 157 n. 2
lucalus 258
lucanus 154 n. 214
lucavus 149, 154
luciculia 176
lumbricus 1–4
lupus 51
lycos 52

milipeda 11, 13–19
minerva 258
mulio 239–40
multipeda 11, 13–19
multipes 16
musca 51, 79, 170, 176, 180, 193, 219–25, 236–9
musca canina 238
muscio 237
musomnium 258
mustio 236–7
myloecus 82

necydallus 142–3
nepa 22

oestrus 226
oniscus 15
ophiomachus 69
opinacus 69

papilio 53, 121–32, 143–4, 186
papiliunculus 126
pediculus 112–20, 160, 178, 253 n. 77
pedis 113
peduclus, peduculus 113, 178
petalis 258
petaurista 61
phalangium 35, 44–56
phryganion 255
piralbus 258
porcellio 15–17
popia 258
prasocoris 243
pseudosphex 190, 198 n. 64
pulex 240–2, 253
pyrallis 248
pyrotocon 248

rauca 254
ricinus 56–60
ruscus 258

salpiga, salpinta, salpuga 50
sanguisuga 5
scabro 188
scarabaeus 16, 79, 81, 157–65, 173, 176, 184, 188 n. 6
sciniphes 104, 235, 246
scolecium 108
scolopendra 11
scorpio 21–34
seps 12
ser 143
serpens 12, 50, 184
sexpedo 258
sfalangium 45
solifuga 50
solipaga 50–1
solipuga 49–52
solipugna 50
sphalangium 45
sphondyle 184
stillo 258
surcularia 92

tabanus 225–9, 238 n. 91
tarmes 151–2, 154, 182, 224
tarmus 224
taurus 160, 164
teredo 131, 137, 151–2, 180, 183, 186, 224
termes see *tarmes*
terrae intestinum 1
tetragnathion 49
thrips 183
timulus 107
tinea, tinia 5, 83 n. 105, 84, 113, 122, 126, 131, 136–40, 149, 178, 181–4, 186, 224, 230 n. 56
tippula 106–7

uruca 122, 124–5, 127–8, 141, see also *eruca*
usia 119

vappo 127 n. 36
vermiculus 1, 26, 38, 48, 96, 105 n. 72, 108–9, 122, 125–7, 136, 141, 145, 150, 152, 156, 167, 171, 180, 183, 221, 227, 229, 237
vermis 1–2, 16–17, 27, 38, 107, 115, 122, 131, 137, 139, 141–2, 147 n. 162, 151–2, 154, 167, 170, 178, 180, 221, 223–4, 248 n. 46, 254
vespa 187–95
volucre 59, 61, 134

zinyphes 200
zinzala 231

3. English

Abraxas grossulariata 123
Acanthoscelides obtectus 180
Acari 56–60, *see also* mite; tick
Acarus siro 60
Acherontia atropos 127 n. 40, 129 n. 53
Acheta domesticus *see* cricket, house
Achroia grisella 129, *see also* moth, wax
Acrididae 62, 70 n. 54, 71
Aedes spp. 229
Agaonidae 212–6
Agelana spp. 37, 51
Aglais urticae 126
Agriotes spp. 167
Amphimallon solstitialis 167
Anacridium aegyptium 73 n. 66
Andrenidae 198 n. 63
Andricus kollari 217
Androctonus spp. 21, 23–4
Anobiidae 140 n. 115, 177, 181
Anobium punctatum 181
Anopheles spp. 229, 232 n. 70, 234, 235 n. 77
Anoplura 112–20
Annelida 1–10
ant 39, 48, 50, 56, 80, 99, 102, 177, 178 n. 132, 191, 195, 198–209, 245, 249–52, 258 n. 133
ant, gold-digging 209–11, 249
ant, velvet 49
Antheraea spp. 141
Anthocoridae 104
ant-lion 200, 249–51
aphid 216–7, 246, 253
Apis dorsata 196
Araneus diadematus 37
Argiopidae 37
Armadillidiidae 13, 15
Asilidae 240

bag-worm 149
bed-bug 7–8, 13, 104–6, 112, 240–1, 246
bee 29, 39, 43, 60, 71, 84, 99–100, 129–32, 157 n. 5, 186, 188–9, 191–8, 201, 219 n. 4, 225 n. 34, 239–40, 245
bee, mason 197
bee-louse 60
beetle 16, 79–82, 138, 140 n. 115, 149–86, 198, 227, 249 n. 50, 256
beetle, blister 52–3, 157 n. 2, 166, 168–74
beetle, click 167
beetle, death-watch 79, 181, 184
beetle, dung 16, 157–64, 171 n. 84
beetle, flea 243, 253
beetle, furniture 181, *see also* Anobiidae

beetle, grain 177
beetle, leaf 171
beetle, long-horn 150, 152–5, 215
beetle, oil 173–5, 185
beetle, rove 185
beetle, soldier 81 n. 97, 169
beetle, stag 149–54
Belostomatidae 25
Bembix rostrata 190, 198 n. 63
Biorrhiza pallida 217
Blaps spp. 80, 82, 185
Blastophaga psenes *see* fig-wasp
Blatta orientalis 80 n. 90
Blattella germanica 80 n. 90
bloodworm 230, 232
bluebottle 219–20
Bombus spp. 197–8
Bombyx mori 141, 146–8
booklouse 113, 136–40
Boreus hyemalis 247
Bostrychidae 181
Brachinus spp. 185 n. 175
Bradyporinae 68
Braula coeca 60
Bruchidae 180
Bruchus pisorum 180
bumblebee 195 n. 46, 197–8
Buthacus arenicola 21
Buthus spp. 21, 23
butterfly 121–9, 142, 243

caddis-fly 149, 255
Calliphora spp. 219–20, 224
Calliphoridae 219–25
Campanotus herculeanus 200
Cantharis spp. 81 n. 97, 169
caterpillar 73, 122–9, 132–6, 148, 151, 186, 244, 253–4, 257
Catharsius spp. 158–9
Cecidomyidae 216
centipede 10–13, 105, 184, 257
Cephidae 190
Cerambycidae *see* beetle, long-horn
Cerambyx cerdo 150, 152
Ceratopogonidae 230
Cetonia spp. *see* rose chafer
Chalcidoma spp. 197
Chilopoda *see* centipede
Chiracanthium punctorium 44
Chironomidae 223, 230–2
Chrysomelidae 171, 243, 253
Chrysops spp. 225

cicada   62, 71–2, 78 n. 77, 91–103, 176
Cicada orni   92
Cicadatra atra   92
Cicadetta montana   92
Cimex lectularius   see bed-bug
Cleridae   186
Coccinellidae   *see* ladybird
Coccoidea   108–13
cockchafer   82 n. 105, 100, 157 n. 2, 160 n. 24, 164–8
cockroach   78, 80–5, 107, 157, 159, 169, 185, 256 n. 96
Coleoptera *see* beetle
Collembola   61, 247
Colletidae   198 n. 63
Copris spp.   158, 160–1
Cossus cossus   see moth, goat
cranefly   167
cricket   62, 64, 66–8, 71–2, 78–81, 91–2, 100, 103, 157 nn. 2 & 5, 184
cricket, bush   62, 63 n. 7, 64–5, 68 n. 48, 69, 70 n. 54, 71–2
cricket, field   78–9
cricket, house   78–9, 81, 82 n. 98, 92 n. 8, 256 n. 96
cricket, mole   67, 69, 78–9, 244
Cryptolestes spp.   178 n. 124
Ctenocephalides spp.   240, 242
Cuclotogaster heterographus   120
Culex pipiens   229 n. 55, 230, 232
Culicoidea   229–36
Culiseta spp.   230
Curculionidae   *see* weevil
Cydia pomonella   124
Cynipidae   216–7
Cynips gallae-tinctoriae   216

Dacus oleae   136
Damalinia spp.   199
Dasyphora cyanella   220 n. 12
Dermacentor spp.   57, 59
Dermestes spp.   138
Dictyoptera   80–8
Diplopoda *see* millipede
Diptera   136, 167, 190, 216, 219–40, *see also* fly; maggot
Dociostaurus maroccanus   62, 77
dragonfly   25, 67 n. 36, 89 n. 125, 239–40
Drosophilidae   *see* fly, vinegar
Dytiscidae   256

earthworm   1–4, 14, 30, 189
Ephemeroptera   *see* mayfly
Ephippiger ephippiger   71, 73 n. 66
Ephippigerinae   68
Erannis defoliaria   123
Ergates faber   150, 152, 156

Eumenidae   187, 198 n. 64
Eupoecilia ambiguella   132–4
Euproctis chrysorrhoea   125
Euscorpius spp.   21, 23, 29

false-scorpion   34, 61
Fannia spp.   219, 220 n. 15
fig-wasp   154, 212–6, 245
Filistata insidiatrix   51
firefly   175–7
flea   61, 105, 112, 177, 220 n. 14, 240–2, 246
fly   39, 50–1, 57, 119, 171 n. 81, 184, 190, 193, 198, 219–29, 236, 238–9, 248, 256–7
fly, black   230, 235 n. 79
fly, flesh   219, 221
fly, gad   194, 225–9
fly, horse   166 n. 50, 219, 222–3, 225–9, 238
fly, house   219, 221–4
fly, olive   136
fly, sand   230
fly, soldier   225 n. 34, 227, 239
fly, vinegar   89, 229 n. 55, 230–1, 236–8, 246
froghopper   97

Galeodes spp.   49, 50 n. 87
gall, gall-wasp   170, 214, 216–7
Galleria mellonella   129, *see also* moth, wax
Gastropacha quercifolia   125
Geometridae   122–3
Geotrupes stercorarius   157–8, 165–6
Gerris lacustris   107
Glomeris spp.   13, 15
glow-worm   157, 165, 175–7, 201, 242
Gluvia dorsalis   50
Glycyphagus domesticus   60 n. 120
gnat   222, 228, 229–37, 252
Gnathocerus cornutus   178
Gonepteryx rhamni   126
Goniodes gigas   120
Gordius spp.   3
grasshopper   62–78, 91–2, 94, 100, 103
greenbottle   219
greenfly   246, 253
Gryllidae   *see* cricket
Gryllotalpa spp.   *see* cricket, mole
Gryllus campestris   see cricket, field
Gymnopleuris spp.   158
Gyrinidae   107 n. 75, 227

Haemaphysalis punctata   57–8
Haematopinus spp.   119
Haltica spp.   243, 253
harvestman   51 n. 90
Heliocopris spp.   158–9
Hemilepistus reaumeri   19

Hemiptera   25, 91–112, 216–7, 246, 253
Hesperophanes spp.   150 n. 175, 153
Hippobosca equina   57
Hippoboscidae   57, 119 n. 119
Hirudinea   *see* leech
Hirudo medicinalis   4, 9
hornet   187–95, 239
horntail   150
horsehair worm   3
Hydrometra stagnorum   107
Hyles euphorbiae   125, 129
Hyles lineata   132–4
Hylobius spp.   181 n. 147
Hylotrupes bajulus   150
Hymenoptera   81, 93, 141–2, 150, 187–217
Hypoderma spp.   225
Hypselogenia spp.   158

Inachis io   126
Isopoda   13–19
Isoptera   *see* termite
Isotoma spp.   247
Ixodes ricinus   57–9
Ixodoidea   *see* tick

Kalotermes flavicollis   199
Kermes vermilio   108–11

Laccifer lacca, lac insect   111
ladybird   170–1
Lampyridae   175–7
Lampyris spp.   175–7
Lasius spp.   202
Latrodectus spp.   44–7, 52–3, 55
leech   4–10, 55, 105–6
Leiurus quinquestriatus   21, 24
Lema melanopa   171
Lepidoptera   67, 121–56, 177, 243–4, 246, *see also* caterpillar; moth
Lepisma saccharina   137 n. 96
Ligia spp.   15
Limnatis nilotica   5, 9
Linognathus spp.   119
Lipeurus caponis   120
Lithobius forficatus   10
Lobesia botrana   132–4
locust   24–5, 62–79, 96, 122, 189
Locusta migratoria   62, 63 n. 7, 77
louse   57–8, 105, 112–20, 137 n. 99, 138 n. 103, 214, 220 n. 14, 223, 235, 240–1, 257
Lucanidae   *see* beetle, stag
Lucanus cervus see beetle, stag
Lucilia spp.   219
Luciola spp.   175–7
Lumbricidae *see* earthworm
Lycosa tarentula   42, 52
Lycosidae   36–7, 51–2

Lyctidae   181
Lytta spp.   168–73, 174

maggot   117, 151, 221, 223, 257
Malacosoma neustria   124 n. 31, 125, 136
Mallophaga   112–20
malmignatte   *see Latrodectus* spp.
Mamestra brassicae   121 n. 2
mantis   65, 67, 85–8
Mantis religiosa   85
Margarodes spp.   112
Margarodes hameli   110
mayfly   88–9, 232, 240 n. 107
mealworm   81 n. 92, 178
Megachilidae   197, 198 n. 63
Melanophila spp.   249 n. 50
Melophagus ovinus   57, 59, 119
Meloe spp.   173–5
Meloidae   168–75
Melolontha melolontha   see cockchafer
Melolonthinae   164–6
Menacanthus stramineus   120
Menopon gallinae   120
Messor spp.   202
midge   229–30
millipede   12–19
mite   60, 118 n. 117, 224, 253
mite, cheese   60
mite, harvest   53, 119
mite, scabies   118
mosquito   193, 229–38, 252
moth   53–4, 121–36, 149, 177, 186, 248, *see also* silkmoth
moth, clearwing   149
moth, clothes   34, 61, 84, 113, 133 n. 69, 136–40
moth, goat   149, 151–2
moth, hawk   54, 125, 127 n. 40, 129, 132
moth, leopard   149, 156
moth, processionary   148
moth, wax   43, 129–32, 137 n. 92, 186
Musca autumnalis   220 n. 12
Musca domestica   see fly, house
Muscidae   219–25
Mutilla europea   49
Mutillidae   48–9
Mylabris spp.   53, 168–74
Myrmarachne formicaria   48
Myrmeleontidae   251

Nemapogon granellus   177
Nereidae   11
Neuroptera   251
Noctuidae   131 n. 62
Nymphalidae   126

Ocypus olens   185
Odonata   *see* dragonfly
Odontotermes obesus   199
Oestridae   221, 225
Onitis spp.   158, 160
Onthophagus spp.   158–9
Operophtera brumata   123
Opiliones   51 n. 90
Orthoptera   25, 62–80, 153 n. 202, *see also* cricket; grasshopper; locust
Oryctes nasicornis   156 n. 219, 158 n. 10
Oryzaephilus surinamensis   177, 178 n. 124

Pachypasa otus   140–1, 146–7
Pagiphora annulata   92
Palingenia longicauda   88 n. 122
Palorus spp.   178 n. 124
Palpita unionalis   136 n. 87
Pamphagidae   68
Pandinus spp.   21, 24
Panorpidae   25, 247
pearl, land   111–12
Pediculus humanus   112, 115
Pemphigus spp.   217
Pentatomidae   104
Philaenus spumarius   97
Philanthus triangulum   194
Philosamia cynthia   141
Philotrypesis caricae   215
Phlebotomidae   230
Pholcus phalangioides   37–8
Phthirus pubis   112, 114–5
Phyllotreta spp.   253
Pieris spp.   121, 126–9, 243
Polistes gallicus   187, 189, 192
Pollenia rudis   220 n. 12
Polyphylla fullo   160 n. 24, 164, 166
Pompilidae   187, 189
pond-skater   107
Prays oleae   136
Pseudoscorpiones   34
Psocoptera   *see* booklouse
Psoroptes spp.   118 n. 117
Psychidae   149, 255
Pulex irritans   240

ragworm   11
Reticulitermes lucifugus   199
Rhagonycha spp.   169
Rhynchites betuleti   132
Rhynchophorus ferrugineus   156
rose chafer   83 n. 105, 107, 164–8, 170
Rutelinae   164–6

Saga viridis   69
Salticidae   46

Sarcophaga spp.   219, 220 n. 8, 221
Sarcoptes scabiei   118
Sarinda spp.   48 n. 72
Saturnia pyri   141
sawfly   190
scale insect   108–13
scarab   157–64
Scarabaeidae   157–8, 164
Scarabaeus sacer   157–9, 161, 163–4
Sceliphron spp.   189
Schistocerca gregaria   62, 77
Scolia spp. 48 n. 72
Scolopendra spp.   10–11
Scolytidae   181, 184
scorpion   4, 10, 21–34, 49, 55, 68, 77, 173, 193–4
scorpion, winged   22, 24–6, 33, 120
scorpion-fly   25, 247
Segestria spp.   46 n. 62, 51
Sesiidae   149
sheep-ked   57
shield-bug   104
silkmoth   140–9, 158 n. 15, 197
silverfish   137 n. 96
Simuliidae   230, 235 n. 79
Siphonaptera   *see* flea
Siricidae   150
Sitophilus granarius   177, 178 n. 124
Sitotroga cerealella   177
Solifugae   46 n. 62, 49–50
Spanish fly   168
Sparganothis pilleriana   132 n. 66
Sphecidae   187, 189–90
Sphingidae   *see* moth, hawk
spider   34–56, 102, 136, 139, 142 n. 135, 144–5, 173, 186, 189, 257
spider, house   35–7, 43
spider, jumping   46, 48
spider, orb-web   35–7, 39, 41
spider, wolf   36, 51–2
springtail   61, 247
Staphylinidae   185
Stegobium paniceum   177, 178 n. 124
Stenopteryx hirundinis   84 n. 106
Stomoxys calcitrans   219, 222–3, 225, 229 n. 55
Stratiomyidae   *see* fly, soldier
Synagris spp.   198 n. 64

Tabanidae   222–3, 225–9
Tabanus spp.   225
Tegenaria spp.   *see* spider, house
Tenebrio molitor   *see* mealworm
Tenebrionidae   80, 82, 178, 185
Tenebroides mauritanicus   177
termite   199
Tetragnathidae   37
Tetraneura ulmi   217 n. 129

Tetranychus spp. 253
Tettigonia viridissima 65, 69
Tettigoniidae *see* cricket, bush
Thaumetopoea pityocampa 148
Theridium spp. 51
Thysanoptera 201
Tibicen plebejus 92
tick 56–60, 112, 119, 160, 224, 238, 257
Tinea spp. 137
Tineola biselliella 137
Tipulidae 167
Torticidae 132–4
Torymidae 215
Tribolium spp. 178
Trichodectes canis 119
Trichodes spp. 186
Trichophaga tapetzella 137
Trichoptera *see* caddis-fly
Trogium pulsatorium 137
Troglophilus spp. 66
Trombicula autumnalis see mite, harvest
Tryxalis spp. 68
Typhaeus typhoeus 158–60

Velia caprai 107
Vespa crabro 187, 189, *see also* hornet
Vespa orientalis 194 n. 37
Vespidae *see* hornet; wasp
Vespula spp. 187, 189

wasp 29, 48, 100, 127, 141–2, 161 n. 33, 187–98, 200–1, 239, 257
wasp, hunting 68 n. 48, 189–90, 194
water-scorpion 25
weevil 132, 156, 177, 179–80, 243
Wohlfahrtia spp. 219
woodlouse 12, 13–19
woodlouse, pill 13, 15, 17–18
woodwasp 150
woodworm 181–4, *see also* Anobiidae

Xestobium rufovillosum see beetle, death-watch
Xylocopidae 197 n. 57
Zabrus spp. 167
Zeuzera pyrina see moth, leopard
Zygaena spp. 129 n. 48